Technology-Enhanced Formative Assessment Practices in Higher Education

Christopher Ewart Dann
University of Southern Queensland, Australia

Shirley O'Neill
University of Southern Queensland, Australia

A volume in the Advances in Higher Education and Professional Development (AHEPD) Book Series

IGI Global
DISSEMINATOR OF KNOWLEDGE

Published in the United States of America by
IGI Global
Information Science Reference (an imprint of IGI Global)
701 E. Chocolate Avenue
Hershey PA, USA 17033
Tel: 717-533-8845
Fax: 717-533-8661
E-mail: cust@igi-global.com
Web site: http://www.igi-global.com

Library of Congress Cataloging-in-Publication Data

Names: Dann, Christopher, editor. | O'Neill, Shirley, 1954- editor.
Title: Technology-enhanced formative assessment practices in higher education / Christopher Ewart Dann and Shirley O'Neill.
Description: Hershey, PA : Information Science Reference, [2020] | Includes bibliographical references and index. | Summary: "This book explore the current impact of mobile technologies and the use of video capture via mobile devices on the learning and assessment of students in higher education, particularly where practical performance examples of their work are required as evidence of attaining competence"-- Provided by publisher.
Identifiers: LCCN 2019023044 (print) | LCCN 2019023045 (ebook) | ISBN 9781799804260 (hardcover) | ISBN 9781799804291 (paperback) | ISBN 9781799804277 (ebook)
Subjects: LCSH: Mobile communication systems in education. | Education, Higher--Effect of technological innovations on. | Computer-assisted instruction. | Educational tests and measurements.
Classification: LCC LB1044.84 .T42 2020 (print) | LCC LB1044.84 (ebook) | DDC 371.260285--dc23
LC record available at https://lccn.loc.gov/2019023044
LC ebook record available at https://lccn.loc.gov/2019023045

This book is published in the IGI Global book series Advances in Higher Education and Professional Development (AHEPD) (ISSN: 2327-6983; eISSN: 2327-6991)

British Cataloguing in Publication Data
A Cataloguing in Publication record for this book is available from the British Library.

For electronic access to this publication, please contact: eresources@igi-global.com.

Advances in Higher Education and Professional Development (AHEPD) Book Series

Jared Keengwe
University of North Dakota, USA

ISSN:2327-6983
EISSN:2327-6991

Mission

As world economies continue to shift and change in response to global financial situations, job markets have begun to demand a more highly-skilled workforce. In many industries a college degree is the minimum requirement and further educational development is expected to advance. With these current trends in mind, the **Advances in Higher Education & Professional Development (AHEPD) Book Series** provides an outlet for researchers and academics to publish their research in these areas and to distribute these works to practitioners and other researchers.

AHEPD encompasses all research dealing with higher education pedagogy, development, and curriculum design, as well as all areas of professional development, regardless of focus.

Coverage

- Adult Education
- Assessment in Higher Education
- Career Training
- Coaching and Mentoring
- Continuing Professional Development
- Governance in Higher Education
- Higher Education Policy
- Pedagogy of Teaching Higher Education
- Vocational Education

IGI Global is currently accepting manuscripts for publication within this series. To submit a proposal for a volume in this series, please contact our Acquisition Editors at Acquisitions@igi-global.com or visit: http://www.igi-global.com/publish/.

The Advances in Higher Education and Professional Development (AHEPD) Book Series (ISSN 2327-6983) is published by IGI Global, 701 E. Chocolate Avenue, Hershey, PA 17033-1240, USA, www.igi-global.com. This series is composed of titles available for purchase individually; each title is edited to be contextually exclusive from any other title within the series. For pricing and ordering information please visit http://www.igi-global.com/book-series/advances-higher-education-professional-development/73681. Postmaster: Send all address changes to above address.

Titles in this Series

For a list of additional titles in this series, please visit: www.igi-global.com/book-series

Preparing the Higher Education Space for Gen Z
Heidi Lee Schnackenberg (State University of New York at Plattsburgh, USA) and Christine Johnson (University of Western Kentucky, USA)
Information Science Reference • copyright 2019 • 253pp • H/C (ISBN: 9781522577638) • US $175.00 (our price)

Competency-Based and Social-Situational Approaches for Facilitating Learning in Higher Education
Gabriele I.E. Strohschen (DePaul University, USA) and Kim Lewis (DePaul University, USA)
Information Science Reference • copyright 2019 • 317pp • H/C (ISBN: 9781522584889) • US $185.00 (our price)

Autoethnography and Heuristic Inquiry for Doctoral-Level Researchers Emerging Research and Opportunities
Robin Throne (Northcentral University, USA)
Information Science Reference • copyright 2019 • 163pp • H/C (ISBN: 9781522593652) • US $135.00 (our price)

Global Perspectives on Fostering Problem-Based Learning in Chinese Universities
Zhiliang Zhu (Northeastern University, China) and Chunfang Zhou (Aalborg University, Denmark)
Information Science Reference • copyright 2019 • 350pp • H/C (ISBN: 9781522599616) • US $195.00 (our price)

Prevention and Detection of Academic Misconduct in Higher Education
Donna M. Velliaris (Independent Researcher, Singapore)
Information Science Reference • copyright 2019 • 324pp • H/C (ISBN: 9781522575313) • US $185.00 (our price)

Educational Technology and Resources for Synchronous Learning in Higher Education
Jiyoon Yoon (The University of Texas at Arlington, USA) and Peggy Semingson (The University of Texas at Arlington, USA)
Information Science Reference • copyright 2019 • 356pp • H/C (ISBN: 9781522575672) • US $185.00 (our price)

Handbook of Research on Critical Thinking and Teacher Education Pedagogy
Sandra P.A. Robinson (The University of the West Indies – Cave Hill Campus, Barbados) and Verna Knight (The University of the West Indies – Cave Hill Campus, Barbados)
Information Science Reference • copyright 2019 • 446pp • H/C (ISBN: 9781522578291) • US $265.00 (our price)

Self-Directed Learning Strategies in Adult Educational Contexts
Francesco G. Giuseffi (Lindenwood University, USA)
Information Science Reference • copyright 2019 • 296pp • H/C (ISBN: 9781522580188) • US $195.00 (our price)

IGI Global
DISSEMINATOR OF KNOWLEDGE

701 East Chocolate Avenue, Hershey, PA 17033, USA
Tel: 717-533-8845 x100 • Fax: 717-533-8661
E-Mail: cust@igi-global.com • www.igi-global.com

Table of Contents

Section 2
Video Feedback and Feedforward

Section 3
Approaches to Dialogue and Dialogic Practices

Detailed Table of Contents

As educators, we aim for students to seek, identify, and utilize a range of feedback to gain an understanding of their present performance in relation to learning goals, and ultimately to identify and use tools to close the gap between present and desired performance. We strive for all students to be their "own first assessors"—intelligent deciders—and develop the independence to self-assess the quality of their own work when they leave higher education institutions and enter the workforce. For students with a print disability such as vision impairment or blindness, traditional forms of feedback may not be successful in providing the information they need to close the gap. The most important issue for these students is access to feedback and agency in the feedback conversation. It is incumbent on higher education educators to find ways to provide equity of access to the provision and reception of feedback for all students. As such, this chapter explores ways for providing feedback to students with a vision impairment to ensure they are able to contextualize and utilize the feedback to improve learning outcomes. This is achieved by aligning the use of mobile technologies and audio feedback with the key principles of connectivism—autonomy, connectedness, diversity, and openness—to provide educators with recommendations.

Evaluating the impact on the student experience of the integrated model of video-enhanced learning, assessment, and feedback discussed in the previous chapter, qualitative data collection employed anonymous online questionnaires, semi-structured interviews, and dialogic interviewing techniques, drawing on summative results data to inform methodological triangulation of the findings. Data analysis combined thematic analysis, constant comparison, and direct interpretation within a grounded theory framework. Illustrative cases present the findings as thick descriptions of the influence of video-based interventions on the experience of six purposively and representatively selected participants. The chapter concludes that an integrated model of asynchronous video-enhanced learning, assessment, and feedback can promote increased reflexivity, enhance learner autonomy, and encourage meta-cognitive self-awareness, while affording greater inclusivity for students affected by dyslexia or Asperger's Syndrome.

In higher education learning environments, there is a dual need for educators to use supportive strategies to motivate students throughout the course, while also aiming to increase the capacity of students to self-regulate their learning. Using instantaneous tools to deliver formative or summative feedback through digital technology has been shown to lead to higher achievement and retention rates. This chapter shows how digital badges can provide instantaneous feedback to support students to feel belonging in the online community and develop self-regulation skills. Instantaneous feedback tools can be used to provide teacher presence throughout higher education courses to increase student engagement, retention, and achievement.

Advancements in digital communication technology (DCT) have significant potential to impact on language teaching and learning by providing new pedagogical opportunities accessible via the internet. Resources that support the design of English language learning and formative assessment have been seen as a fundamental component of the English language learning environments in English language intensive courses for overseas students (ELICOS) and the like. While its importance is universally recognized, the facilitation of a move from the traditional approach to languages learning to a more contemporary approach in keeping with social constructivist theory remains a challenge for both teachers and learners. Part of this challenge involves the need for teachers to reconceptualize the way that they teach to move away from the teacher as instructor to that of facilitator whose pedagogical approach ensures students have opportunities to use English for meaningful purposes. Since this demands an understanding of students' different proficiency levels and the ability to monitor improvement to ensure learning experiences are at the optimum level to facilitate growth, effective formative assessment practices are a vital monitoring tool. Thus, this chapter examines how internet resources can move pedagogy towards a social constructivist approach to improving English language learning bearing in mind the need for a framework to support formative assessment in the monitoring of improvements in proficiency.

Section 2
Video Feedback and Feedforward

In recognizing that effective feedback and pedagogical approaches that support social interaction in learning are vital to students' achievement, this chapter focuses on students' need to be actively involved in their learning and assessment. It reports research into how undergraduate education students utilized assessment feedback. The students provided their perceptions of the feedback strategies employed by completing a survey. This included an investigation into whether the students used the feedback to "feed forward," that is, to answer the question as to whether it was used to enhance their responses in their next assessments. The results revealed a contrast between the students' approach to their learning compared with that of the pedagogy underpinning the course, which sought to empower them in their learning through teaching and assessment practices that engaged them in critical thinking. The students appeared to lack assessment literacy since they treated the feedback as corrective of the assignment being marked and not relevant to them taking-action to improve their future work. It is recommended that this disconnect be further explored since without students' understanding of the purposes of assessment and feedback, in keeping with their educators' intent they are limited in their ability to learn.

This chapter discusses a two-year project that explored the impact of video-enhanced learning, assessment, and feedback on undergraduate first-year students in higher education. Underpinned by a pragmatist epistemology, and arguing the case for a design-based methodological approach within a theoretical framework embracing the cognitive theory of multimedia learning, the community of inquiry, and the conversational framework, the chapter explores contemporary research into assessment and feedback, uses of technology-enhanced learning to promote inclusivity, and educational applications of asynchronous video.

This chapter discusses the implementation of research designed to investigate how video-enhanced learning, assessment, and feedback involves a dialogic approach and contributes to the concept of an integrated model of practice. It builds on the theoretical framework underpinned by both cognitive and social forms of constructivism. Developing design exemplars for video-enhanced learning, assessment, and feedback, three research cycles explored learner-artefact and learner-tutor interactions. The first cycle introduces instructional tutorial videos, freeing up time for formative feedback; the second cycle sees formative video-feedback situated within a conversational framework and video-enhanced assessment activities replace documentation tasks; while the third cycle introduces refinements to the frequency of feedback and regular video diaries, forming an integrated model of video-enhanced learning, assessment, and feedback.

This chapter discusses the impact of feedback as dialogue and video use in a new science methods course that was created to meet new governmental requirements. National demands increased the evidence required by preservice teachers to demonstrate quality teaching and learning in initial teacher education programs impacting on program and course design. This led to increasing reflective opportunities and demonstrating knowledge of subject content for preservice teachers in one small metropolitan university. The ability of 32 preservice teachers to reflect on teaching strategies and their subject knowledge as part of a video teaching assignment with peers in a university classroom is described and discussed. Results indicated high participation in dialogue and dialogic feedback. Personal reflections revealed preservice teachers' understanding of teaching strategies, pedagogies, and subject knowledge improved with support. This has implications for final year preservice teacher progress where they need to demonstrate graduate requirements to transition into the profession.

This chapter evaluates the use of screencast video feedback for summative assessment tasks in the creative arts and analyzes the advantages and disadvantages of such a format when compared to traditional feedback techniques. In 2017, in the second-year course Narrative Animation at the University of South Australia, video feedback was trialed for summative assessment tasks, in an attempt to improve students' understanding of their academic performance. Thirty-seven students participated in the course and received a five-minute feedback video for each of their three submissions. The video feedback provided to students during the course was evaluated at the end of the semester in the form of two online surveys, allowing participating students with the opportunity to critically reflect on the learning experience. The findings of the study disseminate the learning benefits afforded by the video feedback model and provide insight into the varying attitudes of both students and staff.

The use of mobile devices such as mobile phones and tablets in education is a problematic field of research that fits within the scope of assessment, mobile technologies, dialogic practices, and more broadly, feedback. This small pilot study investigated how supervising teachers incorporated a mobile device in the form of a video-enabled app into practicums to promote feedback in the form of dialogue

Foreword

TECHNOLOGY-ENHANCED TEACHING, LEARNING, AND FORMATIVE ASSESSMENT: TRANSFORMING PRACTICES IN HIGHER EDUCATION ACROSS DISCIPLINES AND BEYOND

It can be said that the world of learning has changed for those with access to the Internet, since it has spawned new virtual learning spaces and provided access to almost an infinite number of resources of rich and varied types. It is no longer new to find references made to multimedia and multimodal texts or indeed, to screen technology, as the mediator of communication today. In fact, communication through screens goes hand-in-hand with the growth in the use of smart phones, iPads and Tablets, and the wearable technologies. When combined with the support of free video-conferencing software, these devices allow everyone to connect with each other no matter where they are and what the time is, so long as the Internet is available. Compared with traditional communication, today's digital technological opportunities are at the fingertips of those with access, enabling them to make meaning through multimedia and multimodal texts. In the past, the penning of monomodal texts such as letters and postcards, when not monologic, required much time for a complete exchange of information; and the traditional landline telephone system had limited availability because of financial cost or lack of infrastructure, and was not mobile. In contrast, today's digital technological devices encourage interactive, dialogic communication in real time, 24/7, as well as asynchronous They facilitate group connectivity, in addition to the traditional 'one-on-one'. Thus, technology enhanced communication includes the facility to deepen the message or argument using images or other media. Mobile phones, for instance, allow explanatory photographs to be taken and scanned information to be attached in seconds. Meaning can be constructed and brought to bear in the optimum moment, and can be directed to individuals or groups and shared a multiplicity of times, depending on format, purpose and audience, using social media channels, such as Facebook. It provides support for daily living e.g. google maps provides directions whether one is walking or being transported; the various devices allow users to compare prices of items via the Internet before purchasing, and purchases can be made from almost anywhere in the world. This 'new' context for communication has been transformative in not only changing people's means, modes and frequency of communicative transactions for social and business purposes, but simultaneously creating a virtual world to provide education and support learning.

In this technology-rich context, users of all ages are deeply engaged and seemingly acquire the knowledge and tools involved as quickly as they emerge in the myriad of applications available. It is not surprising that people can be seen staring at the screen of their mobile phone and manipulating the tiny keys to 'text', 'search', and 'talk', when alone and in supposed social and business situations. This is an emergent phenomenon that suggests users can become dependent in relation to their communication needs and even anxious in their anticipation of being in touch (Lopez-Fernandez et al., 2017). However, this same phenomenon can reflect people's deep engagement as learners, off-campus, participating in a virtual learning world. These various devices represent the opportunity for learning online and mobile learning. This opportunity is well promoted, since universities with online learning management systems advertise their many degrees globally, but caution that students are expected to have a computer and access to the Internet to successfully participate. This 'boom' in learning online has taken over two decades to arrive at today's general acceptance, although depending on how developed a country might be there remains some skepticism or misunderstanding that this presents a legitimate option, comparable with traditional, physical on-campus, face-to-face learning.

The use of digital communication technologies via the Internet to enhance teaching, learning and assessment can transform students' educational experience in the way they can connect and communicate both synchronously and asynchronously. While this may seem somewhat simplistic, over the past two decades, opportunities for communication and learning have been developed and enhanced. They provide significant sophistication in terms of the creation of online learning spaces or virtual worlds for learning, particularly in higher education, the focus of this book.

There are many advantages for those who choose to learn online in such virtual spaces. This context for learning provides a space in the CLOUD where a course of study can be presented in terms of its learning objectives, timeframe, structure, learning resources, opportunities for participants' interactions, online tutorials, discussions and questions, learning support strategies and assessment. It acts as a central focus like a 'community center' where learners can participate in a range of ways and have input into their learning. It ideally sets up expectations that learning is an ongoing process that involves critical reflection and dialogue. It needs to have strategies to inform students of their progress, and identify and bridge any learning gaps along the way. The underpinning pedagogical approach is typically in keeping with a social constructivist view. Learners are part of a learning community, working collaboratively, where they have agency in their own learning and construction of knowledge. This contrasts with the traditional approach in higher education where students, in the absence of the Internet, engage in weekly lectures, note-taking, and tutorials, but discussion and connecting between-time is not necessarily planned. Today's online learners are immersed in a wealth of highly relevant resources, tools and strategies that foster their participation and continuity and scaffolding of their learning. However, regardless of how a university course is offered, be it on-campus or online, the design of 'assessment for learning' is vital if students are to actively improve their learning outcomes along the way. Thus, the challenge for those who teach courses in higher education is to design a learning journey where the learning experiences and assessment provide data that students can use to improve in the short term. This means, as a result of reflecting on such data, in conjunction with the opportunity to discuss with their tutor, students should be in a better position to succeed. Sadler (1998) addressed the role and importance of such assessment for learning or 'formative assessment', and recognized the need for learners to have assessment literacy, i.e. the ability to grasp the purposes of feedback and understand how to use it to improve. He emphasized:

Formative assessment does make a difference, and it is the quality, not just the quantity, of feedback that merits our closest attention. By quality of feedback, we now realize we have to understand not just the technical structure of the feedback (such as its accuracy, comprehensiveness and appropriateness) but also its accessibility to the learner (as a communication), its catalytic and coaching value, and its ability to inspire confidence and hope. (p. 84)

It is not surprising, therefore, that technological advancement's virtual learning spaces has dramatically changed the way higher education can be delivered. The application of digital communication technologies, including benefits of artificial intelligence and machine learning, has introduced both improvements to the way students learn and the underpinning learning management systems, besides the way business is conducted. Developing countries seek to adopt these new technologies and introduce 'e-learning' (Kanwal & Rehman, 2017), given their potentially transformational effect on the country as a whole. Smyth (2006) explains:

The availability of broadband services at affordable price levels will have significant impact on gross domestic product (GDP) and attract new investment, create jobs and a larger more qualified labour pool. In addition, greater availability will increase productivity through infrastructure creation and access to new improved services. (p. 409)

The uptake of digital communication technologies to improve higher education learning outcomes and education in general appears to be the major dynamic for change. Much of what is achieved in daily routines and other communicative transactions rely upon the creativity of 'high tech' companies such as Google, that along with Amazon, Facebook, Microsoft and Apple, outlay millions of dollars in ongoing development. From a global perspective, this change is well evidenced. The approach to online education in China, for instance, has grown dramatically in a decade with huge investment in technology for 'quality oriented education' (Cheng, 2018). The adoption of online learning or e-learning is recognized as essential for developing countries to grow. For example, in Pakistan online learning opportunities provide a major solution to reach out to both urban and rural populations to improve the quality of their education. This requires substantial investment, which needs to be far more than the 2% of their GDP that they reportedly recently spent (Kanwal & Rehman, 2017).

The importance of societies having access to the Internet for educative purposes was advocated by Bindé (2005), who recommends education should also be free for all. He highlights the existence of a new digital divide across the globe between those who have access and those who do not. Those 'without' are seen as being deprived of a quality education, because of their lack of access to the wealth of advantages the Internet provides. These include learning resources, funds of knowledge, transformative pedagogies and acquisition of digital literacies. This applies to approximately forty-three percent of the world's population based on the number of Internet users and mobile phone penetration data worldwide (Internet World Stats, 2019). Africa maintains the least penetration at thirty-seven percent of the population. This divide amounts to much more than the lack of reading and writing skills that have traditionally set people apart for learning and work. It deprives people from being exposed to the wealth of resources for learning, including multimedia and multimodal texts, and access to digital forms of communication.

Moreover, by not experiencing learning online and the pedagogies and resources involved, another serious disadvantage is the lack of opportunity to become more active, empowered and autonomous in their own learning. In alluding to only these few examples of the potential of technological advancements to positively impact education, it becomes clear that their adoption is also critical to countries' achieving their major goals.

Technological enhancements may also apply to improving pedagogies and assessment across disciplines, as the authors of this book demonstrate in relation to higher education. This helps ensure that graduates are equipped with the most current knowledge and skills that allow them to maximize their contribution to their fields. I commend the editors of this book for their foresight in recognizing the need to raise awareness of the impact of digital communication technologies in education, and particularly the focus on technology enhanced formative assessment in higher education. This is a learning space that is indeed ripe for transformation in terms of moving away from traditional pedagogy to an approach that is appropriate for living in the 21st century work. There is a need to develop the social, linguistic and cultural knowledges, sensitivities, empathy and collaborative skills, required for living and working in a globalized world, and enabling active citizenship. I also congratulate the authors for making this book possible, for their dedication to their various fields, and for their exciting creative works that demonstrate current practices in the use of technology enhanced formative assessment in higher education. Collectively, these chapters breathe new life into the importance of tertiary educators' understanding of what constitutes learning, how students learn, and the role of the learner in the context of contemporary pedagogy and technology enhanced practices. They clarify how formative assessments need to spark critical reflective dialogue about the learning objectives in focus, to be assistive for students to improve prior to their summative assessment. They also raise awareness of the importance of authenticity of practices to build trust and maintain humanistic attributes in online learning environments, where there is the potential for depersonalisation through the screen mediated- and asynchronous- communication.

In particular, these chapters demonstrate how academics in higher education across three continents, and working in a wide range of different disciplines, are eminently rising to the challenge of pedagogical change in the new world of digital communication technologies. Their research provides deep insights into technology enhanced formative assessment and adds to the knowledge in the field through discussions of practice, providing leadership in this vital but neglected space of 'assessment for learning'. So, this book is most beneficial for the Academy in its presentation of cross-disciplinary insights into current practice that provide stimulating opportunities for the cross-fertilisation of ideas for transforming practices in higher education and beyond.

Professor Xiang Long
Guilin University of Electronic Technology, China

REFERENCES

Bindé, J. (Coord.). (2005). *Towards knowledge societies: UNESCO world report. UNESCO reference works series*. Paris: UNESCO. Retrieved from http://unesdoc.unesco.org/images/0014/001418/141843e.pdf

Cheng, Y. (2018, October 17). *China Daily*. Retrieved from http://global.chinadaily.com.cn/a/201810/17/WS5bc6a025a310eff303282da3.html

Internet World Stats. (2017). *Usage and population studies*. Retrieved from www.internetworldstats.com/stats.htm

Kanwal, F., & Rehman, M. (2017). Factors affecting e-Learning adoption in developing countries: Empirical evidence from Pakistan's higher education sector. *IEEE Access: Practical Innovations, Open Solutions*, *5*, 10968–10978. doi:10.1109/ACCESS.2017.2714379

Lopez-Fernandez, O., Kuss, D. J., Romo, L., Morvan, Y., Kern, L., Graziani, P., ... Rousseau, A. (2017). Self-reported dependence on mobile phones in young adults: A European cross-cultural empirical survey. *Journal of Behavioral Addictions*, *6*(2), 168–177. doi:10.1556/2006.6.2017.020 PMID:28425777

Sadler, D. R. (1998). Formative assessment: Revisiting the territory. *Assessment in Education: Principles, Policy & Practice*, *5*(1), 77–84. doi:10.1080/0969595980050104

Smyth, G. (2005). Wireless technologies bridging the digital divide in education. In *ECEL2006-5th, European Conference on e-learning: ECEL2006* (pp. 407-415). Reading, UK: Academic Conferences Ltd.

Preface

Technology in all its forms is available to educators in higher education to use to enhance teaching, learning and assessment practices that form the important cycle of academic success. Yet, at the same time, because of the vast choices in resources, strategies and continued innovative developments of today it presents a challenge with increasing conundrums for educators to decide what constitutes its most effective use. These conundrums present both opportunities and threats to student learning depending on the technology related decisions made; however, the effectiveness of both the educator's teaching and the student's learning have the potential to be magnified when technology is utilized to enhance the formative assessment of learning.

New technologies have presented themselves to educators via personal, social or institutional encounters and consequently have gained some form of use in the teaching and learning process. This process has resulted in numerous technologies being implemented across the spectrum of disciplines in higher education. Within these applications of technology, and given the diversity of programs offered by higher education institutions, this is no surprise. Such diversity has resulted in some educators being aware of a range of technologies that are possibly valuable for educative purposes. As a result, some early adopters attempted to use some of the more well-established technologies in their teaching and learning with growing success and innovation. With the ongoing explosion of resources available via the Internet and improved learning management systems to support online learning opportunities, virtual learning spaces and blended learning this mode of study has fast become the norm, not least of all the recognition that it provides an opportunity to offer short micro-courses as tasters to inform and support would be online learners. Moreover, within this rich context for teaching and learning, the innovative technologies of today have the capacity to personalize and transform students' learning through their ability to enhance formative assessment feedback processes.

While not all technologies are specifically designed to enhance formative assessment or even the learning process, the creativity, innovation and professional commitment of academics to improve students' learning and success is reflected in the reports of the authors who have contributed to this text. For example, technologies such as Eye Tracking and Second Life's conceptions of Avatars, as well as collaborative learning in simulated online Lab work are examples described in these chapters as ways of enhancing formative assessment feedback processes to support the learning of students in higher education. To ensure students are being prepared for the future and acquire the skills and knowledge that can help them be autonomous learners and transformative in their personal practice, there is a need to redefine and reconceptualise formative assessment as the trigger to closely monitor and support students' learning in keeping with their Zone of Proximal Development (Vygotsky, 1978). Technologies of today can be utilized to model a measured and considered approach that will ensure the much-needed innovation in the field of formative assessment that focuses on progressing students' learning (Henderson et al., 2018). Each of the authors in this text have considered the literature around their intentional formative feedback practice and laid out a platform for consideration. They have shown how technology can contribute to the answer to the challenges of improving student learning outcomes through a focus on reconceptualizing formative feedback as 'feed' that may be one-way or two-way, either monologic or dialogic, respectively (Dann, Dann, & O'Neill, 2018). Importantly, they illuminate the kind of changes to formative assessment processes that will facilitate a more learner-centered approach.

Thus, the technologies and processes outlined in the chapters in this text provide a much needed insight into current research in technology enhanced formative assessment across several disciplines that may not be found in many jurisdictions. The authors recognize that there are global forces that encourage the use of technologies for learning but recognize that they do not necessarily create a level playing field for all learners, despite the fact that they are shown to be enabling in the teaching of diverse learner groups. Nevertheless, teachers and students in developing nations may not have access to these technologies or Internet connections may be unreliable, or they may require professional development to come to grips with the pedagogical underpinnings of using technology to learn online compared with a traditional approach to learning. Depending on limitations of use some exposure to the way technologies can impact positively on learning would assist in their motivation to adopt.

The student experience in higher education has become a central focus for Australian universities in the light of the fact that prospective students exert their opportunity to select their preferred choice. The student experience is monitored and evaluated nationally through annual surveys to gain feedback. This provides comparative data that is reported online and may be used in the funding of universities, thus initiatives that have the capacity to improve and transform student learning are vital to universities' raison d'etre. These data are impacted by the learning and assessment experiences of students. While assessment processes are a key stress for students, they do have the ability to provide their feedback not only on their course experience to academics but on their overall university learning experience based on their journey towards their critical qualification. This often forms the basis of students' perception of the effectiveness of their university along with whether they were prepared for work and acquired a job. Thus, the main subject of this book, in terms of informing and encouraging technology enhanced formative assessment processes that can improve learning bodes well with the identified needs of the higher education sector.

Formative assessment practices may have existed in some form throughout history but in modern terms it can be traced back to 1967 when Scriven coined the term "formative evaluation" (Allal & Lopez, 2005). But most recently, the term "mid-stream", noted by Kahl (2005, p. 38), was used to position the formative assessment process as a tool for measuring students' grasp of the topic they were learning. Sadler (2010) followed Scriven's work by consolidating the idea that formative assessment requires feedback to be clearly understood by both the students and their teachers. This appears to be the earliest indication that parity between the teacher and the student is essential to the learning and assessment process. It was not until the research of Black and Wiliam (1998) that it was established that formative assessment could actually improve learning, which was based on their meta-analysis of 250 studies. In earlier research by Kluger and DeNisi (1996), where meta-analysis was also conducted, the results highlighted that in a third of the studies feedback had worsened performance. But, this identified problems with the feedback design because the evaluation was found to have focused on the person as opposed to the task. The discrepancies identified in this meta-analysis have also contributed to the focus of the work reported in the chapters of this text, and its structure, in demonstrating how formative assessment has moved forward to not only focus on the task but for teachers to have analyzed the task and the course of learning to identify the feedback markers to feed forward to monitor and progress a student's learning. Thus, the question of what actions are effective in the formative assessment space and how they are involved in communicating to students and teachers where a student is positioned in their ZPD is an essential addition its reconceptualization.

This book depicts innovative approaches to supporting student learning and the student experience through technological adaptations. These approaches range from those tested to ones that are proposed or have application to formative assessment practices in general. All told, they comprise an important collection that is critically reflective of the current status of technological innovations in the formative assessment space of higher education. It is hoped that these examples and innovations will be helpful to academics and help encourage enhancements to practice as described. These may include trialing of new approaches and attempts to facilitate and increase the opportunity for critical dialogue between students and their educational leaders in pedagogy and the underpinning formative assessment. However, essentially, formative assessment processes need to support students' attempts to progress through their course of learning.

Establishing formative, technologically enhanced, assessment processes is a time-consuming practice and a difficult undertaking. It is problematized by the need for higher education to ensure the appropriate authority's rigorous assessment policies are applied. New approaches to the formative space, while often attractive to individual early adopters of technology, are ultimately considered by administrators, who need to establish stable, and efficient learning and teaching approaches, in the light of their scalability and reliability. In addition, in the scaling of processes, institutions need to consider the diversity of their programs and cohorts, and the adaptability and appropriateness of the technology and or process to meet students' needs. However, in the consideration of the diversity of degree programs the challenge of them becoming fractured by the introduction of variations in course structures has been recognized as a new variable to contend. Originally, using Australia as an example, higher education offered programs with a standard number of courses taken over a two-, three- or four- year period, and these courses culminated in the completion of a diploma, bachelor degree or other qualification. The traditional view was to have standalone assessment items in each course that would provide students with learning to support their future endeavors, including work and learning goals. This traditional approach to the design of higher education programs that require long-term sustained commitment, unless governed by a regulation authority, such as teaching, nursing and engineering, may be seen as less attractive and necessary in the light of the development of alternative, technologically enhanced learning online. These short skill-based courses provide a new layer of qualifications typically acknowledged through merits provided by digital badges. They may be in the guise of a Massive Open Online Course (MOOC) that is typically free, or Micro and Mini courses that attract a fee and may be designed to meet specific needs and can be designed to be stackable to provide credit towards a full university course. These provide the learner with a shorter learning experience and commitment, that is seen as appealing for their short time frame and adaptable to the busy learners of today, who include many in full-time work. Thus, these new strategies, that may be economically driven on the part of both providers and those who take them up, when placed on top of the traditional program/degree structures, create new expectations on time to complete and assessment processes and techniques. To enhance and personalize teaching and learning, and provide 24/7 support to all students, and administrative support as well, only the adoption of 'smart technologies', including the application of artificial intelligence and machine learning can suffice (Fahimirad & Kotamjani, 2018).

This book represents the current state of thinking and practice about technologically enhanced processes for formative assessment processes. Formative assessment is positioned as the provision of information that is understood by students about their attempts to learn some form of content together with the processes of developing students' understanding of the information provided. We see this process as more than one instance of providing written information about a task. Each of the authors in this text also demonstrate how they enact this communication process and in some cases propose how this process can be enhanced. It is hoped that the chapters found here can challenge existing ideas of the formative assessment space for higher education, trigger new ideas that place students at the center of the process and promote the ongoing communication of ideas between the student and the academic. The editors hope that as readers approach each chapter they will consider new ways of developing dialogue within their courses and beyond, and that, such dialogue becomes the key component of the design of formative assessment techniques and the learning experiences of students.

The text is separated into three distinct sections although there is continuity between since the first section explores the ways that the use of technology can enhance the formative assessment processes for all learners, acknowledging the growth in student diversity. It particularly includes ways that technology has enhanced the learning and assessment experiences of students that present with a print disability, autism/Asperger's Syndrome, and emotional stress as well as those who have English as an added/second or foreign language. The second section focuses on the use of video in the formative assessment process for learners. It showcases the work around managing the video process, how video can be built into course design and finally how video can be used in professional practice learning situated in field practice. The final section presents ideas that position the importance of incorporating dialogue within the formative practices in higher education more broadly. It presents how dialogic practices can be incorporated in online Lab work through a collaborative learning approach in an engineering program, how a university support team can build their activities within a course to become part of the process with academics. The section finishes with an example of how eye-tracking software may be used to support the processes of formative assessment, in addition to the final chapter that identifies the key actions and activities of academics and students in the formative assessment process proposing action to increase the focus on the humanistic characteristics of the process to balance the technological influence being felt.

Section 1: Enhanced Formative Assessment for ALL learners

This section presents enhanced formative assessment as applied to diverse groups of learners. It presents four chapters that explore four separate student groups and how technology can support their learning and assessment. The first chapter asks academics to "Talk to Me". The authors take a personal view of how we might cater for students with a print disability and explore ways to provide feedback to students with such a disability. They recommend that key principles presented by Tschofen and Mackness (2012), which are connectivism, autonomy, connectedness, diversity and openness, be used in feedback processes when mobile technologies are used in the audio feedback process. They propose this in answer to the driving question of "How can we close the gap in the provision of feedback to students with a print disability or visual disability?" Technology-enhanced practices are shown to play a key role in removing barriers for students with visual disabilities but points out that their full access in not yet the norm. This may be because of the low incidence of such a disability but this chapter demonstrates how the idea of technologically enhance feedback needs to be dialogic rather than one-way process, which is shown to be concerning. This chapter also concludes that there is a lack of research depth around this topic and proposes further exploration of the research problem with a particular need to understand how learning management systems can be better equipped through advances in technology to enhance the formative process for students with print disabilities.

Chapter 2 focuses on the student experience where the author evaluates the impact of the process of integrating video enhanced feedback in to learning and formative assessment. He presents thick descriptions of how video has influenced six students affected by Autism. He concludes that an integrated model of asynchronous wide enhanced learning and assessment can promote increased reflexivity and enhance learner autonomy. He also concludes that the process encourages meta-cognitive self-awareness and affords greater inclusivity for students. This chapter is one of a trilogy of chapters by this author, where in combination provide an in depth critical insight into evidence-based practice that examines the work of an academic in research, learning experience design and practice to improve student learning outcomes. This chapter on Autism builds the concept that increased use of formative assessment that is enhanced by technologies, such as providing video feedback, can impact on a broader range of learners. It is concluded that the context within which the technology is positioned and the relationship between its modes of use are significant in realizing the potential of technologically enhanced processes in higher education. The findings indicate that the use of video enhanced assessment and feedback can help students with Asperger's Syndrome to overcome social communication issues that impact on the ability to engage with the dialogue around the feedback.

The third chapter focuses on the utilization of instantaneous feedback to promote self-regulated learning in online higher education courses. It addresses the growing practice of rewarding students with digital badges. Importantly, for the current expectations of the provision of opportunities for personalized learning, it addresses the need for academics to promote and develop students' self-regulated learning. In a higher education sector that is increasingly delivering courses via online modes this chapter shows how digital badges can provide feedback to support students who may feel disconnected through the process of online study. The authors positions digital badges within the field of teacher presence and show how they can be used to increase student engagement retention and build their achievement. The authors make five insightful recommendations to implement effectively digital badges within the higher education context, and conclude that digital badges are an authentic pedagogical assessment tool that can effectively engage learners in the feedback processes. They argue how the use of digital badges have the capacity positively contribute to student motivation, and be used to manage and track students' learning and performance in the formative and summative assessment space. The final chapter in this section presents a detailed account of how the Lab learning and assessment experience in engineering can be improved to meet well established learning design features. It describes and explains how the availability of Remote Access Labs (RALs) have the potential to change conventional practices to adopt a collaborative learning approach that is dialogic and able to better address students' learning needs and provide flexibility of access to the learning and assessment experience online. In doing this it raises the issue of how 'markers' need to be identified in the learning design to facilitate formative assessment practices that are able to 'feed forward' and inform both teachers and learners on how learning can be improved.

SECTION 2: VIDEO FEEDBACK AND FEED FORWARD APPROACHES

This second section of the text explores the use of video and how it is applied and theorized in the formative and feedback/feedforward processes. The section includes five chapters, which range from feedback and feedforward approaches to the theoretical position and study of the phenomenon of video use for students in higher education who identify as autistic. The final two chapters present the application of video in the course development and delivery of a science course and to a small case study of how a mobile device can collect video for formative feedback.

The first chapter, Chapter 5, entitled "Feedback to Feedforward: Promoting Student Reflection and Learning Through Teacher Evaluation," examines how undergraduate students used their feedback from graded assessments through the administration of a survey. The survey sought to identify if students were using their feedback to make improvements in their subsequent assessments. The results indicated that the students did not feed forward productively. The author proposes further research into the disconnect between educators and their students in an effort to find strategies that can more effectively harness the use of academics' time, energy and expertise in relation to teaching, learning and formative assessment.

The next chapter, Chapter 6, in this section, focuses on a theory-practice research framework for video-enhanced learning, assessment and feedback. As part of the trilogy linking research, curriculum design and students' learning experience this chapter sets the research base and explores the educational implications of asynchronous video. The author presents a global view of the literature on video enabled formative assessment illuminates practices in the UK. He establishes an epistemological foundation in the Deweyan tradition and highlights the gap in literature around the learner experience of video enhanced assessment and feedback for students with Autism. Chapter 7 by the same author describes the three cycles undertaken in his research that provide a strong methodological understanding of his examination of the use of video enhanced learning, assessment, and feedback for Autistic students. The chapter highlights ways that video can reduce or eliminate disadvantages encountered through traditional more monologic approaches to assessment. His approach seeks to enable the meaningful engagement of students and academics through detailed and personalized feedback in ongoing dialogic practices. The chapter presents detailed data that supports the need for increasing critical dialogic practices with all students.

Chapter 8, "Course Development: Feedback and Video to Improve Teaching Capacity," discusses the impact of feedback dialogue and video use in an undergraduate initial teacher education science methods course. It explores the abilities of thirty-two preservice teachers in a small metropolitan university in Australia where the results showed a high level of engagement in the feedback dialogue. Personal reflections of preservice teachers revealed improved understandings of content though the dialogic approach. The author concludes that course design was pivotal in the creation of a learning environment that could incorporate dialogic practices. She also concludes that the use of a video App enabled students to develop and practice their metacognition skills through the ongoing attempts that were built into the course learning design. Chapter 9 reports on the use of video in course assessment in undergraduate media arts. The author's use of screencast video feedback for assessing tasks in the creative arts where students were studying narrative animation is evaluated through student surveys. This technology enhanced approach is compared with traditional feedback techniques, taking into account the response from students' opportunity to critically reflect on their learning experience.

Chapter 10 concludes this section with a discussion on the use of a video-enabled App as a foundational piece to improving supervising teacher and preservice teacher critical reflective practices through dialogue during practicum experiences. It describes how the use of this App enabled clearer alignment of the practicum conversations and teaching exemplars to the assessment criteria for the practicum placement. The chapter concludes that the use of this video-enabled App acted as a vehicle that enabled 'focused dialogue' that enhanced preservice teachers' pedagogical growth and developmental. It also identified the need for institutional changes in order for sustained use of video-enabled apps to become a standard part of teaching, formative assessment and learning.

SECTION 3: APPROACHES TO DIALOGUE AND DIALOGIC PRACTICES IN HIGHER EDUCATION

Chapter 11 contextualizes dialogic practice involving the use of technology enhanced practices within the discipline of Engineering, in the delivery of learning in Lab work. The author contrasts the conventional learning experience with that involved in a small case study where students are involved in collaborative approach to learning. In addition, the approach involves learning online through Remote Access Labs, where the task can be simulated and accessed from anywhere anytime subject to a reliable Internet connection. The design of tasks is explored in relation to the need to incorporate formative assessment that is able to feed forward. The chapter provides a detailed discussion of well-established underpinning, task-design, principles, from the perspective of students' progressing within their Zone of Proximal Development.

Chapter 12 advocates for the use of eye-tracking software and its advantages for improving pedagogy in the higher education context. This is a challenging perspective and shows how new technologies can utilized for purposes other than their originally design intent. The authors explore the literature around eye-tracking and explain the processes used in its application, together with how the data produced may be analyzed to improve practice. They provide a short history of the concept of eye-tracking, and how the process is set up in plain English. It is then contextualized in terms of how it might be applied within the context of formative assessment. The current and past systems available are described, and then they provide a case study of its application to the provision of formative feedback. The chapter proposes that eye-tracking can be used to enhance the learning and assessment process because it can be used to build a profile of a learner's characteristics when learning off a screen. It suggests that future iterations of screen technology will have eye-tracking capabilities that will change the way computers or related devices interact with users because of the data provide. This may be a new form of data to be made available in the learning and assessment space that has yet to be fully explored.

Chapter 13 investigates the mechanisms in place provided by higher education support personnel to assist students in their learning journey. It presents the case for increased collaboration between student support professionals and the course teacher/leader. The authors, who are experienced in such provision of support from the perspective of a regional, remote and rural student base as well as delivery nationally and worldwide in an online university in Australia, focus on the importance of the dialogic. They present an approach that enables dialogic practices within courses to be expanded beyond the student, to the students' peers and the course teachers/academics. Their view challenges the limitations of conventional practices and thinking in terms of who should be involved in facilitating a dialogic approach, and how that inclusion might occur. They discuss how students engage with their feedback and what is available for students in this area within the Australian higher education sector. They present related research findings and describe how they developed a three-step process to support students in understanding the importance of the feedback they receive, and how they can productively engage with it. Their three-step process was introduced in first year courses as part of a suite of resources developed specifically to support students as they encountered assessment demands. This chapter also explores the potential for resources developed outside of a course to also impact the development of students' feedback literacy, as well as the students' learning and development within the authors' collaborative, cross-professional approach. In drawing attention to active teaching feedback literacies, this chapter builds a new perspective on dialogic practices and shows that technology is not and should not be the driver of feedback and formative actions.

Chapter 14 presents a theories-view of the concept of feedback situated within higher education research. It presents the conceptions of feedback mark two from Boud and Molley (2013) and then aligns it to the conditions of success described in Henderson et al.'s (2018) report. Both these constructs align to a proposed framework that is argued to be representative of the influence on the enacted curriculum. The chapter argues for the provision of professional development of both staff and students to ensure a more detailed understanding of the importance the dialogic approach as a necessary way to ensure authentic critical reflective practices to transform learning in the formative space. Although a humanistic view may seem contrary to the technological focus of this text both authors seek to engage with technology to bring the various nuanced behaviors to the fore in the best interests of students' learning. A call is made for greater examination of the key actions and activities involved in technology enhanced formative assessment and how it can best trigger improvements in learning prior to the end of a course. Thus, the need for a baseline understanding of what actually happens in effective formative assessment practices, including additional exemplars across disciplines, is recommended for future research and publication.

The concluding remarks brings together the central points of each of these chapters. It builds a picture of the current research and practice, and provides a road map to help 'feed forward' to inform those involved in the higher education sectors of how learning can be improved. It advocates for development of a systematic approach to support all students' learning that better utilizes the technology enhanced practices as outlined, although not exhaustive. In essence the text is prompting consideration of a paradigm shift in pedagogical approach in the recognition of the need for a cognitive shift. This shift requires an awareness of the conventional narrative surrounding assessment as 'summative' reflecting the days of examinations, and built on a traditional monologue approach, in keeping with the engrained usage of 'feedBACK'. Contrary to this, a dialogic approach is cognizent with a social constructivist approach to learning, where knowledge is seen as being co-constructed among peers with their educators in two-way conversation, and students have voice and agency in the process. The authors do not seek to pander to those voices that may be informed by limited assessment literacies but call for change. They seek to assist in contributing to a much needed catalyst for change to building a deeper understanding of a dialogic approach to formative assessment that values all actors in the process with parity within a vision for learning improvement being possible for all.

Shirley O'Neill
University of Southern Queensland, Australia

Christopher Dann
University of Southern Queensland, Australia

REFERENCES

Allal, L., & López, L. M. (2015). Formative assessment of learning. A review of publications in French. In OECD (Ed.), *Formative assessment: Improving learning in secondary classrooms* (pp. 241–264). Paris: OECD Publishing.

Black, P., & Wiliam, D. (1998). *Inside the black box: Raising standards through classroom assessment.* London: Department of Education and Professional Studies, King's College.

Boud, D., & Molloy, E. (2013). Rethinking models of feedback for learning: The challenge of design. *Assessment & Evaluation in Higher Education*, *38*(6), 698–712. doi:10.1080/02602938.2012.691462

Dann, B., Dann, C., & O'Neill, S. (2018). Formative assessment via video feedback on practicum: Implications for higher education and professional teacher accreditation bodies. In T. Richardson, B. Dann, C. Dann, & S. O'Neill (Eds.), *Formative assessment practices for pre-service teacher practicum feedback: Emerging research and opportunities* (pp. 158–183). Hershey, PA: IGI Global. doi:10.4018/978-1-5225-2630-8.ch006

Fahimirad, M., & Kotamjani, S. S. (2018). A review of application of artificial intelligence in teaching and learning in educational contexts. *International Journal of Learning and Development*, *8*(4), 106–118. doi:10.5296/ijld.v8i4.14057

Henderson, M., Boud, D., Molloy, E., Dawson, P., Phillips, M., Ryan, T., & Mahoney, P. (2018). *Feedback for learning: Closing the assessment loop*. Australian Government. Department of Education and Training.

Kahl, S. (2005). Where in the world are formative tests? Right under your nose. *Education Week*, *25*(9), 38.

Kluger, A. N., & DeNisi, A. (1996). The effects of feedback interventions on performance: A historical review, a meta-analysis, and a preliminary feedback intervention theory. *Psychological Bulletin*, *119*(2), 254–284. doi:10.1037/0033-2909.119.2.254

Sadler, D. (2010). Beyond feedback: Developing student capability in complex appraisal. *Assessment & Evaluation in Higher Education*, *35*(5), 535–550. doi:10.1080/02602930903541015

Scriven, M. (1967). The methodology of evaluation. In R. Tyler, R. Gagne, & M. Scriven (Eds.), Perspectives on curriculum evaluation (pp. 39-83). Chicago: Rand McNally.

Tschofen, C., & Mackness, J. (2012). Connectivism and dimensions of individual experience. *International Review of Research in Open and Distance Learning*, *13*(1), 124–143. doi:10.19173/irrodl.v13i1.1143

Vygotsky, L. S. (1978). *Mind in society: The development of higher mental process*. Cambridge, MA: Harvard University Press.

Section 1

Enhanced Formative Assessment for All Learners

Chapter 1
"Talk to Me!": Empowering Students With a Vision Impairment Through Audio E-Assessment Feedback

Melissa Cain
Australian Catholic University, Australia

Melissa Fanshawe
University of Southern Queensland, Australia

ABSTRACT

As educators, we aim for students to seek, identify, and utilize a range of feedback to gain an understanding of their present performance in relation to learning goals, and ultimately to identify and use tools to close the gap between present and desired performance. We strive for all students to be their "own first assessors"—intelligent deciders—and develop the independence to self-assess the quality of their own work when they leave higher education institutions and enter the workforce. For students with a print disability such as vision impairment or blindness, traditional forms of feedback may not be successful in providing the information they need to close the gap. The most important issue for these students is access to feedback and agency in the feedback conversation. It is incumbent on higher education educators to find ways to provide equity of access to the provision and reception of feedback for all students. As such, this chapter explores ways for providing feedback to students with a vision impairment to ensure they are able to contextualize and utilize the feedback to improve learning outcomes. This is achieved by aligning the use of mobile technologies and audio feedback with the key principles of connectivism—autonomy, connectedness, diversity, and openness—to provide educators with recommendations.

INTRODUCTION

The role of feeding up, feeding back, and feeding forward in higher education student achievement (commonly recognized as the feedback 'loop') has been well documented and rationalized over the past 15 years (Sadler 2010, 2013, 2015; Nicol & Macfarlane-Dick, 2006; Wiggins 2012; Harks, Rakoczy,

DOI: 10.4018/978-1-7998-0426-0.ch001

Hattie, Besser, & Klieme, 2014). Feedback is "the lynchpin to students' effective decision-making, and the basis of improved learning outcomes" (Henderson, Boud, Molloy, Dawson, Phillips, Ryan & Mahoney, 2018, p. 1). The feedback debate revolves around the qualities of good feedback, including the form, timing, and mode of such feedback. Additionally, the role of tacit knowledge in students acquiring an implied understanding of quality, as compared to goals and standards, and the impact of motivation and self-regulation in students' independent learning and progression, is important to becoming their own 'first marker'. As the use of e-learning and the use of mobile technologies increases in the higher education space, the design and delivery of feedback to students about the progression of their learning continues to evolve. This evolution is building evaluative data, upon which educators can evaluate the effectiveness of their teaching.

The age of e-learning has resulted in advances in higher education accessibility and removal of traditional barriers of access to print for students who are blind or have a severe vision impairment (Harpur & Loudoun, 2011; Akcil, 2018). Mobile learning promises more equitable access to education and social participation through the use of tools such as voice recognition, screen reading, voice recording, podcasts, e-books, speech-based interaction with mobile phones, digital pens for electronic note taking, optical character recognition, and digital braille devices (Boris, 2013; Buzzi, Buzzi, & Leporini, 2012; Hullen, 2012). This form of learning connection to a range of online information sources and knowledge sharing embodies the constructivist learning theory of 'connectivism'; "a new learning theory for a digital age", characterized by four key principles: autonomy, connectedness, diversity, and openness (Tschofen & Mackness, 2012, p. 124).

Despite these advances, full accessibility through a connectivist model is not yet a reality. The research of Payne, Kirkpatrick, Goodacre, and McClean (2006) reports that students and disability practitioners in higher education "frequently reported difficulties in timely access to course materials in accessible formats" (p. 5). In addition, most university staff "do not consider themselves experts with technology and thus rely on disability liaison services and IT support when making pedagogical adjustments" (Payne et al., 2006, p. 5). When entering university, those who are braille dependent (for example) require content to be formatted in 'word', rather than PDF, to ensure accessibility for digital braille devices. Similarly, digital accessibility may be provided through screen-reading software such as 'Non-Visual Desktop Access' (NVDA). If NVDA access is hindered students are often unable to 'read' their assessments, marking criteria, and feedback due to inaccessible formats in use. Character recognition software "may fail to correctly include footnotes, and page numbers often are not easily identifiable on pages of flowing texts, including tables, graphs or other graphical representations" (Harpur & Loudoun 2011, p. 156).

It is critically important that students with print disabilities (such as blindness, dyslexia, and motor disabilities) have the same opportunities to succeed academically and achieve their goals as their peers. 92 countries including Australia are signatories to the Salamanca Statement and Framework for Action on Special Needs Education (United Nations Educational, Scientific and Cultural Organization, UNESCO, 1994) and are therefore legally bound to recognize and respond to diverse needs of all students making accommodations and adjustments to suit different learning styles and rates. Harpur & Suzor (2013) note that the inability of people with print disabilities to access knowledge in all its forms in higher education institutions has "great flow-on effects in hindering their full participation in society" (p. 748). According to the Disability Discrimination Act (Commonwealth of Australia, 1992) indirect discrimination is, "the application of facially non-discriminatory policies which do not have the intent of discriminating but have a discriminatory effect in practice" (Harpur & Loudoun, 2011, p. 155). This may occur when higher education institutions do not provide equitable access to learning and assessment materials to students with a print disability.

International effort to address this issue is building. The United Nations General Assembly (2007) convened the Convention on the Rights of Persons with Disabilities and the Marrakesh Treaty (World Intellectual Property Organisation, 2012) was established to facilitate access to published works for persons who are blind, visually impaired, or otherwise print disabled. These are in addition to the Guidelines on information access for students with print disabilities (Australian Vice Chancellors' Committee, 2004) to promote and represent best practice for universities in Australia. Examples from the latter include that "it is standard practice for all University departments to be able to provide materials in accessible format", and that "staff have access to advice and assistance on the creation of documents which can be made accessible to students" (Australian Vice-Chancellors' Committee, 2004, p. 3). It is intriguing then, that none of the recommendations from these international developments address the provision and reception of feedback.

The tenets of inclusive education focus on the individual needs of learners. This chapter aims to answer the following question: 'how can mobile learning provide quality and personalized feedback in a timely manner so that every student can successfully achieve their learning goals?' This question is pertinent, as to the knowledge of the authors; there is no focused research on how feedback is best provided to students with a print disability in higher education. Most notably, large research projects that examine feedback in higher education contexts (Henderson et al., 2018; Lebler, Harrison, Carey, & Cain, 2015; Tomitsch, Curwood, Thomson, & Hendry, 2016), have not included an examination of how students with a disability receive or engage with feedback. The vast majority of sources in the literature focusing on accessibility address how best to provide students textbook and course notes in accessible form.

Harpur (2010) suggests that the cost to universities of ensuring students are provided with accessible feedback is "negligible" but "the potential educational impact is great" (p. 122). As such, this chapter seeks to provide a fuller picture of how feedback is delivered and received by students with a print disability (specifically those who are blind or have a vision impairment) in higher education through the recent advances in mobile learning. This chapter will consider the principles of effective feedback, the role of self-determination on successful implementation of feedback, equity of access to feedback and engagement in the feedback conversation, the importance of the social dimension of feedback, and tacit understandings about quality and comparability. An exploration of the types of mobile tools that can enhance feedback and the advantages of e-assessment feedback (EAF) through audio files will provide a practical presentation of current practices in higher education institutions in Australia.

LITERATURE REVIEW

Principles of Effective Feedback

According to Rust, O'Donovan, and Price (2005) assessment constitutes "probably the single biggest influence on how students approach their learning" (p. 231). Sadler (2013) concurs and adds that feedback is "often regarded as the most critical element in enabling learning from an assessment event" (p. 55). However, the mere provision of feedback does not, necessarily lead to improvement in performance (Cain, 2014; Fanshawe, 2015). Sadler states that although feedback should play a critical role in improvement, in practice it often "seems to have no or minimal effect" (2013, p. 55). Most feedback given to learners is rarely utilized and as such, educators need to focus on both the nature of effective feedback and most importantly, how feedback is provided and received (Henderson et al., 2018). Hattie & Timperley (2007) advocate that feedback should address three levels–the task level (that is, the learning intent and the specific requirements of the task); the process level (the processes, skills, and strategies required to complete the task); and the self-regulatory level (relational thinking and self-monitoring to compare and adjust their work in relation to the required standards, criteria or intent) (p. 86). This should in turn address three questions answerable by the student— where am I going? (feeding up); how am I going? (feeding back); and, where to next? (feeding forward) (Hattie & Timperley 2007, p. 86).

Several researchers have worked to summarize the features of effective feedback. Nicol and Macfarlane-Dick (2006) offer that effective feedback

helps clarify what good performance is; facilitates the development of self-assessment (reflection) in learning; delivers high quality information to students about their learning; encourages teacher and peer dialogue around learning; encourages positive motivational beliefs and self-esteem; provides opportunities to close the gap between current and desired performance; and provides information to teachers that can be used to help shape teaching (p. 205).

Wiggins (2012, p. 10-16) adds to these observations and suggests that feedback should also "be goal referenced; tangible and transparent; provide actionable information; user friendly; delivered in a timely manner; ongoing; and consistent (stable, accurate and trustworthy)". The research of Evans (2013) provides a further 12 principles of effective feedback. We add two of these to the observations of Nicol and Macfarlane-Dick (2006) and Wiggins (2012) as being most pertinent for students who are blind or have a vision impairment. They are "clarifying with students the different forms and sources of feedback available including e-learning opportunities" and "clarifying the role of the student in the feedback process as an active participant and not as purely receiver of feedback and with sufficient knowledge to engage in feedback" (Evans 2013, p. 79).

The National Council of Assessment and Feedback (2015) suggest that students view traditional models of assessment as something done to them, and not with them, and that feedback has essentially a teacher-centered concept. Recent research, (cf Henderson et al., 2018) however, points to the importance of the social dimension of feedback within "all feedback exchanges generated within assessment design" (Evans 2013, p. 71). Feedback should be an active and learner-centered 'participatory process', as the feedback loop can only be considered complete when students have engaged with and utilized feedback provided, thus the central role of the student in the success of feedback process (Henderson et al., 2018). Personalized feedback tailored to the individual, task, and context assists to build relationships between teachers and students as part of the social and constructivist dimension of learning.

Self-Regulation

Lebler (2008) suggests that ideally, assessment processes should reflect the kinds of evaluations students should be able to employ after graduation, and that students need to develop "the inclination and ability to be their own first marker if they are to continue to learn independently and effectively" (2008, p. 4).

This 'self' level (Hattie & Timperley, 2007) engages the student in personal evaluation, which is where we want our learners to be. At this level, the teacher is minimally involved as the learner engages in self-directed error detection and strategic thinking for deep learning. It is important to note, however, that feedback given at the 'task' level is often interpreted at the 'self' level, resulting in negative emotional responses which divert attention from the task, instead "focusing it upon the self where it is perceived as a generalized criticism leading to negative feelings like self-doubt, anger, or frustration" (Evans 2013, p. 96). The ultimate aim of assessment and feedback, therefore, should be to empower students to become self-regulated learners who understand and can apply feedback effectively in all its forms (Boud, 2000). In this chapter, we note the importance of feedback that is personalized and dynamic–'adaptive formative feedback'–to increase self-motivation and enable students to be judges of their own learning. Boud and Soler (2016) highlight the notion of feedback as a "co-productive process in which both students and others have key roles to play" (p. 403).

Sadler (2013) suggests that there are three basic requirements for learners to become proficient and self-regulated learners. They must be able to

acquire a concept of high quality and can recognize it when they see it; with considerable accuracy judge the quality of their works-in progress, and connect this overall appraisal with particular weaknesses and strengths; and choose from their own inventories of potential moves those that merit further exploration for improving quality (p. 54)

We look to Deci et al.'s (2008) 'self-determination theory' which emphasizes that effective feedback will enable students and promote intrinsic motivation towards their learning. They describe autonomously motivated learners as those who value and experience self-endorsement of their actions, and note that for autonomous motivation to be present, three needs must be in place. Students must have a level of competence, connect with others, and have a sense of autonomy in their goals. As such, effective feedback should also allow students opportunities to engage in discussion that acknowledge motivations and beliefs about self (Nicol & Macfarlane-Dick, 2006).

The Role of Tacit Knowledge in the Giving and Receiving of Feedback

Students need to be inducted into the ways that "complex qualitative judgements" (Sadler 2010, p. 546) can be made about their work, and for this to happen, they must possess some of the tacit evaluative skills of their teachers. Sadler (2013) refers to this as a type of 'knowing to' or having developed an "awareness and responsiveness so they can detect anomalies or problems for themselves" (p. 58). Feedback can only be effective, therefore, when students possess this knowledge base. For students to successfully apply and adapt knowledge, Sadler (2013) stresses that they must be able to "interact with information and skills"; "incorporate them into their existing knowledge bases and structures" and "build knowledge that will serve them as adaptable personal capital" (p. 56). When teachers transmit feedback, there is the often an incorrect assumption that messages are clearly received by students and are easily decoded and translated into action, although there is considerable evidence that feedback is "invariably complex and difficult to decipher, and that students require opportunities to construct actively an understanding of them before they can be used to regulate performance" (Nicol & Macfarlane-Dick, 2006, p. 201).

Henderson et al. (2018) note that students naturally possess "less complex understandings of the purpose of feedback than staff", and that this disparity "could hint at mismatches between implicit feedback cues that staff may consider obvious, but which students may fail to recognize and implement" (p. 9). Previously, Sadler (2013, 2010) has similarly identified the critical role that tacit knowledge plays in students fully understanding and applying assessment criteria and feedback. For students to have a clear understanding of the acceptable level of quality on a task, students must be able to assess their work in the same manner their teachers do. This requires not only being familiar with set criteria, but with more salient criteria which influence a teacher's qualitative and holistic interpretation of a piece of work (Sadler, 2010). Such knowledge and skills cannot be imparted through explicit teaching alone, and thus students must experience and be inducted into non-overt methods of making judgements about quality. Rust et al., (2005) stress the difficulty in developing such tacit knowledge as this involves years of "observation, imitation, dialogue and practice" (p. 152). This involves being able to attend to two facets of their work: a holistic evaluation of its quality and smaller separate aspects of quality as gained through both explicit and tacit knowledge of assessment criteria. For feedback to be effective, therefore, students need a sound working knowledge of three concepts: task compliance, quality, and criteria. These assessment concepts must be understood "not as abstractions but as core concepts that are internalized, operationalized and applied to concrete productions" (Sadler, 2010, p. 548). Participation in the feedback loop, opportunities to take part in continual dialogue about feedback, as well as invitations to negotiate feedback, assist students to gain an understanding of the types of tacit understandings that educators take for granted. We might compare and relate this to the way in which students who with a vision impairment miss out on vital non-verbal clues and cues such as gestures and facial expressions that hold nuanced contextual meaning and information.

The Impact of Vision Impairment on Learning for Students in Higher Education

The work of Pacheeco, Lips, and Yoong (2018) found that for students with disabilities, participating in higher education is a "more challenging and stressful experience due to the barriers imposed by their impairment" (p. 1). Vision impairment creates a unique challenge to accessing information in higher education "as the content and assessment is designed for those who can see" (Armstrong 2009; Cain & Fanshawe, 2019). Learning outcomes are written with the assumption that all students are sighted and students with a vision impairment "are expected to attain the same learning outcomes to succeed in the course" (Permvattana, Armstrong, & Murray 2013, p. 18). Sighted students are able to access printed images, diagrams, and tables and engage and interpret easily however, engagement with these printed images is not possible for students with a vision impairment (Rule, Stefanich, Boody, & Peiffer, 2011). In addition, any e-learning materials "are not frequently designed to integrate with the range of assistive technologies" (Permvattana, Armstrong, & Murray 2013, p. 18), resulting in inaccurate or inaccessible information. Interpretation bias can occur when images or pictures are described for a student and hidden information inferred by the pictures cannot be interpreted or easily conceptualized (Prosser & Loxley, 2007). Online Learning Management Systems (LMS) can be difficult for students to navigate due to the differences between the ways that courses are designed (Bennett, Lockyer & Agostinho, 2018), and the need to utilize inaccessible drop-down menus and icons.

Mobile learning has increased access to higher education for many learners, including those who are blind or vision impaired (Kukulska-Hulme, 2012), creating the delivery of learning to students "anytime and anywhere" (Wang, Wu & Wang, 2009, p. 93). The increasing accessibility functions of mainstream devices and vision specific adaptive technologies can enable students who are blind or vision impaired more equitable access to higher education (Pieri & Diamantini, 2012). Crossland, Silva, and Macedo (2014) found that a high percentage of people with vision impairment have access to mobile phones, tablets, and/or laptops. Software products such as 'screen readers' use a synthesized voice to read out documents and what appears on the screen. Students may use free inbuilt products, such as Voiceover® (Apple), Text-to-speech (Android), iMax (Mac), Thunder® (Windows) or downloadable products specifically designed for non-visual readers, such as eBooks, Nonvisual Desktop Access (NVDA), or JAWS® (license required) or digital braille readers that produce print information on a braille device. Jiang et al. (2017) explains how Optical Character Recognition (OCR) software used in conjunction with screen readers can assist people with vision impairment to "not only recognize signs, but also reconstruct sentences and convert them to speech".

Speech recognition software or virtual assistants, such as Siri®, Google Home®, or Cortana® "simplify the complex, multistep things we're all tired of doing via drop down menus, complicated workflows, and hopscotching from app to app" (Pierce, 2015, p. 1). Sankalpani et al. (2018) promote the use of virtual assistants for people with a vision impairment as they are available at low cost and assist many people to simplify their lives. Smartwatches, a combination of screen readers and virtual assistants, can be used in accessibility mode to employ screen readers such as Voiceover, virtual assistant Siri® along with hauptics, vibration to ensure full access to notifications to communication (Galdon, et al., 2013).

Mobile technologies in particular, have provided inclusion for students through ability to access information on the go, creating a sense of belonging through increased access to the learning environment and participation in social interactions between learners (Akcil, 2018). The principles of connectivism are important here in the potential for networked learning "to accommodate the psychological diversity of participating individuals" (Tschofen & Mackness 2012, p. 124). In their study of students with vision impairment, Bishop and Rhind's (2011) work highlights that social inclusion plays a major role in developing a positive self-concept. De Carlo et al. (2012) agrees that social inclusion is essential for belonging in education and found that learning environments that provided opportunity for students to participate equally meant students had a better quality of life overall.

Despite these challenges, inclusion for students with vision impairment is possible when conditions are created to access "learning spaces on an equal footing" (Thurston, 2014, p. 105). For all learners within the mobile learning environment, access to education is facilitated through digital means; students participate in online discourse and develop their social identity within the group through online interactions. Ihrig (2013) recognizes that many students with vision impairment naturally do not like to be perceived as different. Whilst online, other students cannot determine if a student is accessing the information by braille or by print. The need to identify faces, body language, and other social cues is difficult for students with vision impairment (Cain & Fanshawe, 2019) but in an online environment this need is eliminated. Knowledge about identities are created without a visual presence and other users are likely unaware the student has a vision impairment. Whitburn (2014) talks of the 'alienating legacy' (p. 2) for students with vision impairment in navigating their way to and within education facilities which can be difficult due to navigate through the spaces.

Electronic Feedback in Higher Education

There has been significant growth in the number of articles focusing on electronic assessment feedback (EAF) possibilities within the past 10 years, as 'e-assessment' is playing "an increasingly important role in the transformation of higher education" (Lafuente, Remesal, & Álvarez Valdivia, 2013, p. 443). This encompasses both formative and summative feedback delivered or conducted through information communication technology of any kind, including CD-ROM, television, interactive multimedia, mobile phones, and the Internet (Gikandi, Morrow, & Davis, 2011). EAF is considered wide-ranging in that it can be a/synchronous, face-to-face, or at a distance, involve automated or personal feedback and different mediums, and be used to support individual and group learning. One of the much-lauded benefits of EAF is the provision for automated marking, reducing the marking workload for academics. It is important here to make the distinction between computer-generated scoring and feedback, where the former provides a mark but no feedback guidance (Ware, 2011). Lafuente, Remesal, and Álvarez Valdivia (2013) support EAF as an excellent medium "not only to track students' learning, but also to support it" (p. 443). Advocates of EAF argue that such approaches encourage students to adopt deeper approaches to, and greater self-regulation of learning. Frequently cited benefits of EAF for students include higher achievement and retention rates (Ibabe & Jauregizar, 2010) through the provision of more relevant and authentic assessment feedback experiences and collaborative learning opportunities (Gilbert et al., 2011). Such technology affords immediacy and "anytime, anywhere, anyhow approaches" and is suitable for use with large numbers of students" (Gikandi et al., 2011).

A key factor in the efficacy of e-feedback technologies is "the nature of the interaction between students and their lecturers within the process" (Evans 2012, pp. 86-87). Henderson et al. (2018) believe this interaction creates a learning environment with a culture for interactive and dialogic feedback, a design inclusive of the individual needs of the students, and one that assists students to gain capacity to understand and use feedback as self-regulated learners. EAF can be both official and unofficial (Henderson, et al., 2018). Official feedback includes feedback from academics, support staff, and automatic feedback derived from tools such as digital badges and submission receipts within the Learning Management System (LMS). Students can also contribute to the unofficial feedback discussion through the use of audience response systems (e.g. clickers) and embedded students' response activities. Many students use feedback from their peers in LMS chats and forums, as well as social media to gain information and clarify assessment requirements. Henderson et al. (2018) found that 55% students also used unofficial feedback from family members, friends, peers, private tutors prior to submitting assessment. This suggests students' desire to receive regular and intermittent feedback prior to submitting assignments and should be a consideration within feedback planning.

TALK TO ME! RECOMMENDATIONS FOR ACCESSIBLE, EQUITABLE, AND MEANINGFUL FEEDBACK THROUGH AUDIO ENGAGEMENT

Meaningful feedback "is one of the most powerful means of empowering students, as well as enhancing and transforming their learning" (Rasi & Vuojarvi, 2017, p. 292). In order to design effective accessible feedback for students with a vision impairment, educators need to be proactive in gaining a practical understanding of the needs of students as individuals and understanding their students' disabilities.

Reflecting on the principles of effective feedback discussed in this chapter and the possibilities for enhanced inclusivity and removal of barriers for students with a vision impairment, we address the research question to provide recommendations for differentiation and specific adjustments for higher education students with print disabilities. These recommendations aim to provide agency and choice to learners, to promote an active role in the feedback dialogue, to inculcate self-regulatory skills, and independence in evaluation of work, with the ultimate goal of students progressing to becoming their own 'first marker'. The principles of effective feedback that are directly addressed here include that feedback is user friendly and timely, utilizes a variety of forms including e-learning opportunities, promotes dialogue around learning, and encourages active participation in the feedback process. Of central importance, and one that has not been included by any of the experts in the review of literature, is that effective feedback should be accessible to all students.

Given Crompton & Burke's (2018) findings that "the majority of adults own more than one mobile device" (p. 53), the authors propose that the use of a variety of audio options would reduce identified barriers and provide access to the feedback loop for students with vision impairment. A common observation of various adjustments for differentiation is that not only do the adjustments support the targeted student or group of students, but very often benefit a wider range of learners (Foreman & Arthur-Kelly, 2017). For example, the use of audio files also benefits students with other print disabilities such as dyslexia and some motor disabilities (Emery & Atkinson, 2009) and students without disabilities who may find processing print material complex (Hapur & Loudoun, 2011). Audio feedback can provide improved learning outcomes for these students as it is convenient and timely, builds emotional connectivity within the learning environment (Woodcock, 2016), and helps develop self-regulated and independent learners connected to knowledge and other learners.

Audio Feedback Is Convenient and Timely

Providing feedback through audio means, allows students to listen to feedback on mobile devices in any location (Lunt & Curran, 2010) including at home, on public transport, and even when travelling overseas. Cann (2015) found that most students preferred to receive feedback on their mobile phone, allowing students to engage with feedback in privacy and enabling them to respond to their feedback "in different ways and at different times" (Dixon, 2015, p. 98). From the teacher's perspective, audio feedback can also be recorded in any location, allowing teachers the ability to provide feedback in a timely manner on mobile devices. Learning analytics can play a part in this conversational feedback loop by providing teachers with information about students' activities and engagement with learning tasks so they can reflect on student learning processes.

The timing of feedback delivery is crucial. Feedback is most effective when it is provided as soon as possible after the work has been submitted but also at time when students are likely to engage with the feedback and teachers are available to respond to questions and engage actively in the feedback loop (Cain, 2014; Cann 2015). Most LMS have analytics inbuilt or can be available through external open-source online application, such as The Loop Tool. This information can be used by teachers to provide feedback to students or as evidence to inform changes and improvements to learning resources and/or learning design. It can be used to identify individual students who are not engaging with course materials and to connect with students who may be at risk of failing to provide intervention. Lecturers can use information from the learning analytics as feedback for their own pedagogy and course design, by noting student engagement. By checking analytics for students with vision impairment, lecturers can examine which add-ins are not being accessed and start conversations with students to understand what students need to assist their learning and implement strategies to support this. It is also important to offer opportunities for students to clarify their feedback, either by phone or built-in dialogue tools, which will enable students to fully participate in the feedback conversation.

Audio Feedback Builds Emotional Connectivity and Personal Connections

Ensuring emotional engagement to learning and personal connectivity with students in higher education remains a major challenge, particularly for courses delivered online (Dixon, 2015). Evans' (2013) research found that feedback is a powerful generator of emotions, and successful assessment feedback is attributed to "student access, retention, completion, and satisfaction within college and university contexts" (p. 70). Dixon (2015) too believes that educators should consider the importance of attending to the emotional dimension of learning, and "the roles of feedback for learners' motivational beliefs and self-esteem" (Nicol & Macfarlane-Dick, 2006, p. 199). These aspects can be accommodated through the use of audio feedback (Rowe, Fitness & Wood, 2015). Woodcock (2016) found that students who received audio feedback individually catered to their assessment and learning goals found constructive criticism easier to receive. Recent research demonstrates that audio feedback should be considered as an option to promote emotional connectivity between both students and teachers, and students and their peers (Rasi & Voujarvi, 2017). This is true for all students but particularly those for whom print is inaccessible and whom experience barriers to important social connections in the learning space.

Online chats assist students to take ownership of their learning by reflecting on their contributions and those of their peers, and the personal advice and feedback provided through these forums (Joosten, 2012). This form of formative EAF provides students with an opportunity to connect with peers, analyze and synthesize course content, and connect through teacher and peer dialogue. Within this space, students with a vision impairment are inconspicuous in their disability (Pieri & Diamantini, 2012) and are able to access such dialogue 'on the same basis' as their peers (Crossland, Silva, & Macedo, 2014). This is a key expectation of inclusive education. This social dimension to learning assists students to feel confident in their abilities and develop positive beliefs about themselves as learners (Nicol & Macfarlane-Dick, 2006).

Audio Feedback Through Mobile Technologies

There are a number of mobile tools that support asynchronous audio feedback in mobile learning. As higher education educators are 'time jealous' and have limited time to examine new tools to provide audio feedback, a number of technological tools that are easy to use and readily available have been considered here.

Smartphones are regularly used in higher education to record audio feedback for students and according to Nortcliffe and Middleton (2011), success has been due to their handy size, accessible functions, and affordability. Recordings can be made as voice memos and sent as by email, text message, messenger, or 'air dropped' through Bluetooth. Ice et.al., (2007) suggests that email is not a preferred method of receiving text files, and students prefer the feedback to be housed within the course LMS. Sending audio files through Google Drive® and DropBox® (Cann, 2015) or uploading to the online hosting sites such as SoundCloud® are other methods to reduce concerns over download size and speed (Alleyne, 2013).

The literature suggests the online technological tools within the LMS can "impact on teaching presence and sense of community" within the course (London 2018, p. 1). London (2018) used voice recorders and trialed embedding audio files in discussion boards to allow students to hear verbal feedback from the teacher, however, the process of feedback distribution in this trial was determined to be arduous by the teachers. Race & Williams (2018) prefer inbuilt synchronous and asynchronous tools within the LMS such as online chats and Zoom®, that can embed audio within discussion forums, and which can be easily be accessed on mobile devices. Inbuilt LMS feedback tools such as GradeMark® or Turnitin® for iPad can be used to provide summative audio feedback. Fawcett and Oldfield (2015) found the inbuilt tools took less teacher effort to upload and valuable as a means to create teacher presence within the learning environment. Audio feedback within these systems was found to be received as more personal, conveyed greater nuance in the message, and a more accurate interpretation of the intended meaning (Ice, Curtis, Phillips, & Wells, 2007). Students with vision impairments found these tools to be inclusive as they enabled them to access the feedback and discussion in the same method as their peers (Hewett et al., 2017). An important part of the feedback loop is that students are active participants within the feedback. This can be facilitated in the LMS through technological tools such as a discussion thread, feedback dialogues, or plug-ins such as Zoom® or Voicethread®.

Audio EAF can also contribute to tacit understandings about quality that may be inferred or misinterpreted when written (Rasi & Voujarvi, 2017). Students may interpret written feedback incorrectly, or interpret feedback aimed at 'task' level as feedback at the 'self' level, and therefore the feedback is not utilized effectively to positively impact learning (Lunt & Curran, 2010). Audio feedback allows the students to hear the feedback in the lecturer's 'voice'; listening to the "instructor's strain and intonation, excitement or monotone provide the student receiving the feedback additional information about what the instructor means" (Rasi & Vuojarvi, 2017, p. 299). These important nuances of communication are missing with written or computer generated audio descriptions. Lunt and Curran (2010) found that only 5% of students opened written feedback as compared to 50% when audio files were provided. In their study using audio feedback, 62% respondents had listened to the feedback more than once (McGarvey & Haxton, 2011) primarily to identify 'feed forward' suggestions. Participants felt they were able to develop shared understandings of feedback and use it to meet their goals (McGarvey & Haxton, 2011).

Audio Feedback: Important Considerations

When using audio descriptions and feedback, the selection of language is very important in detailing a concept, image, diagram, or map to a person with vision impairment. When labelling objects the tendency is to use visual descriptions that may not be understood by all people, given varied prior knowledge and experience. A good example lies in the following assessment. Figure 1 was provided to the students. The lecturer included a description in the 'alternate text' section of 'classroom'. The assessment task stated: 'You need to rearrange the learning environment to ensure collaborative learning for the students in the class. Explain how you would do this'. As a classroom can be so varied in layout and could be primary, secondary or tertiary education, the user has missed a lot of information about the image, that the alternate text did not provide and therefore does not allow the same access as sighted peers.

Carefully detailing descriptions in content and feedback will assist students to access the information that is important in this learning context. In the example above, an alternative text might be 'This image is of a classroom in a secondary school in Australia. The room is approximately 7 m by 7 m,

Figure 1. 'A classroom'
(Image retrieved from https://www.pexels.com/photo/chairs-classroom-college-desks-289740/ under Creative Commons license)

with wooden floors. It contains singular tables, approximately 80 cm long and 60cm deep. There are 6 rows of tables with 5 in each row. Each table has a four-legged chair behind it, with a wire shelf under the chair to hold books. The room is set up for the students to undergo a test'. Additional feedback or dialogue between the teacher and students may be required to ensure understanding. As expressed by Murray (2008) "students with a vision impairment possess the same abilities to process and understand spatial concepts with differences explained by intervening variables" (p. 85) such as audio description or hands on experiences.

Audio Feedback: Possible Challenges

Despite technology for audio feedback being available for over a decade, the adoption has been slow. Rasi and Vuojarvi (2017) acknowledge that educators generally believe the process of giving audio feedback takes longer than written feedback. Research conducted on the time taken to provide audio versus written feedback (Allen & Bentley, 2012; Borup, West, & Thomas, 2015; Lunt & Curran, 2010) found, however, that audio feedback was a more time effective option. Allen and Bentley (2012) relate that the average time taken to mark assignments when providing audio feedback is three minutes less, compared to written comments provided through Gradebook in Blackboard® (Allen & Bentley, 2012). Lunt & Curran's (2010) research notes that one minute of speech takes approximately six minutes to write, illustrating that audio feedback "has the potential to actually save a significant proportion of staff time" (Fawcett & Oldfield, 2016, p. 80). Despite providing more equitable accessibility, audio feedback has also been criticized by Rasi and Voujarvi (2017) for "the lengthy time required to listen to the feedback, and difficulty skimming the feedback and returning to specific points at a later time" (p. 295).

Although some studies (cf Bourgault, Mundy, & Joshua, 2013) did not find significant difference in results for students using written or audio feedback, for students who are vision impaired, using audio EAF on familiar digital platforms provides equitable access which may also relieve anxiety in engaging with feedback (Datta & Talukdar, 2016). However, if the goal is to create feedback that is accessible to all, the authors acknowledge that audio EAF is not an option for students who are deaf or who have a hearing impairment. With the aim to promote belonging and a sense of inclusion for all students, and with a connectivist belief that students need to be actively connected to all stakeholders in the feedback loop, a number of inbuilt mainstream accessibility and disability specific software options are capable of recording audio and text simultaneously. The Smartphone is capable of recording in voice memos and dictating into notes. Products such as Dragon® can be trained with intelligent speech recognition which also includes grammar correction and on-screen editing. VoiceThread® and OneNote® for Android® and Virtual Assistants, also have the capability to turn to speech to text, which can be used to provide written feedback where required.

To be effective, any tools used for delivering feedback require staff and student training. Staff need knowledge of pedagogical tools that will assist them to trial audio feedback and an understanding of the language required to convey conceptual understanding within the feedback (Gould & Day, 2013). Students need to become accustomed to the culture in which to receive feedback, the mobile technology being used, and how to retrieve the audio files (Kukulska-Hulme, 2012).

CONCLUSION AND IMPLICATIONS FOR FUTURE RESEARCH

All students must be able to seek, identify, and utilize a range of feedback to gain an understanding of the present state of their performance in relation to learning goals and to identify the tools to close the gap between present and desired performance. Effective feedback allows students to be active participants in feedback conversations and play an important role in closing the feedback loop. For students with a print disability such as students with a vision impairment or blindness, full access to learning materials including feedback is often reduced or impossible due to the preference for presenting learning materials and course site design to suit those who can see. This has significant implications for many areas of learning including social connectivity, autonomy, and academic success. The increased use of digital technologies in higher education has played a significant role in removing some barriers, however, full access is not the norm. This chapter has reviewed the possibilities for the use of mobile devices and specifically audio feedback in providing EAF.

Vision impairment is a low incidence disability and many teachers will be unaware of how to cater for their needs in the online learning environment (Cain & Fanshawe, 2019). Due to technological advances, there are a range of adaptive software that are inbuilt in mainstream devices which enable students to 'read' the screens. As it is important for students to be active participants in their feedback, the access to feedback through these adaptive technologies allow students to receive and provide feedback to gain tacit understanding of quality the teacher possesses (Akcil, 2018). This chapter has demonstrated the importance of a pedagogical approach to inclusion through mobile technology, but also the necessity of educators' awareness of equitable access to content, assessment, and feedback, bringing the inclusive education agenda to the fore. Despite the possibilities of audio feedback options for students with a vision impairment to access higher education in an inclusive manner, it is important that audio feedback is not used as 'one way' feedback from the teacher to the student. Rather as part of the connectivist paradigm, where students have an equally important role to play in the self-regulation or their learning, opportunities to actively engage by discussing feedback will therefore close the feedback loop.

As there is an absence of research on feedback for students with a print disability and tools that will suit all learners, the authors recommend further investigations concerning the individual elements of LMS that may be analysed for their accessibility. In addition, training modules should be developed for higher education teachers and students on accessible LMS design and pedagogy, with a focus on feedback as "one of the most powerful means of empowering students, as well as enhancing and transforming their learning" (Rasi & Vuojarvi, 2017, p. 292).

REFERENCES

Akcil, U. (2018). The use of mobile learning for visually impaired learners school in tolerance education contents. *Quality & Quantity*, *52*(2), 969–982. doi:10.100711135-017-0552-1

Allen, R., & Bentley, R. (2012). *Feedback mechanisms: Efficient and effective use of technology or waste of time and effort*. London, UK: *STEM Annual Conference*.

Alleyne, L. (2013). Breaking sound barriers: Custom integrated audio distribution is taking off in the cloud. *Residential System*, *14*(1), 2–5.

Armstrong, H. (2009). Advanced IT education for the vision impaired via e-learning. *Journal of Information Technology Education*, *8*, 243–256. doi:10.28945/691

Australian Vice-Chancellors' Committee. (2004). *Guidelines on information access for students with print disabilities*. Retrieved from https://www.adcet.edu.au/download/attachment/5352-1/

Bennett, S., Lockyer, L., & Agostinho, S. (2018). Towards sustainable technology-enhanced innovation in higher education: Advancing learning design by understanding and supporting teacher design practice. *British Journal of Educational Technology*, *49*(6), 1014–1026. doi:10.1111/bjet.12683

Bishop, D., & Rhind, D. (2011). Barriers and enablers for visually impaired students at a UK Higher Education Institution. *British Journal of Visual Impairment*, *29*(3), 177–195. doi:10.1177/0264619611415329

Boris, C. (2013). 3 mobile apps for converting voice to text. *Entrepreneur Asia Pacific*. Retrieved from https://www.entrepreneur.com/article/225584

Borup, J., West, R. E., & Thomas, R. (2015). The impact of text versus video communication on instructor feedback in blended courses. *Educational Technology Research and Development*, *63*(2), 161–184. doi:10.100711423-015-9367-8

Boud, D. (2000). Sustainable assessment: Rethinking assessment for the learning society. *Studies in Continuing Education*, *22*(2), 151–167. doi:10.1080/713695728

Boud, D., & Soler, S. (2016). Sustainable assessment revisited. *Assessment & Evaluation in Higher Education*, *41*(3), 400–413. doi:10.1080/02602938.2015.1018133

Bourgault, A., Mundy, C., & Joshua, T. (2013). Comparison of audio vs. written feedback on clinical assignments of nursing students. *Nursing Education Perspectives*, *34*(1), 43–46. doi:10.5480/1536-5026-34.1.43 PMID:23586205

Buzzi, M., & Leporini. (2012). Investigating mobile learning and accessibility for blind learners. In M. Allegra, M. Arrigo, V. Dal Grande, P. Denaro, D. La Guardia, S. Ottaviano, & G. Todaro (Eds.), Mobile learning for visually impaired people (pp. 26-40). Consiglio Nazionale delle Ricerche – Istituto per le Tecnologie, Didattiche.

Cain, M. (2014). Participants' perceptions of fair and valid assessment practices in higher music education. In D. Lebler, S. Harrison, & G. Carey (Eds.), *Assessment in music: Principles and Practices* (pp. 87–106). UK: Springer.

Cain, M., & Fanshawe, M. (2019). Opening eyes to vision impairment: Inclusion is just another way of seeing. In S. Carter (Ed.), *Opening eyes onto inclusion and diversity*. Pressbooks.

Cann, A. (2015). Engaging students with audio feedback. *Bioscience Education, 22*(1), 31-41. doi-org.ezproxy.usq.edu.au/10.11120/beej.2014.00027

Commonwealth of Australia. (1992). *Disability discrimination act*. Canberra: Author.

Crompton, H., & Burke, D. (2018). The use of mobile learning in higher education: A systematic review. *Computers & Education*, *123*, 53–64. doi:10.1016/j.compedu.2018.04.007

Crossland, M., Silva, S., & Macedo, A. (2014). Smartphone, tablet computer and e-reader use by people with vision impairment (report). *Ophthalmic & Physiological Optics*, *34*(5), 552–557. doi:10.1111/opo.12136 PMID:25070703

Datta, P., & Talukdar, J. (2016). The impact of support services on students' test anxiety and/or their ability to submit assignments: A focus on vision impairment and intellectual disability. *International Journal of Inclusive Education*, *21*(2), 160–171. doi:10.1080/13603116.2016.1193561

DeCarlo, D., McGwin, G., Bixler, M., Wallander, J., & Owsley, C. (2012). Impact of paediatric vision impairment on daily life: Results of focus groups. *Optometry and Vision Science: Official Publication of the American Academy of Optometry*, *89*(9), 1409–1416. doi:10.1097/OPX.0b013e318264f1dc PMID:22863790

Deci, E., Ryan, R., Hunsley, J., Pelletier, L., Vallerand, R., & Koestner, R. (2008). Self-Determination theory: A macrotheory of human motivation, development, and health. *Canadian Psychology*, *49*(3), 182–185. doi:10.1037/a0012801

Dixon, S. (2015). The pastoral potential of audio feedback: A review of the literature. *Pastoral Care in Education*, *33*(2), 96–104. doi:10.1080/02643944.2015.1035317

Emery, R., & Atkinson, A. (2009). Group assessment feedback: The good, the bad and the ugly. In *A word in your ear conference proceedings* (pp. 1–7). Retrieved from http://research.shu.ac.uk/lti/awordinyourear2009/docs/emery-atkinson-Solent_Audio_Feedback_paper.pdf

Evans, C. (2013). Making sense of assessment feedback in Higher Education. *Review of Educational Research*, *83*(1), 70–120. doi:10.3102/0034654312474350

Fanshawe, M. (2015). The implications of monocular vision on orientation and mobility. *Journal of the South Pacific Educators in Vision Impairment*, *8*(1), 77–86.

Fawcett, H., & Oldfield, J. (2018). Investigating expectations and experiences of audio and written assignment feedback in first-year undergraduate students. *Teaching in Higher Education, 28*, 1–15.

Foreman, P., & Arthur-Kelly, M. (2017). *Inclusion in action* (5th ed.). Cengage.

Galdon, P., Madrid, R., Rubia, E., Cuestas, E., Diaz, A., Estrella, A., & Gonzalez, L. (2013). Enhancing mobile phones for people with visual impairments through haptic icons, the effect of learning processes. *Assistive Technology, 25*(2), 80–87. doi:10.1080/10400435.2012.715112 PMID:23923690

Gikandi, J., Morrow, D., & Davis, N. (2011). Online Formative Assessment in higher education: A review of the literature. *Computers & Education, 57*(4), 2333–2351. doi:10.1016/j.compedu.2011.06.004

Gilbert, L., Whitelock, D., & Gale, V. (2011). *Synthesis report on assessment and feedback with technology enhancement*. Southampton, UK: University of Southampton.

Gould, J., & Day, P. (2013). Hearing you loud and clear: Student perspectives of audio feedback in higher education. *Assessment & Evaluation in Higher Education, 38*(5), 554–566. doi:10.1080/02602938.2012.660131

Harks, B., Rakoczy, K., Hattie, J., Besser, M., & Klieme, E. (2014). The effects of feedback on achievement, interest and self-evaluation: The role of feedback's perceived usefulness. *Educational Psychology, 34*(3), 269–290. doi:10.1080/01443410.2013.785384

Harpur, P. (2010). Accessible or not? Academics' hand written comments on assessment items written by students with print disabilities. *Disability & Society, 25*(1), 121–122. doi:10.1080/09687590903363472

Harpur, P., & Loudoun, R. (2011). The barrier of the written word: Analysing universities' policies to students with print disabilities. *Journal of Higher Education Policy and Management, 33*(2), 153–167. doi:10.1080/1360080X.2011.550088

Harpur, P., & Suzor, N. (2013). Copyright protections and disability rights: Turning the page to a new international paradigms. *The University of New South Wales Law Journal, 36*(3), 745–778.

Hattie, J., & Timperley, H. (2007). The power of feedback. *Review of Educational Research, 77*(1), 81–112. doi:10.3102/003465430298487

Henderson, M., Boud, D., Molloy, E., Dawson, P., Phillips, M., Ryan, T., & Mahoney, P. (2018). *Feedback for learning: Closing the assessment loop: Final report*. Victoria: Monash University.

Hewett, R., Douglas, G., McLinden, M., & Keil, S. (2017). Developing an inclusive learning environment for students with visual impairment in higher education: Progressive mutual accommodation and learner experiences in the United Kingdom. *Journal European Journal of Special Needs Education, 32*(1), 89–109. doi:10.1080/08856257.2016.1254971

Hullen, J. (2012). The relevance of e-learning-accessibility for people with vision impairment. In M. Allegra, M. Arrigo, V. Dal Grande, P. Denaro, D. La Guardia, S. Ottaviano, & G. Todaro (Eds.), *Mobile learning for visually impaired people* (pp. 72–79). Consiglio Nazionale delle Ricerche – Istituto per le Tecnologie, Didattiche.

Ibabe, I., & Jauregizar, J. (2010). Online self-assessment with feedback and metacognitive knowledge. *Higher Education: The International Journal of Higher Education and Educational Planning*, *59*(2), 243–258. doi:10.100710734-009-9245-6

Ice, P., Curtis, R., Phillips, P., & Wells, J. (2007). Using asynchronous audio feedback to enhance teaching presence and students' sense of community. *Journal of Asynchronous Learning Networks*, *11*(2), 3–25.

Ihrig, C. (2013). Clinical pearls. *Optometry and Vision Science*, *90*(3), 302–303. doi:10.1097/OPX.0b013e31828b8189 PMID:23400025

Jiang, H., Gonnot, T., Yi, W., & Saniie, J. (2017). Computer vision and text recognition for assisting visually impaired people using android smartphone. *2017 IEEE International Conference on Electro Information Technology*. 10.1109/EIT.2017.8053384

Joosten, T. (2012). *Social media for educators' strategies and best practices*. San Francisco, CA: John Wiley & Sons.

Kermit, P., & Holiman, S. (2018). Inclusion in Norwegian higher education: Deaf students' experiences with lecturers. *Social Inclusion*, *6*(4), 158–167. doi:10.17645i.v6i4.1656

Kukulska-Hulme, A. (2012). Prospects for inclusive mobile learning. In M. Allegra, M. Arrigo, V. Dal Grande, P. Denaro, D. La Guardia, S. Ottaviano, & G. Todaro (Eds.), *Mobile learning for visually impaired people* (pp. 13–25). Consiglio Nazionale delle Ricerche-Istituto per le Tecnologie, Didattiche.

Lafuente, M., Remesal, A., & Álvarez Valdivia, I. (2013). Assisting learning in e-assessment: A closer look at educational supports. *Assessment & Evaluation in Higher Education*, *39*(4), 443–460. doi:10.1080/02602938.2013.848835

Lebler, D. (2008). Perspectives on assessment in the learning of music. In D. Bennett & M. Hannan (Eds.), *Inside, outside, downside up: Conservatoire training and musicians' work* (pp. 181–193). Perth, Australia: Black Swan Press.

Lebler, D., Harrison, S., Carey, G., & Cain, M. (2014). *Assessment in music: An approach to aligning assessment with learning outcomes in the higher music education*. Queensland: Griffith University.

London, J. E. (2018). Use of synchronous audio feedback in discussion boards with online RN-BSN Students. *Nurse Educator*, *1-4*. doi:10.1097/NNE.0000000000000633 PMID:30585883

Lunt, T., & Curran, J. (2010). "Are you listening please?" The advantages of electronic audio feedback compared to written feedback. *Assessment & Evaluation in Higher Education*, *35*(7), 759–769. doi:10.1080/02602930902977772

McGarvey, D., & Haxton, K. (2011). Using audio for feedback on assessments: Tutor and student experiences. *New Directions*, *7*(7), 5–9. doi:10.29311/ndtps.v0i7.459

Murray, I. (2008). Instructional eLearning technologies for the vision impaired. *Curtin University espace*. http://hdl.handle.net/20.500.11937/645

National Council of Assessment and Feedback. (2015). *Focus on learning; formative feedback*. Retrieved from https://www.ncca.ie/media/1925/assessment-booklet-3_en.pdf

Nicol, D., & Macfarlane-Dick, D. (2006). Formative assessment and self-regulated learning: A model and seven principles of good feedback practice. *Studies in Higher Education, 31*(2), 199–218. doi:10.1080/03075070600572090

Nortcliffe, A., & Middleton, A. (2011). Smartphone feedback: Using an iPhone to improve the distribution of audio feedback. *International Journal of Electrical Engineering Education, 48*(3), 280–293. doi:10.7227/IJEEE.48.3.6

Pacheeco, E., Lips, M., & Yoong, P. (2018). Transition 2.0. Digital technologies, higher education and vision impairment. *The Internet and Higher Education, 37*, 1–10. doi:10.1016/j.iheduc.2017.11.001

Payne, T., Kirkpatrick, D., Goodacre, C., & McClean, P. (2006). *Creating accessible teaching and support for students with disabilities. Carrick Institute for Learning and Teaching in Higher Education* [Report]. University of Tasmania.

Permvattana, R., Armstrong, H., & Murray, I. (2013). eLearning for the vision impaired: A holistic perspective. *International Journal of Cyber Society and Education, 6*(1), 15–30. doi:10.7903/ijcse.1029

Pierce, D. (2015). Say what? *Wired, 23*(10), 86.

Pieri, M., & Diamantini, D. (2012). Approaches to mobile learning accessibility. In M. Allegra, M. Arrigo, V. Dal Grande, P. Denaro, D. La Guardia, S. Ottaviano, & G. Todaro (Eds.), *Mobile learning for visually impaired people* (pp. 52–63). Consiglio Nazionale delle Ricerche – Istituto per le Tecnologie, Didattiche.

Prosser, J., & Loxley, A. (2007). Enhancing the contribution of visual methods to inclusive education. *Journal of Research in Special Education, 7*(1), 55–68.

Race, A., & Williams, M. (2018). Digital audio feedback technology: Nursing students' perspectives. *Computers, Informatics, Nursing, 36*(2), 63–67. doi:10.1097/CIN.0000000000000418 PMID:29401070

Rasi, P., & Vuojarvi, H. (2017). Toward personal and emotional connectivity in mobile higher education through asynchronous formative audio feedback. *British Journal of Educational Technology, 49*(2), 292–304. doi:10.1111/bjet.12587

Rowe, A., Fitness, J., & Wood, L. (2014). The role and functionality of emotions in feedback at university: A qualitative study. *Australian Educational Researcher, 41*(3), 283–309. doi:10.100713384-013-0135-7

Rule, A., Stefanich, G., Boody, R., & Peiffer, B. (2011). Impact of adaptive materials on teachers and their students with visual impairments in secondary science and mathematics classes. *International Journal of Science Education, 33*(6), 865–887. doi:10.1080/09500693.2010.506619

Rust, C., O'Donovan, B., & Price, M. (2005). A social constructivist assessment process model: How the research literature shows us this could be best practice. *Assessment & Evaluation in Higher Education, 30*(3), 231–240. doi:10.1080/02602930500063819

Sadler, D. R. (2010). Beyond feedback: Developing student capability in complex appraisal. *Assessment & Evaluation in Higher Education, 35*(5), 535–550. doi:10.1080/02602930903541015

Sadler, D. R. (2013). Opening up feedback: Teaching learners to see. In S. Merry, M. Price, D. Carless, & M. Taras (Eds.), *Reconceptualising feedback in higher education: Developing dialogue with students* (pp. 54–63). London: Routledge.

Sadler, D. R. (2015). Backwards assessment explanations: Implications for teaching and assessment practice. In Assessment in music: From policy to practice (pp. 9-19). Cham, Switzerland: Springer.

Sankalpani, P., Wijesinghe, I., Jeewani, I., Anooj, R., Mahadikaara, M., & Jayakody, J. (2018). Smart assistant: A solution to facilitate vision impaired individuals. *2018 National Information Technology Conference (NITC)*. 10.1109/NITC.2018.8550059

Thurston, A. (2014). Disability, power and equality in the school. *International Journal of Disability Development and Education, 61*(2), 105–107. doi:10.1080/1034912X.2014.905048

Tomitsch, M., Curwood, J., Thomson, K., & Hendry, G. (2016). *Professional development for academics: Evaluating a personalised mobile website for learning from student feedback. Final report*. Sydney, NSW: University of Sydney.

Tschofen, C., & Mackness, J. (2012). Connectivism and dimensions on individual experiences. *International Review of Research in Open and Distance Learning, 13*(1), 124–143. doi:10.19173/irrodl.v13i1.1143

United Nations Educational, Scientific and Cultural Organization (UNESCO). (1994). *The Salamanca statement and framework for action on special needs education*. Adopted by the World Conference on Special Needs Education: Access and Quality, Salamanca, Spain.

United Nations General Assembly (2007). *Convention on the Rights of Persons with Disabilities: Resolution adopted by the General Assembly*. A/RES/61/106. Author.

Wang, S., Wu, C., & Wang, Y. (2009). Investigating the determinants and age and gender differences in the acceptance of mobile learning. *British Journal of Educational Technology, 40*(1), 92–118. doi:10.1111/j.1467-8535.2007.00809.x

Ware, P. (2011). Computer generated feedback in student writing. *TESOL Quarterly, 45*, 769-774.

Whitburn, B. (2014). The 'inclusion' of students with vision impairments: Generational perspectives in Australia. *International Journal of Whole Schooling, 10*(1), 1–18. Retrieved from http://hdl.handle.net/10536/DRO/DU:30062408

Wiggins, G. (2012). 7 Keys to Effective feedback. *Educational Leadership, 70*(1), 10–16.

Woodcock, P. (2016). Towards dialogue: Audio feedback on politics essays. *European Political Science, 16*(2), 193–205. doi:10.1057/eps.2015.101

World Intellectual Property Organisation. (2012). *Marrakesh treaty to facilitate access to published works for persons who are blind, visually impaired, or otherwise print disabled*. Geneva: Author.

Chapter 2
The Student Experience of Video-Enhanced Learning, Assessment, and Feedback

James McDowell
Lancaster University, UK

ABSTRACT

Evaluating the impact on the student experience of the integrated model of video-enhanced learning, assessment, and feedback discussed in the previous chapter, qualitative data collection employed anonymous online questionnaires, semi-structured interviews, and dialogic interviewing techniques, drawing on summative results data to inform methodological triangulation of the findings. Data analysis combined thematic analysis, constant comparison, and direct interpretation within a grounded theory framework. Illustrative cases present the findings as thick descriptions of the influence of video-based interventions on the experience of six purposively and representatively selected participants. The chapter concludes that an integrated model of asynchronous video-enhanced learning, assessment, and feedback can promote increased reflexivity, enhance learner autonomy, and encourage meta-cognitive self-awareness, while affording greater inclusivity for students affected by dyslexia or Asperger's Syndrome.

INTRODUCTION

Background and Context

This chapter discusses the student experience of the video-enhanced learning, assessment and feedback interventions developed and integrated over the three cycles described in Chapter 7. Employing the data collection and analysis strategies outlined in Design-Based Research Cycles 2 (DBRC2) and 3 (DBRC3), purposive sampling led to the development of a series of illustrative cases which highlight key findings emerging from the investigation, with emphasis placed on the presentation of raw data derived from semi-structured and dialogic interviews conducted with participants towards the later stages of the study. The participant group was comprised of undergraduate first-year students, all of whom were male in the

DOI: 10.4018/978-1-7998-0426-0.ch002

18-25 age range, and it is important to note that the six illustrative cases found in this study therefore describe the experience of male participants only. Section 2 introduces the cases of Robert and Keith, two students affected by dyslexia whose experience of the initial video-enhanced assessment (VEA) and video-enhanced feedback (VEF) activities in DBRC2 provided strong direction to the development of the integrated model of video-enhanced learning, assessment and feedback developed in DBRC3.

In Section 3, a further four illustrative cases highlight the impact of this integrated model, with the presentation of interesting and unusual cases including those of one student affected by dyslexia, one student affected by an autistic spectrum condition, and two of their neurotypical peers.

It should be noted that undergraduate students in the UK enter higher education following completion of either A-Levels, which are predominantly assessed by examination, or alternative qualifications approved by the Business and Technology Education Council (BTEC), which are predominantly assessed through coursework; the latter are referred to using the acronym BTEC.

SECTION 2 ILLUSTRATIVE CASES FROM DBRC2

1. Illustrative Case 1: Robert (VEF)

Robert, a 21-year-old male design-stream student from a BTEC background, self-reported the impact of dyslexia as mid-range, although expert witness accounts from the School's Disabilities Coordinator suggested that the impact of Robert's dyslexia placed him in the upper range of the mid-range spectrum. Although Robert's dyslexia had been formally recognised following clinical assessment aged 5, he had struggled with even simple spelling throughout school, and it had taken him an additional two years to complete the qualifications required for university entrance.

Despite having no previous experience of learning using video, Robert accessed all the instructional tutorial videos (ITVs) developed in DBRC1, commenting: "*Found it much easier to pick up - would have struggled a lot without the tutorials*". He also displayed a keen enthusiasm for using video in the assessment and feedback process when the conceptualisation of the Video-Feedback Loop (VFL) was outlined to the cohort at the start of DBRC2. Pre-empting the implementation of the Video-Enhanced Assessment and Feedback (VEAF) interventions, Robert conducted self-directed research into cost-free options that might enable students to engage in VEF activities outside the studio environment. Finding that *screencast-o-matic.com* offered web-based screen capture functionality that facilitated the production of learner generated content, Robert employed his personal *YouTube®* account to host the output.

Thus, Robert became the first participant to engage in the learner-tutor VFL, proactively creating a video that highlighted an issue causing problems in a programming exercise, and emailing a private link to this content on *YouTube®* with a request for feedback. This short video (01m:31s) highlighted the area of concern with a visual demonstration of both the code and its compiled output, and briefly outlined attempts to overcome the problem. This provided sufficient detail from which to identify the source of the problem, and, using *Camtasia®* to capture the screen as the *YouTube®* video was played, a short video-feedback (01m:34s) artefact incorporating webcam video of the tutor's head and shoulders as a picture-in-picture was recorded which pinpointed the area of the code requiring attention.

In order to sustain dialogue around the issue as an asynchronous video-based conversation, this video-feedback artefact ended with a request to provide follow-up video evidence of how the feedback had been used, and to confirm whether these actions had solved the problem. The video-feedback artefact was then uploaded to *Mahara®*, and made available to Robert through a private 'View', to ensure that the feedback was shared only with the intended recipient. Along with the video-feedback, instructions highlighting how to embed content hosted on *YouTube®* within pages in *Mahara®* were included on this private View. This enabled Robert to situate his original video in closer proximity to the feedback for future iterations of the VFL.

Robert's initiation of a VFL led to an asynchronous conversation of six cycles as he worked through a series of problems that he encountered during his programming development work. In the semi-structured interview, Robert confirmed that he had used all the ITVs, and had repeat-viewed many of these in order to consolidate his learning. On the subject of VEF, Robert described it as *a very good method*, highlighting the specificity of the feedback as particularly useful given his dyslexia *Could see what was going on more – combination* [of audio and video] *worked best. Let you know what line you needed to be on exactly* (Robert). Robert also praised the practice of making the video-feedback available on *Mahara®*, adding that it was *"more handy - it's just there"*, before going on to express concerns for privacy should this have been made publicly available on *YouTube®*.

Talking specifically about the first video-feedback artefact received in response to his initiation of the VFL which had included the picture-in-picture feature, Robert commented that seeing the tutor in the video had been a distraction, and that it was *a bit spooky seeing you at the bottom of the video - bit strange. Just your audio would be better, but it's a really good way to get feedback about coding problems* (Robert).

When asked to compare his experience of feedback through video and through text, Robert suggested that video had made a significant difference for him, suggesting that he *wouldn't have done as well without it*, ultimately enabling him to develop a much more polished product than expected. Citing the specificity of the feedback in combination with the opportunity to repeat-view the video, Robert appreciated the sense of control afforded by video-feedback, commenting that he had watched pieces of video-feedback on his mobile phone while on public transport, and suggesting that the experience was *more relaxing - you know exactly what's going on, and you can re-play it to double-check you're sorting out everything you need to* (Robert).

On his experience of the VEA activities and the principle of using video to document development work, Robert confirmed that the experience of producing a summative self-assessment video in VEA Activity 3 had been *better for me. It's much easier to explain what you've done, and better than using screenshots . . . a much easier way of documenting your work* (Robert)

The interview also explored how VEA might be integrated within the structure of the module in future iterations. Confirming as someone affected by dyslexia that the frequency of video-feedback received through the VFL had made a positive impact on his experience, Robert indicated that a focal point for accessing video offered the potential for stronger engagement *'It would be good to have it in Mahara® each week too so you can see where you've got to, and it's more monitored* (Robert).

2. Illustrative Case 2: Keith

Keith, a 19-year-old male programming-stream student from an A-level Mathematics and ICT background, self-reported the effects of his dyslexia as 'mild'. Both his elder siblings had been recognised as affected by dyslexia, and consequently Keith underwent early assessment at school, receiving support throughout secondary education.

In-studio observations made by the researcher over the course of the academic year, identified Keith as a strong student who clearly possessed a good grasp of the programming techniques needed to produce the relevant aspects of the portfolio, but whose design skills initially required further development. Evidence from Keith's blog entries indicated that his use of the ITVs was systematic, working through the two series of tutorials in a linear fashion, although in the semi-structured interview he acknowledged having needed to return to certain ITVs, confirming that the purpose had been to consolidate his appreciation of the finer technical details associated with key techniques:

The hardest part I found was remembering what all the different icons mean, and sometimes I just get plain confused how to do something I might have only done a few minutes ago. Because of this I had to keep re-watching videos to remind myself how to do different aspects within the Editor (Keith).

For VEA Activity 3, Keith produced a 6-minute video-summary reflecting on the development of the design and programming work he had produced over the academic year. Electing to capture the video using *Camtasia®*, Keith created a well-structured and clearly presented overview of how the ITVs had contributed to the development of skills and techniques used in the production of the software assets comprising the portfolio. Employing a 'back-and-forth' approach which demonstrated how the assets had evolved through iterative cycles, the presentation combined voiceover-enabled screen-captures of media-rich blog entries with walkthrough demonstrations of the finished products, culminating in 'live' demonstrations of these artefacts.

Keith's video summary demonstrated a high level of reflexivity, while also exhibiting evidence of meta-cognitive self-awareness through explanations highlighting how the tutorials had facilitated the development of knowledge and skills. From informal discussions initiated by colleagues with no stakeholder connection with the investigation, it emerged that Keith had successfully transferred aspects of his knowledge and skills into other areas, producing digital assets for assessment in concurrent modules which these other tutors confirmed were of an extremely high quality.

On VEA, Keith suggested that using video *felt more dynamic and exciting* than using text to present an account of how the development work had progressed during the year had been:

A lot easier - using Camtasia® was also fun as I was learning how to use that as well . . . Made me think about how I'd done things, had to plan out what I was going to say; it was good to go back and look at how my work's progressed (Keith).

In connection with video-feedback, while Keith was sufficiently confident in his programming work to have not initiated a VFL under VEF Activity 1, he had received video-feedback on his work under the umbrella of VEF Activity 2 *absolutely amazing to get feedback like that - a lot more detailed than what I'm used to. Seems to work really well because of the detail. Very difficult to get that if just text* (Keith). In an earlier semi-structured interview, Robert had described the positive impact of receiving regular video-feedback in the course of a VFL sustained over six-weeks. In this interview, Keith was therefore asked to consider whether an increased frequency of short activities might work with VEA. Although concerned that he *"might have found it intimidating to start with"*, primarily due to being self-conscious around a newly formed peer group, Keith pointed to patterns of activity on *YouTube®* in which asynchronous conversations took place between those posting and commenting on videos as potentially worthy of emulation in an educational context.

SECTION 3 ILLUSTRATIVE CASES FROM DBRC3

1. Illustrative Case 1: Neil

Neil, a design-stream student from an art and design background, arrived at university with a prior clinical assessment of dyslexia, and with a history of dyslexia having impacted negatively on his academic performance since primary school. In contrast to those dyslexic participants interviewed as part of DBRC2, Neil self-reported the impact of his dyslexia as severe; this was corroborated by an expert witness account from the School's Disabilities Coordinator, who suggested that Neil's dyslexia had been assessed as severe-to-extreme, and confirmed that outside timetabled classes he was supported in the production of written work in other modules by a specialist Dyslexia Support Tutor (DST) employed by the institution's Disability Support Services. Against this background, the extent of Neil's dyslexia was a key factor in his selection as a case intended to illustrate the impact of the video-enhanced approach for students affected by the condition.

Engagement with ITVs, VEF and VEA

Neil engaged with the ITVs from the outset, and in dialogic interviews strongly welcomed the opportunity to engage with learning materials which did not rely on the written word. Neil began producing video diaries from the third week of the first semester in response to a problem-focused trigger event (cf. Garrison, Anderson, & Archer, 2001) connected with the use of the software development environment. While his early diary entries highlighted how he had used the ITVs, Neil frequently used the diary entries to raise problems encountered in the subsequent application of the skills developed in order to engage in dialogue around the problem by initiating a conversation within the VFL, thereby eliciting a video-feedback response.

The work in the first semester was predominantly of a design nature, and Neil highlighted his engagement with video-feedback through further weekly video-diary entries which demonstrated his progress beyond troublesome areas (Mayer & Land, 2005). While the impact of dyslexia on Neil's design-based work was noticeable, the impact in the Semester 2 programming exercises within the development environment was immediately observable as extreme and debilitating. In addition to being unable to distinguish between higher and lower case alphabetic characters, Neil could not distinguish between mirrored non-alphanumeric characters including '{ ' and '}', and '(' and ')' commonly used to open and close key programming constructs such as methods and classes.

In Neil's case, videos aiming to highlight problems with development work and video-diary entries documenting the development process quickly became indistinguishable, and the VFL served to facilitate a process of continuous asynchronous dialogue around the programming exercises. Over the course of the academic year, Neil produced 34 videos with a combined duration of 01h:41m:36s, and during the twelve-week period in which the focus was on programming, he initiated a total of seventeen online learner-tutor loops in addition to the weekly face-to-face dialogic interactions in the studio.

Learner Evaluation

On ITVs: Neil described the ITVs as being *very useful*, citing several technical examples of where these had helped, and, as noted earlier, he welcomed the opportunity to learn using materials that did not rely on him to interpret text-based content. On VEA: Neil found the VDEs made a highly significant difference to his capacity to complete the documentation process, affording him the opportunity to approach the task in a manner in which he felt comfortable and in control. Qualitative analysis of the many VDEs and semi-structured interview data highlighted numerous categories including SPEED, EASE OF USE, UTILITY, DYSLEXIA, and MOTIVATION clustered throughout both his work and his evaluation of the interventions. With dyslexia being the single largest factor affecting his academic performance since entering secondary education, Neil spoke from experience when he commented that:

If I had to write up my progress, I'd spend a week writing what I'd done the previous week. Pointless. Using video is a massive improvement, and doing video blogs also means you're more likely to do it on a weekly basis, as your program will change. With text version, might be tempted to do it all at the end (Neil).

On VEF: Neil indicated that he much preferred creating a video to illustrate the problems he was encountering than to use text, citing SPEED and UTILITY as the main factors. He also suggested that video-feedback had aided his comprehension of the subject matter substantially, exemplifying INTELLIGIBILITY, as video-feedback pinpointed exactly where the error was while explaining the problem, highlighting the SPECIFICITY category identified in the analysis of questionnaire responses in the previous chapter:

I'd sooner send a video saying I've done this wrong than type it all down and then have to sit there and correct it all and have to read it through about five times to make sure it's right . . . [and] I liked getting feedback in that format, as wouldn't have understood if it was a PDF, with circles around the errors I still wouldn't get it. Because you showed exactly where the error was, while explaining it, I knew where to go (Neil).

Moreover, the final sentence in the quote above also underlines the significance for Neil of a link between the visual, on-screen demonstration and the explanatory voiceover, echoing the *Modality Principle* of the *Cognitive Theory of Multimedia Learning* (Mayer, 2001). In summing up his experience of VEF, Neil also indicated the value he placed on the dialogic aspect of the VFL, commenting on how this impacted on his engagement with the module: *Video feedback feels more like a conversation. Makes you engage with entire course more* (Neil).

Outcomes

Neil's final mark for the module was 50%, having achieved identical marks in both the design and programming elements. Despite theory-informed (e.g. Woodfine, Nunes, & Wright, 2008; Asbjørnsen, Obrzut, Eikeland, & Manger, 2010) predictions that Neil's performance in this module should have exceeded his attainment levels in modules where no video-enhanced interventions were introduced, he achieved a score below his personal mean for the year (53.3%). Analysis of purely quantitative data might have suggested that the video-based approach had therefore been unsuccessful in this case, however in the semi-structured interview conducted following the release of the results, Neil was adamant that: *there's no way I would have passed this module without video. I would have failed programming for sure, and it would have taken weeks to explain my work if I'd had to do it as text* (Neil).

Corroboration by DSC

The impact of documenting work using VDEs was confirmed by the DSC as having had a significant impact on Neil's ability to complete the module. While his mark was slightly below that of his personal mean score for the year, the DSC indicated in a semi-structured interview that this "almost certainly, was significantly higher than it would have been" in the absence of the intervention.

Reflections

Neil's case serves not only to illustrate how traditional assessment models which require learners to produce written reports can seriously disadvantage those students whose dyslexia is severe, but also to highlight how video can be used to scaffold the development of fundamental programming skills which require the acquisition of techniques based in the lexical paradigm. With regard to the programming exercises in particular, in-class observations and dialogic interviews highlighted that Neil's dyslexia meant he required significant support from the DST to accomplish even simple tasks in the development environment, and that the DST's skill-set limited the help which could be provided with programming constructs. Having identified that visual problems normally requiring the use of blue film over the screen were markedly diminished when viewing the ITVs, the use of video-feedback allowed Neil to see exactly where errors could be found, and the opportunity to engage in VEA enabled his participation in an asynchronous conversation using video. When viewed against the background of the cases of dyslexic participants in DBRC2, Neil's case adds weight to the suggestion that video can help to minimise the disadvantages in assessment which disabilities legislation is intended to mitigate.

2. Illustrative Case 2: Graeme

Graeme, an 18-year-old male, was a design-stream student from a BTEC background. Graeme did not display any indications of being affected by any specific learning difficulties, and gave no declaration of such on entry.

Engagement with ITVs, VEF and VEA

In-class observations and dialogic interviews highlighted Graeme's steady engagement with the ITVs, rarely getting ahead of or slipping behind the weekly schedule throughout Semester 1. Despite the appearance of being an outgoing and socially confident individual, Graeme was hesitant to engage with VEA in the studio due to peer-group inhibitions, and while he began to do so on a fortnightly basis from the fourth week of Semester 1, early entries did not include a voiceover. All Graeme's VDEs were produced outside the studio environment, often late at night prior to the studio class; initially, these VDEs were very short (c.30-45s), and the content usually restricted to demonstrations of having used the ITVs.

Over the course of the academic year, Graeme produced 15 videos with a combined duration of 00h:31m:27s. Graeme's engagement with video-feedback was limited during Semester 1, with little or no attempt made to request assistance, or to engage in conversation around feedback, either face-to-face or through the medium of video. Having expressed reservations regarding his capacity to succeed in the programming component of the module, a trigger event (cf. Garrison, Anderson & Archer, 2001) early in Semester 2 saw Graeme initiate a VFL around a seemingly intractable problem encountered in the development environment. While it transpired that the problem was relatively minor and the solution trivial, the provision of video-feedback pinpointing the root of the problem was enthusiastically welcomed by Graeme, leading him to enter into ongoing dialogue on his programming work within a VFL which persisted for six weeks, in which he produced weekly VDEs each of 2-3 minutes' duration. During this time, Graeme's progression through the ITVs became accelerated. He gained the confidence to experiment and apply the newly-learned techniques in different areas, quickly overcoming threshold concepts and transgressing those areas identified as posing troublesome knowledge (Mayer & Land, 2005) noticeably faster than his design-stream counterparts.

While Graeme's Semester 1 design-based work was of a standard that warranted an average grade, his work for the Semester 2 programming exercise was of a standard that exceeded expectations. Deductive thematic analysis of the dialogic exchanges within the VFL for instances of autonomy revealed that Graeme was the first of the design-stream participants to show evidence of autonomy within the programming component, and notably, Graeme was the only design-stream student to receive a higher mark for his programming work than for his design-based work.

Learner Evaluation

On ITVs: In the semi-structured interview, Graeme spoke positively regarding the experience of using video-based learning materials, with categories exemplified including EASE OF USE, UTILITY, and INTELLIGIBILITY:

They are very easy to learn from and understandable in what they do. I find it the easiest way of learning, video tutorials. Considering I didn't know anything beforehand, these videos are brilliant . . . I thought the [design] videos were good at the introductory level, but the programming tutorial videos really stand out as the most useful for me as I expected to find coding the hardest part (Graeme).

On VEA: Graeme revealed correlations with categories described above, including SIMPLICITY:

A lot better than doing a long document - I do it closer to when I've done it, so it doesn't change and I remember what I've done. Reminds me later of what I've done, rather than getting a backlog of stuff. A lot more reliable to do videos (Graeme).

On VEF: Given the background of his experience described above, Graeme's thoughts on video-feedback were contextualised against and framed as contingent upon the production of VDEs as *Much better than written feedback. Could point out what the issues were - far easier than writing it down. That were good - easier for you to explain in video than in emails* (Graeme).

Outcomes

Graeme's final mark for the module was 63%, which as highlighted above was comprised of a higher mark for the programming and documentation components than the design element. This mark was marginally above Graeme's mean score for the year (59%).

Reflections

While inductive thematic analysis of Graeme's evaluation of the intervention highlighted his focus on the ease and utility he associated with using video, his case illustrates how the opportunity to combine visual specificity with high frequency dialogue around formative assessment and feedback can enable the establishment of teaching presence, and divert a student away from potential disengagement. Graeme had chosen to take the design-stream route in part due to his concerns around programming, however his engagement through video enabled him to discover not only that his fears around programming were unfounded, but also that he was capable of developing his skills in that area to a level which exceeded those displayed in his original chosen field.

3. Illustrative Case 3: David

David, a 19-year-old male, was a programming-stream student from a BTEC background. David gave no declaration of any learning difficulties on entry, and did not exhibit any indications of being affected by either dyslexia or Asperger's Syndrome (AS). While already competent at scripting websites using HTML and JavaScript, David had no previous experience in programming compiled languages.

Engagement with ITVs, VEF and VEA

In-class observations and VDEs showed that David engaged with the ITVs from Week 1, and in dialogic interviews David spoke positively regarding the experience of using video-based learning materials. David began producing video diaries from the second week of the first semester, although these were initially screencasts-only and did not include a voiceover. Once the availability of microphones had been emphasised, David began to record voiceover-enabled screencasts, but was observed to take two weeks to overcome inhibitions in vocal delivery later confirmed as associated with immersion within the new peer group.

In addition to using screencasts to create video-diary entries, David also used the technology to capture evidence of a problem in the configuration of the development environment that he identified as causing shutdown errors, making this available first within *Mahara®* and then to the relevant technical support team by email. David's VDEs demonstrated rapid progress through the ITVs, but he was also observed repeat-viewing materials both within and across weeks. While his VDEs were initially descriptive, qualitative analysis using the categories emerging through the Key Pedagogic Aims identified in DBRC2 revealed that David was the first of the DBRC3 participants to show evidence of reflexivity and autonomy.

The standard of work produced by David in Semester 1 was very high, and conversational video-feedback within the VFL was primarily of a constructive rather than corrective nature, with the emphasis on highlighting those areas where the work could be perfected, rather than needing to assist David in transcending threshold concepts (Mayer & Land, 2005) or overcoming a grade boundary. In Semester 2, David excelled in the programming exercise, and his approach to the VDEs became one of a truly reflexive learner who consolidated his learning through deep explanations linking conceptual and practical aspects of the work in exemplar materials that could be used to teach other students.

Over the course of the academic year, David produced 22 videos with a combined duration of 01h:09m:03s. Notably, David had encountered a problem late in the academic year that he had not solved at the time of producing his weekly VDE. In the process of explaining the problem, he identified a possible solution and implemented it successfully before making a reflexive commentary on his own cognitive and meta-cognitive processing in relation to the problem. It is fortunate for this study that David elected to continue recording this process and to share it within *Mahara®*. It is highly unlikely that this event would or could have been captured in the process of producing a written, text-based report to document his progress.

This particular incident, within the case, serves to highlight the importance of establishing an informal, conversational nature to asynchronous communication between learner and tutor using video, exemplifying both Mayer's (2001) *Personalisation Effect* at work on the student's side within a conversational framework (Laurillard, 2002). The processes of software development and documentation had previously required students to either complete or break off from the development task in order to engage in the reporting task. Inserting distance between the two tasks enabled the production of a VDE to allow students to remain focused on aspects of the development task as they report on them (Cheng & Chau, 2009; O'Donoghue & Cochrane, 2010; Jordan, 2012). In this study, the proximity of the development and reporting tasks was facilitated by the coincidence and convergence of the development environment and the reporting medium.

Learner Evaluation

On ITVs: David described the design-focused ITVs as *very effective*, and *really good at an introductory level*, but indicated that he *would prefer a book for programming*, and, citing a preference for working directly with sample code, suggested that the lexical nature of the development environment made this less appropriate for learning a programming language. On VEA: David gave only positive expression to his experience of using video diaries to document development work throughout the year, citing *Mahara®* as a central focus for hosting and collating the various video materials, and highlighting aspects of the experience which revealed correlations with categories described above, including SIMPLICITY, EASE OF USE:

It allowed me to see my progress through the module in semester 1 from a skills point of view and a [software] development point of view. Mahara® has also been useful for storing different versions of my work, which makes editing them at home and at uni a lot easier . . .

. . . overall, it's much simpler than print screening a program and pasting it into a word document. Very easy to see what you did each week thanks to the Mahara® views and a great method of personal review and a way to get peer feedback (David).

On VEF: Despite much of the video-feedback received on his work being constructive rather than corrective, as indicated above, David made specific reference to overcoming problems in his evaluation of VEF, and in common with Neil, made a link between the visual demonstration and voiceover, further underlining the relevance of Mayer's *Modality Principle* for video-feedback: *I could see what the problems were with my work and fix them straight away . . . it's just so much simpler to view a video and hear someone speak than it is to read* (David).

Outcomes

David's final mark for the module was 90%, comprised of an even balance between the marks for the design, programming and documentation elements. This mark stood out over David's personal mean for the year (66%), and initially gave rise to concerns that David may have focused overly on the module to the detriment of other modules. In the semi-structured interview conducted following the release of the results, David acknowledged that he had put more effort into this module than other modules, however he insisted that he expended no more or less effort on those modules than he otherwise would have done, regardless of the primary medium employed in the delivery of this module, and that due to his inherent laziness he simply found the teaching methods employed in the other modules to have been "less inspiring".

Reflections

David was a highly independent and autonomous learner. Analysis of the evidence found in David's development work and his documented activities over the year yielded results which suggest exceptionally high levels of problem-solving, reflexivity, and meta-cognitive awareness across all aspects of his performance on the module. Notably, David was the only student to make any substantial mention of *Mahara®* as having played a role in relation to the storage and presentation of the VDEs and the VEF artefacts received, suggesting a broader meta-cognitive awareness than was evidenced by most of his peers. That he was not motivated to engage more substantially with other modules, yet prepared to engage so deeply with the module and to acknowledge the video-enhanced system as the trigger for him having done so is also worthy of note.

Against this background, David's case adds further weight to those indications arising from that of Graeme that video-enhancement of teaching, assessment and feedback methods can offer a vehicle through which to engage students, by offering an opportunity to maximise academic potential through dialogue around formative feedback, by establishing and enhancing teaching presence within the online learner-tutor relationship, and by encouraging reflexivity, and autonomy.

4. Illustrative Case 4: William

The final case in the series explores the experience of William, a programming-stream student from an A-level mathematics, sciences and ICT background, who arrived at university with a prior clinical diagnosis of an autistic spectrum condition. In common with both Neil and the other dyslexic participants interviewed in DBRC2, William's SpLD had impacted negatively on his academic performance in secondary education, and his AS had necessitated multiple changes of school.

Although described by the DSC as 'mid-range', William's AS initially left him unable to make direct eye contact; while this improved over the year, he remained unable to maintain eye contact for more than a few seconds, and he was observed to display anxiety when confronted by sudden, unexpected noises. William was supported by an institutional Personal Learning Support Plan which entitled him to two weeks' extension on assignment deadlines 'on demand', and which excused him of any requirement to give presentations to the class. While extensions to assignment deadlines were never requested, the nature of the VEA activities meant that students presented their work through video rather than face-to-face, and it was evident from questions he asked during the first session of the academic year that William was eager to engage through this medium.

Engagement with ITVs, VEF and VEA

From the outset, William engaged extensively with all video-enhanced activities, devouring the ITVs with an exceptionally strong appetite. Initially, dialogic interviews were conducted face-to-screen in order to minimise any anxiety associated with face-to-face contact, and it was evident from these exchanges that the opportunity to engage with learning materials that did not rely on face-to-face communication was strongly welcomed. William recorded VDEs evidencing his work from the outset, with a trademark pattern to the introductions: *Welcome to week* [X] *of my* [core module] *learning experience. This week I have . . ..*

Other students were initially hesitant to engage in learner-tutor loops, primarily due to social inhibitions as in David's case, but ultimately did so in response to some trigger event (Garrison, Anderson & Archer, 2001) such as an unresolved problem or threshold (Mayer & Land, 2005), and frequently in the manner described in Neil's case. In stark contrast, William's inability to recognise social barriers left him unfettered by the norms and expectations of the new peer group, and he launched into the production of between one and three video diaries per week of up to seven minutes, duration in full anticipation of an asynchronous video-feedback response, fluently referring back and forth within and between his VDEs in a manner of which he was incapable in face-to-face settings. Both Attwood (2000) and Baron-Cohen (2008) have previously reported on a disconnect between the sophistication of written and spoken expression in persons diagnosed with AS, and William's case highlighted a direct parallel with these observations in the differences between his *in-video* and *in-vivo* expression.

In common with Alex, the student with AS described in the previous chapter who experienced positive results in a group-work exercise through using computer-mediated communications, evidence of William's growing confidence in his ability to communicate through the medium emerged over the course of the academic year. In a further parallel with Alex's case, after spending time communicating asynchronously using video, William not only sought regular feedback on his work, but also appeared significantly more comfortable in both dialogic interviews and face-to-face interactions with peers. In a further parallel with Neil's case, the nature of William's VDEs collapsed any distinction between problem-highlighting within the development work and documentation of the development process, and the VFL again facilitated continuous asynchronous dialogue around module activities. Over the 24-week academic year, William produced 37 videos with a combined duration of 03h:32m:12s. Early in the first semester, his VDEs were of only 3-5 minutes, duration, however he later posted up to three 7-10 minute VDEs per week, sustaining an ongoing online learner-tutor loop throughout this time in addition to the weekly face-to-face dialogic interactions.

William also found it necessary to report an unsolved problem at the time of producing one of his weekly VDEs, and in a parallel with David's case, identified and implemented a possible solution in the process of explaining the problem. Having previously stated in a dialogic interview that *"if you find a logical error you should be able to fix it yourself"*, it was notable that William was prepared to share this apparent shortcoming in his approach to a programming problem, and to move immediately into self-analysis to explain how a cognitive shift in his conceptual understanding had enabled him to solve the problem. As suggested, in connection with David's case, it is unlikely that this event would have been reported on in the process of producing a written, text-based report to document his progress.

Perhaps the most significant observation emerging from William's case was a pattern which emerged at the peak of the frequency and duration of his weekly VDEs. Having developed an appreciation of the programming language which exceeded that expected of students using only the ITVs, William introduced a new technique into his development work that involved the application of advanced calculus. In the process of demonstrating how this worked, the manner of his explanation began to resemble that of a tutor rather than a student reporting on his progress, and this consolidation of his own learning by teaching others evidenced in the video diaries underlined the high level of autonomous learning that William achieved.

Learner Evaluation

On ITVs: In the semi-structured interview, William responded quickly to an initial question regarding his use of the ITVs before moving the interview on towards VEA:

Because the tutorials were very easy to understand and comprehend, I can get through them very quickly, so especially there were some weeks when it was very easy to learn new stuff and demonstrate. It's definitely been very useful (William).

Later in the interview however, he returned to the ITVs, acknowledging that the tutorials had been a particularly useful introduction to the subject area, while demonstrating autonomous learning by indicating that he had accessed more advanced written materials in order to extend his knowledge and understanding.

On VEA: Giving prior contextualisation to his response to the first question on VEA, William indicated that one of his A-level reports had run to around 100 pages, adding that in this module it was:

. . . a relief not to have to write so many documents . . . having to write so many pages [at A level] was rather overwhelming . . . so moving on to here, a lot of this has been alleviated by making videos instead of say writing in text how I analysed what was the best possible solution (William).

William shared details of the different technical approaches he had used to produce VDEs, suggesting that, contrary to advice given in the studio, he had started with a scripted approach, but had found it more effective to record both audio and visual components synchronously *before getting into making voiceovers for the videos, I would just import the video before the voiceover ... but then it just made more sense to overlay audio on top of that and talk about it* (William).

He went on to suggest that using VDEs had forced him to be more concise in what he had to say about his work, while later responses highlighted examples of categories used in the qualitative data analysis, including SPEED, SIMPLICITY, INTELLIGIBILITY:

It really has made things easier from past projects. It is definitely easier to document that way, than it is to write many pages of text . . . the videos give more insight into why I chose to implement it the way I have ... yes, audio and video have made things much easier (William).

On VEF: William also found video-feedback to be preferable to text or text-plus-graphics, and citing the co-incidence of spoken audio and visual demonstration as a key in requesting help from a tutor, highlighted examples of the MIXED MEDIA, and SPECIFICITY categories:

Video really does help as you can explain and show at the same time. In my opinion, I felt I was able to both show the problem and explain why I think it's happening or possible solutions that I want another opinion on (William).

Personalisation was another key aspect of VEF for William:

[VEF] is very appropriate. Most notably, having a sort of personalised feedback, is, for me, better than having a more general feedback. If the class-size was large enough, I think a generalised feedback saying, oh, 'so and so many people have done this wrong' . . . could still be useful but I would prefer it personalised (William).

Finally, William appeared to associate REFLEXIVITY with VEF, exemplifying a further category from the earlier qualitative analysis:

It seems the delay between the giving video and the receiving feedback, I guess . . .it gives the student the chance to try and sort out the problem themselves or otherwise try and better explain the problem . . . so, I think the delay is to the benefit of the system (William).

Outcomes

William's final mark for the module was 81%, although this was comprised of lower marks for the design element than the programming element, with the highest marks awarded for the documentation of the process. While William achieved a summative mark for the module that was markedly above his personal mean for the year (73%), this was not atypical as a profile for an overall A grade student, narrowly placing him as the student with the highest personal mean across the group, with another programming-stream student following closely on 72%.

Corroboration by DSC

The same DSC had been in post during the stays of both Alex and William at the institution, and although initial scepticism had been slow to break down in Alex's case, this was more readily suspended in order to explore the evidence in William's case. Before examining any evidence, the DSC explored the basis for those preconceptions that preceded it, offering as initial hypotheses the suggestions that any apparent engagement might be (1) superficial, and/or (2) designed to disguise an inability to make any meaningful social connection.

The DSC recognised the duration and frequency of the VDEs produced as consistent with patterns of behavior that might be expected of a student with AS, while acknowledging the parallels between the sophistication of expression demonstrated within the VDEs and that seen in written work. Of surprise to the DSC was the apparent ease with which William entered into learner-tutor interactions, and both the reflexivity shown in the process of correcting errors while recording, and the consolidated learning in which William began to switch emphasis from that of learner to tutor. While aware of the original research in which primary school children with AS were characterised as "little professors" (Asperger, 1944), the DSC suggested that the switch in role was nonetheless significant for William, confirming perceptions of a change in self-confidence around peers since his arrival in Semester 1.

Reflections

William's case serves not only to illustrate how traditional assessment models that require learners to deliver face-to-face presentations can seriously disadvantage students with an autistic spectrum condition, but also to highlight how video can be used to scaffold the development of skills which require face-to-face engagement in a process of evidencing the acquisition of skills and techniques in software development. The use of the VFL enabled William to participate in a high frequency, formative, feed-forward themed asynchronous conversation around assessment and feedback using video.

When viewed against the background of the illustrative cases highlighting the experience of dyslexic participants in this study, William's case contributes further evidence that video can help students with certain specific learning difficulties to overcome potential disadvantages presented by traditional assessment methods.

Breakdown of Illustrative Case Participants' Use of VDEs

Enabling comparison of the illustrative case participants, Table 1 offers a breakdown of the number and average duration of VDEs created by these participants, and a brief comment on each.

Formative video-feedback which focused on specific aspects of the work documented was created in response to each VDE. These MP4 files were uploaded to students' feedback area in Mahara within 24 hours, triggering an email notification of its availability to the student.

Table 1. Breakdown of VDE production in cases of four illustrative case participants across the 24-week teaching schedule

Teaching Schedule Week		1-6	7-12	13-18	19-24
Neil	*No. of VDEs*	8	9	9	8
	Mean Duration	02m:11s	03m:08s	03m:32s	03m:01s
	Comment	Neil engaged strongly from the outset, initiating VFLs early in each semester. Highlighting dyslexia-related issues around coding, Neil showed clear evidence of progress in response to video-feedback.			
Graeme	*No. of VDEs*	3	3	5	4
	Mean Duration	00m:41s	01m:52s	02m:48s	02m:27s
	Comment	Graeme's early reports on using ITVs were short and succinct, and there was no voiceover in the first two VDEs. His initiation of a VFL on a coding issue led to evidence of deeper engagement and reflexivity.			
David	*No. of VDEs*	3	6	8	5
	Mean Duration	01m:03s	02m:36s	03m:33s	04m:23s
	Comment	Initially hesitant to use voiceover, David's VDEs became lengthier and increasingly fluent as the year progressed. David displayed meta-cognitive awareness when he overcame a problem while recording.			
William	*No. of VDEs*	6	9	13	9
	Mean Duration	03m:02s	05m:37s	07m:02s	05m:48s
	Comment	Engaging strongly from the outset, William created the most VDEs. His initiation of VFLs led to ongoing conversations each semester, and displayed meta-cognitive awareness around a programming problem.			

SUMMARY OF POINTS OF RELEVANCE FOR PRACTITIONERS IN HIGHER EDUCATION

On Video-Enhanced Learning

This research found that while the introduction of instructional tutorial videos helped to address the requirement for more time to spend on formative feedback, two participants initially felt short-changed by the intervention, expressing a preference for face-to-face lectures, although these same students later indicated that they had learned a lot from having engaged with the ITVs.

On Video-Enhanced Assessment

As seen in the previous chapter, this research highlighted inhibition around initial engagement with the production of video-diary entries, with students reluctant to accompany a screencast demonstration of work completed or a problem encountered with a voiceover. This appeared to be associated with immersion in a new peer group, with students shedding these inhibitions within two to three weeks.

On Video-Enhanced Feedback

As highlighted by the cases of Robert (Section 2-1), Neil (Section 3-1) and William (Section 3-4), this research has illustrated how video can be used to provide feedback that is both highly specific and useable, that is personalised to the individual student, and that, as part of an integrated, formative and conversational model, can be provided in a manner that is both timely and which enables students to make improvements to their work. Feedback emerged as the use of video most consistently valued by students in both cohorts of participants.

On Supporting the Deployment of a System for Video-Enhanced Learning, Assessment and Feedback

This research suggests that the use of an e-portfolio system enhances the value of a system for video-enhanced learning, assessment and feedback. As highlighted by the cases of Robert (Section 2-1) and David (Section 3-3), *Mahara®* offered a focal point for both students and tutor. Providing a single point of contact for students to access instructional tutorial videos, to store and share video-diary entries, and to house video-enhanced feedback artefacts, the proximity of these artefacts appeared to encourage repeat viewing of video tutorials and video-feedback.

On the Video-Feedback Loop

The cases of Neil (Section 3-1) and Graeme (Section 3-2) illustrate how the video-feedback loop proved an effective tool to catch errors early in the development of students' work, while the opportunity to highlight errors with pinpoint accuracy promoted specificity in feedback. This ensured that students did not expend effort mistakenly by following an erroneous development path, and the timeliness of the feedback also meant that students arrived at the subsequent studio session ready to move on to the next phase of their work, rather than needing to raise issues hanging over from the previous week.

The research also highlighted the development of a repository of video-feedback artefacts and video-diary entries as a by-product of the video-feedback loop. Where students agree to share these with peers and future cohorts, a FAQ database of common errors and examples of good practice covering problems encountered, solutions attempted and problems overcome can be generated through the production of these items as learner-generated content.

As exemplified by Graeme's case (Section 3-2), the research indicated that students showed signs of stronger engagement with the assessment and feedback process when afforded the opportunity to use video-diary entries. Students also reported having adopted the process of producing regular video-diary entries in other modules, indicating that the technique had provided a valuable trail of evidence from which to produce traditional written documentation supported by screenshots. For students with a SpLD, this effect was amplified, as exemplified by Robert (Section 2-1), the dyslexic participant in DBRC2, who elected to use the technique throughout the remainder of his undergraduate studies, negotiating the necessary variation to documentation requirements for individual modules directly with the relevant tutors.

As highlighted by Neil's case (Section 3-1), the research suggests that indicators of the establishment of cognitive presence emerging in response to the introduction of ITVs can be identified in video-diary entries, while teaching presence can be established through a tutor's video-feedback responses to those video-diaries within dialogic exchanges facilitated by the video-feedback loop.

The research found that the integrated system and its constituent components were effective within a specific, studio-based and product development-focused module in the computing discipline. Further research is needed to assess the transferability of both the individual techniques and the integrated approach to other subject areas and to other disciplines. Moreover, this investigation was conducted with two groups of participants in relatively small class sizes, and further research is also necessary to assess how the approach can be scaled to larger classes.

On Inclusivity

The research suggested that students affected by SpLDs benefitted greatly from the introduction of asynchronous video. As illustrated in the cases of Robert (Section 2-1), Keith (Section 2-2) and Neil (Section 3-1), students affected by dyslexia were enabled to engage with learning materials, to understand and act upon feedback on their work, and to produce audio-visual accounts of the development of their work, all without the struggle to interpret textual sources. Further, as exemplified by William's case (Section 3-4), a student with ASwas able to produce detailed presentations of work undertaken, and to engage in high frequency conversation around formative feedback on software development work without being disadvantaged by any requirement to do so in a face-to-face setting.

This research has demonstrated how asynchronous video can offer a viable medium through which to overcome disadvantages inherent in traditional forms of assessment, providing examples of how the requirements of disabilities legislation (e.g. Her Majesty's Stationery Office, HMSO; 2001; 2010) can be fulfilled. Further, additional benefits were experienced by students affected by dyslexia, with indications that video had not only made a significant contribution to improvements in the academic performance of all dyslexic students in the study, but also that the impact for these participants was proportionally larger than for the neurotypical group, suggesting that asynchronous video offers a medium through which to achieve greater equality of opportunity in the area of educational attainment for students with SpLDs. For students with AS, the opportunity to benefit from computer-mediated dialogic exchange without the trauma associated with face-to-face interaction was a key finding in the pre-cursor study outlined in the previous chapter, and this research indicates that asynchronous video adds a further dimension to the use of online text to achieve this benefit.

On Reflexivity and Autonomy

As illustrated in Graeme's case (Section 3-2), this research found that the use of asynchronous video in the assessment and feedback process was consistent with increased frequency of demonstrations of reflexivity in students' reports on their work. In particular, the pause for reflection inserted into a dialogic exchange by asynchronicity created a space in which students, unable to rely on the immediacy of synchronous communication, would ponder how best to describe the problem they wished to highlight, and, as seen in the cases of David (Section 3-3) and William (Section 3-4), in the process of demonstrating and describing those problems, would often solve it. While this pattern was predominantly displayed by students whose learning was of a higher order than that of their peers, the link between encouraging reflexivity and the development of autonomy as a learner appeared to be amplified by the use of asynchronous video as a tool to produce a visual demonstration of a problem, and was associated with meta-cognitive self-analysis on the part of the student.

On Dialogue

This research confirms the value of dialogue in teaching (cf. Dewey, 1944; Laurillard, 2002). In particular, the use of high frequency, dialogic formative feedback, adapted from Oxbridge practices (Gibbs & Simpson; 2004) was found to work well, and was consistent with the model of assessment for learning (Torrance, 2007). As mentioned elsewhere in this thesis, these benefits are clearly not restricted to circumstances such as those described in this investigation where asynchronous video has featured strongly, however the combination of dialogue and video appears to have been synergistic, each amplifying the benefits of the other, as illustrated in Neil's case (Section 3-1). Moreover, the use of dialogue as an interviewing and data collection tool was particularly appropriate against this background, and perhaps owes its success in part to having situated formative feedback within a conversational model.

CONCLUSION

While this investigation has shown that, even when employed in isolation, video offers potential benefits in the areas of learning, assessment and feedback, the research has found that the context within which video is employed and relationship between its modes of use are also of significance if that potential is to be maximised. Notably, the benefits of using asynchronous video in the assessment and feedback process can be amplified in cases where a model of assessment for learning is adopted (Torrance, 2007), where learning, assessment and feedback activities take place in close proximity, and where a video-enabled feedback loop facilitates frequent dialogic exchanges between learner and tutor. In promoting this frequent, high-quality, learner-tutor interaction around formative feedback, the research has also shown that a pedagogic strategy which integrates video-enhanced learning, assessment and feedback can strengthen both cognitive presence and teacher presence within a blended learning community, offering support for key theorisations underpinning the Community of Inquiry Framework.

For students affected by a specific learning difficulty, and for those affected by dyslexia in particular this investigation has consistently demonstrated that using video-enhanced approaches to learning, assessment and feedback affords stronger engagement with, improved outcomes from, and ultimately enhanced self-confidence around participation in higher education. This study has also shown how video-enhanced assessment and feedback problems can help students diagnosed with an autistic spectrum condition can overcome social communication issues that impact on both the ability to engage with feedback and the capacity to engage in dialogue around that feedback. To this end, video-enhanced learning, assessment and feedback offers a clearly defined strategy for institutions seeking to respond to legislative requirements around equality for students affected by unseen learning difficulties, thereby enabling greater participation for students otherwise disadvantaged by traditional forms of assessment and feedback. In summary, this study has highlighted video-enhancement of assessment and feedback practices can help to improve the student experience of the assessment and feedback process, while helping to realise pedagogic ideals of reflexivity, learner autonomy and meta-cognitive self-awareness in the learning process, and in so doing has taken important steps *towards an integrated model of video-enhanced learning, assessment and feedback.*

REFERENCES

Asbjørnsen, A. E., Obrzut, J. E., Eikeland, O.-J., & Manger, T. (2010). Can solving of Wordchains be explained by phonological skills alone? *Dyslexia*, *16*(1), 24–35. doi:10.1002/dys.394 PMID:19562661

Asperger, H. (1991). Autistic psychopathy' in childhood (U. Frith, Trans.). In U. Frith (Ed.), Autism and Asperger syndrome (pp. 37–92). Cambridge, UK: Cambridge University Press. (Original publication 1944)

Attwood, T. (2000). *Asperger's Syndrome: A guide for parents and professionals.* London: Jessica Kingsley.

Baron-Cohen, S. (2008). *Autism and Asperger's Syndrome.* Oxford, UK: Oxford University Press.

Cheng, G., & Chau, J. (2009). Digital video for fostering self-reflection in an ePortfolio environment. *Learning, Media and Technology*, *34*(4), 337–350. doi:10.1080/17439880903338614

Dewey, J. (1944). *Democracy and education.* New York: Free Press.

Garrison, D. R., Anderson, T., & Archer, W. (2001). Critical thinking, cognitive presence, and computer conferencing in distance education. *American Journal of Distance Education*, *15*(1), 7–23. doi:10.1080/08923640109527071

Gibbs, G., & Simpson, C. (2004). Conditions under which assessment supports students' learning. *Learning and Teaching in Higher Education*, *1*(1), 3–31.

Glaser, B. G., & Strauss, A. L. (1967). *The discovery of grounded theory: Strategies for qualitative research.* New York: Aldine de Gruyter.

Her Majesty's Stationery Office (HMSO). (*2001*). *The Special Educational Needs and Disabilities Act.* Retrieved from http://www.legislation.gov.uk/ukpga/2001/10/pdfs/ukpga_20010010_en.pdf

Her Majesty's Stationery Office (HMSO). (2010). *The Equality Act.* Retrieved from http://www.legislation.gov.uk/ukpga/2010/15/pdfs/ukpga_20100015_en.pdf

Jordan, L. (2012). Video for peer feedback and reflection: Embedding mainstream engagement into learning and teaching practice. *Research in Learning Technology*, *20*(sup1), 16–25. doi:10.3402/rlt.v20i0.19192

Laurillard, D. (2002). Rethinking university teaching: A conversational framework for the effective use of learning technologies (2nd ed.). London: Routledge Falmer.

Mayer, J. H., & Land, R. (2005). Threshold concepts and troublesome knowledge (2): Epistemological considerations and a conceptual framework for teaching and learning. *Higher Education*, *49*(3), 373–388. doi:10.100710734-004-6779-5

Mayer, R. E. (2001). *Multimedia learning*. Cambridge University Press. doi:10.1017/CBO9781139164603

McDowell, J. (2015). A black swan in a sea of white noise: Using technology-enhanced learning to afford educational inclusivity for learners with Asperger's Syndrome. *Social Inclusion*, *3*(6), 7–15. doi:10.17645i.v3i6.428

O'Donoghue, M., & Cochrane, T. (2010). The role of live video capture production in the development of student communication skills. *Learning, Media and Technology*, *35*(3), 323–335. doi:10.1080/17439884.2010.509350

Torrance, H. (2007). Assessment as learning? How the use of explicit learning objectives, assessment criteria and feedback in post-secondary education and training can come to dominate learning. *Assessment in Education: Principles, Policy & Practice*, *14*(3), 281–294. doi:10.1080/09695940701591867

Woodfine, B. P., Nunes, B. P., & Wright, D. J. (2008). Texted-based synchronous e-learning and dyslexia: Not necessarily the perfect match! *Computers & Education*, *50*(3), 703–717. doi:10.1016/j.compedu.2006.08.010

Chapter 3
Utilizing Instantaneous Feedback to Promote Self-Regulated Learning in Online Higher Education Courses:
The Case for Digital Badges

Melissa Fanshawe
University of Southern Queensland, Australia

Nicole Delaney
University of Southern Queensland, Australia

Alwyn Powell
University of Southern Queensland, Australia

ABSTRACT

In higher education learning environments, there is a dual need for educators to use supportive strategies to motivate students throughout the course, while also aiming to increase the capacity of students to self-regulate their learning. Using instantaneous tools to deliver formative or summative feedback through digital technology has been shown to lead to higher achievement and retention rates. This chapter shows how digital badges can provide instantaneous feedback to support students to feel belonging in the online community and develop self-regulation skills. Instantaneous feedback tools can be used to provide teacher presence throughout higher education courses to increase student engagement, retention, and achievement.

DOI: 10.4018/978-1-7998-0426-0.ch003

INTRODUCTION

Preparing students for their professions and life-long learning should be important goals in higher education courses (Çakir, Korkmaz, Bacanak, & Arslan, 2016). A student's ability to self-regulate learning is important for academic achievement and a necessary skill for 21st Century learners (Zhu & Mok, 2018). Students' capabilities, understandings and attitudes will enable students to live life and seek employment in a global market (MCEETYA, 2008).

Whilst educators often focus on the course content, delivery and assessment of their courses, Garrison, Anderson, & Archer (2000) propose the online learning environment should be a collaborative process, which is reliant on social presence, teaching presence and cognitive presence to engage learners and develop a sense of belonging (Thomas, 2012; Thompson, Vogler, & Xiu, 2017). This online Community of Inquiry (Garrison, Anderson, & Archer, 2000) can be created through targeted use of digital technologies, to increase self-regulation and afford student learning (Gibson, 1977; Thompson, Vogler & Xiu, 2017).

Digital badges, online representations of an achievement or award, are one pedagogical tool that can assist educators to create a Community of Inquiry. The instantaneous feedback of the digital badge can assist students to track course progression and engage with the course content (Dowling-Hetherington & Glowatz, 2017). Utilising digital badges provides timely acknowledgement of task completion and alongside other forms of feedback, contributes to teaching presence within a course.

BACKGROUND

Online learning has rapidly increased with over 7 million students studying higher education online (Allen & Seaman, 2014). Online delivery has provided access for students who may not have been able to study on campus, which has resulted in a change of student population. Students may be studying part-time, juggling work or family commitments or live in rural or remote areas. Universities have needed to make adjustments from a pedagogy of face-to-face instruction to online delivery to adjust to changing demands and competition in the higher education sector (Starkey, 2010). While initial changes focused on topic content and assessment (Hill & Barber, 2014) to cater for student demographics, delivery format and capabilities of technology in online learning are still being explored (Carey & Stefaniak, 2018; Chiasson, Terras & Smart, 2015).

The widespread global extent of digital technologies, high speed internet and accessible cost of devices means access to information is readily available for more people (Dowling-Hetherington & Glowatz, 2017). Many students entering higher education have engaged with various digital and multimedia platforms, such as smartwatches, smartphones, tablets, laptops, gaming platforms and social media. The devices are usually interactive and users are rewarded for engagement with 'likes', rewards and levels of accomplishment where instantaneous feedback is common upon completion of tasks (Hamari, 2017). Courses using digital badges in higher education (Abramovich, 2016) have shown students are proud of the credentialing of badges and levels and feel satisfaction upon completion and can share these with others in the online community or on social media (Jones, Hope & Adam, 2018).

Gamification within these technologies uses prior behaviors to develop knowledge of the intrinsic and extrinsic motivators of its users (Dowling-Hetherington & Glowatz, 2017) to keep them motivated and retained in the activity and guide behavior (Gamrat, Zimmerman, Dudek, & Peck, 2014). These technologies are also able to tell the levels of their users and provide pathways at their levels, in what McDaniel and Fanfarelli (2016) refers as 'completion logic'; using data from the completed activity to send the student to the next task or level, depending on their responses in the first task. Users are drawn in to wanting to receive badges, have incentive to continue to improve (CasIlli & Hickey, 2016) and take on the behaviors and skills that successfully navigate the game.

Despite the widespread use of rewards and gamification in real-life technology contexts, gamification in education has yet to be fully utilized (Dowling-Hetherington & Glowatz, 2017). Course design that focuses purely on providing content, information and assessment can contribute to students feeling isolated and disengaged in learning (Dossel, 2016). Higher education courses need to adopt an approach that values students as part of the learning process, rather than just receivers of information (Hill & Barber, 2014). Consideration of higher order thinking skills, problem solving, opportunities to relate to each other, as well as the resilience to persist and manage their own learning will help students to develop skills to self-regulate learning and create a sense of engagement and belonging.

Bista, Nepal, Colineau, & Paris, (2012) believe that educators need to use the psychology of gamification to engage and retain students in learning. Using gaming psychology, courses should scaffold learning, create and facilitate students' capabilities to succeed in an online community (Hill & Barber, 2014). Higher order thinking skills and problem solving need to be developed through the process (Ford, Izumi, Lottes, & Richardson, 2015) along with resilience and social attitudes to relate to others in the community (Hill & Barber, 2014).

Use of gamification strategies aligns to the Community of Inquiry Framework (Garrison, Anderson & Archer, 2000) which aims to afford learning in the online environment by focusing on three core elements; social presence, teaching presence and cognitive presence. The framework recognizes that environments that foster social engagement and create a sense of belonging will increase the learning experience (Lehman & Conceicao, 2010; Weidlich & Bastiaens, 2017). Thus, creating an online community with support and feedback from both teacher and peers, assists students to gain a sense of belonging and educators can employ tools to enable students to move towards self-regulated learning.

The three elements within the Community of Inquiry Framework are codependent.

1. *Social presence* is engaged when there is a sense of belonging in the course, between the student-student, and student-educator and a sense of trust within the community (Saville, Lawrence & Jacobson, 2012). Interaction facilitated through avenues such as discussion forums, enables students to share experiences, show their personalities and build common goals.
2. *Teaching presence* is more than the design and organization of the online course. Teacher presence involves pedagogical approaches and sequenced learning opportunities that engage students in learning in numerous modes; such as videos, recordings, face-to-face meetings, discussions and personal communication. Creating a teaching presence, requires ensuring every student online feels like the teacher knows who they are and what they have already completed within the course and students feel a sense of connection to the educator.

Figure 1. Example of two uses of digital badges in higher education
(author's own badges)

3. *Cognitive presence* is the ability of the course material to stimulate students and enable them to analyze and synthesize information that is pertinent to their field of study and make meaning of the context. This can be problematic in online courses as the conversation often lacks the immediacy and individualized feedback that can occur in a face-to-face classroom. Individual personalized feedback has vital importance to provide growth in learning for students (Dowden, Pittaway, Yost & McCarthy, 2013), however this paper will discuss the role instantaneous technology plays in assisting students to feel their teachers are aware of their learning (Casilli & Hickey, 2016).

To maximize the learning experience within a course students must be fully aware of their goals and expectations within the course and utilize feedback from educators and peers to meet these learning goals (Cheng, Watson & Newby, 2018; Hensiek, DeKorver, Harwood, Fish, O Shea, & Towns, 2017; Weidlich & Bastiaens, 2017). Feedback is the communication exchange regarding learning progress which can be provided through a range of sources. It can be covert or overt and to do with content or behaviors shown by students in their learning. Feedback is essential in assessment as it enables students to know their current progress and what they need to do to increase their performance and reach the required learning goals (Cheng, Watson & Newby, 2018).

Learning Management Systems (LMS) used for online courses have capabilities to track and monitor student progression. Educators can strategically select tools to ensure social, teaching and cognitive presence is felt by the students with the aim to provide an engaging learning experience for the students (Hill & Barber, 2014). A digital badge is one such pedagogical tool that can be used within the Learning Management System to provide instantaneous feedback to students, assist students to self-regulate their learning and build a Community of Inquiry within the learning environment. Resembling gamification, the badge "is a representation of an accomplishment, interest or affiliation that is visual, available online, and contains metadata including links that help explain the context, meaning, process and result of an activity" (Gibson, Ostashewski, Flintoff, Grant, & Knight, 2013, p. 1).

The first badge was used to provide instantaneous feedback to students who had met the requirements of assessable forum participation regarding probability in mathematics. The other badges were sent at halfway and upon completion, as motivation and so students knew they had met the submission requirements. The following section of this chapter uses the Community of Inquiry Framework to analyze the usefulness of the digital badge in providing feedback which affords learning in online higher education courses.

THE USE OF DIGITAL BADGES TO ASSIST SELF-REGULATED LEARNING

The use of digital badges in higher education is still quite novel, although studies from courses using digital badges have shown an increase in student participation, achievement and retention (Ibabe & Jauregizar, 2010; Xiong & Suen, 2018). Badges in higher education, have been primarily used for credentialing students, particularly in large Massive Online Open Courses (MOOCs) as they provide an efficient way to digitally record, store and manage completion of tasks (Jirgensons & Kapenieks, 2018). However digital badges can also be used as a pedagogical tool to support learning and assessment (Hills & Hughes, 2016) and provide timely feedback to students (Jirgensons & Kapenieks, 2018).

Creating an Environment that Facilitates Self-Regulated Learning

Self-regulated learning refers to the knowledge and skills students have about their learning; the forethought to plan and set goals, employ task related academic skills and reflect upon and monitor performance (Zimmerman, 1990). Studies (Roth, Ogrin, & Schmitz, 2016) show, students who engage in self-regulated learning, reported higher motivation, course satisfaction and performance. An important function of feedback is that it can be provided not just to feed forward to the next assessments, but can also be to develop the self-regulating skills and attitudes that students require in their employment beyond higher education (Nicol & Macfarlane-Dick, 2006).

A common trait of self-regulated learners, is the ability of students to receive feedback in the way it is intended and to apply this feedback to their skills, knowledge or learning strategies to achieve goals (Cramp, 2011). This requires the learning environment feedback to provide both formative assessment to and from the student about their progression through the course (Zimmerman, 1990) as well as ways to meet the next learning outcomes. Using the socio-constructivist approach, it is essential that communication channels are not just between educator and student, but designed to allow peer feedback between students, and support self-feedback and reflection. Cognitive presence is operationalized through the Practical Inquiry Model (Garrison, Cleveland-Innes, & Fung, 2010; Garrison, & Akyol, 2013), whereby students interact with peer and teacher feedback to develop an external understanding of their development, and reflect on these understanding in regards to their own internal perception.

By creating a learning environment based on the Community of Inquiry Framework, feedback should be an essential part of the course design and used to motivate students to participate, to reinforce desired skills and behaviors (Cano, Martin, Ginns, & Berbén, 2018) and to inform students of their progress (Nelson & Schunn, 2009). The environment should facilitate the students' role as self-regulated learners who ask questions of educators and their peers and seek feedback to ensure understanding (Cassidy, 2011).

The Use of Digital Badges as Feedback

Digital badges have been increasingly popular in online courses in the past ten years, used primarily as a credentialing tool, awarded to students as they meet certain goals or skills. The intention of the digital badge, in a similar way to gamification, is for students to note the levels they have achieved and influence motivation to achieve the next level. Digital badges have gained popularity (Dowling-Hetherington & Glowatz, 2017) due to their instantaneous nature and ability to be programmed to automatically be received by students upon completion of milestones, which is particularly useful in large online classes.

More recently, digital badges have been used in higher education as an authentic pedagogical approach to assist students to move towards self-regulation by providing feedback to motivate, evaluate and regulate skills and behaviors, and establish a Community of Inquiry within the learning environment (Finkelstein, Knight & Manning, 2013).

The Use of Digital Badges to Motivate

Digital badges are an excellent tool to motivate students (Goodyear & Nathan-Roberts, 2017). Finkelstein, Knight, and Manning (2013) stated, a badge "marks a milestone that gives a sense of pride for having achieved something of value and an awareness that new opportunities might be unlocked as a result of achievement" (p. 6). LMS can be set to instantaneously apply badges as soon as a task is complete or set conditions are met. These conditions may include completing a module, replying to a forum post or completing steps in a lesson. Badges can also be tailored to have multiple conditions. Badges are collected and displayed on the student's profile page and are displayed in the same manner as videogames. Figure 2 shows the similarity between an iPhone health awards and the LMS Studydesk.

Receiving a badge from on online community is equivalent to earning a badge for passing a level in a video game, a like on *Facebook®*, or for closing the exercise ring for completing the required steps in a day on a smartwatch. Notifications will be automatically sent to the students' Studydesk/email and can also be set to SMS their mobile phone. A wide range of images can be chosen to be delivered, along with

Figure 2. Awards on iPhone Health App (left) and awards on LMS (right)
from author's own iPhone and LMS.

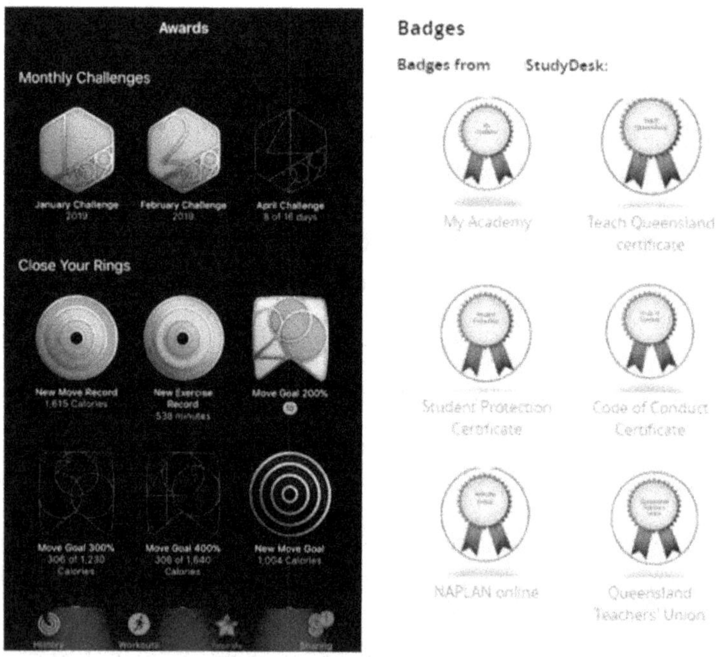

the reason the badge was awarded and the date. Messages can be included with the badge to motivate, give more information and feel personal to students. Although the badge is automated on task completion, the badge will be sent from the educator's email, mimicking teacher presence.

Students studying online can often feel lonely and disconnected from the educators, the course and the University. Breaking down the course into manageable sections, where completion is instantaneously recognized, may assist students through task completion by viewing progression as achievable steps throughout the course (Finkelstein et al., 2013). Receiving personalized emails at the completion of each task with the badge notification, may increase this feeling of connection and teaching presence and becomes a recognition of "a visible marker of a relationship or bond with the issuer" (Finkelstein et al., 2013, p. 6).

In much the same way as gamification psychology has shown that people value receiving badges for completing levels or tasks, studies that have examined digital badges in higher education, have also shown high levels of student satisfaction with badges (Goodyear & Nathan-Roberts, 2017). Evans (2013) believes through digital rewards, the positive impact is on the student's learning self-concept, in which the affective dimension of teacher presence provides students with motivation to complete the next learning goal (Evans, 2013).

The Use of Digital Badges to Evaluate Progress

Using digital badges to plan and evaluate progress is another way badges provide feedback to support self-regulation of learning for students in Higher Education (Iwata, Telloyan, Murphy, Wang, & Clayton, 2017). Once activities are set up in the LMS, students are shown a completion bar on their profile page, outlining the activities that need to be done and the number they have completed. The activity completion showed the students that they had received 5/10 or 50% of the course badges. This is also shown on the LMS as shown in Figure 3.

Seeing the badges and completion tracker on the LMS, may motivate students to map pathways for future activities and timelines (Bowen & Thomas, 2014), which leads towards the goal of self-regulation. Despite concerns that students may try to complete activities quickly to merely receive the badge and then move to the next task, students must complete the required tasks as set up by the educator. Therefore, the task must be completed to receive the badge. Expectations need to be communicated through LMS.

Figure 3. Example of activity completion progress bar
(author's own course)

A key role of the educator in the formation weeks, is to facilitate student navigation through the course and provide feedback on how to succeed within and beyond the course. At the beginning of the course, educators can show students how to plan and set goals (Cheng, Watson & Newby, 2018), organize their study into manageable sections and track their completion using the LMS online tools. Students are aware of expectations and what is needed to complete the tasks. When the badge is received, students are aware that they have completed the requirements, which can reduce anxiety for students (Goodyear & Nathan-Roberts, 2017)

Whilst digital badges instantaneously provide feedback that the task has been completed, they do not sit alone. Digital badges are a part of the complex learning environment and alongside formative, personalized feedback provided through peers or educators within the Community of Inquiry. Formative assessment can be identified through forum discussions (Xiong & Suen, 2018) with formative feedback made though peers and educators monitoring the posts (Facey-Shaw, 2018). Many students have pride in earning and displaying badges within the forums, with one study (Abramovich, 2016) showing 24% of students even posted their badges on external social media sites. Emails to educators from students to check on badges, shows they are using these digital tools to track their progress, which emulate student self-regulatory practice (Evans, 2013). Such effect on self-regulatory behaviors show the positive effects of feedback, which feeds up and beyond into students' own professional practice (Ferguson, 2011), essential in an increasingly competitive job market (Bowen & Thomas, 2014).

The Use of Digital Badges to Regulate Learning

Online learning provides a flexible learning system where learners have autonomy over time and place of learning, depending on the students' circumstances and needs (Iwata et al., 2017). Alongside measures which clarify task and performance requirements, digital badges provide acknowledgement of learner effort as students participate and progress through a course. This may allow students to experience success and in turn influence motivation and persistence.

Iwata et al. (2017) implemented badges to instantaneously send to students as they successfully completed a set of quizzes in the online course. If students received a certain score, they would be awarded the badge immediately, sent via LMS to their profile page and directly by email. If they did not receive a

Figure 4. Examples of badges on student profile pages of Learning Management Systems. The students can collect and display their badges on their profile page and/or share through social media.
(Author's own courses)

badge, they could reattempt the task. Upon completion of all quizzes, they had then successfully completed the course. Students used the badges to receive feedback they had completed the task to a satisfactory standard. As students continue to receive positive feedback for completing activities, it can regulate behavior as students know the expectations to succeed (Facey-Shaw, 2018; Fanfarelly & McDaniel, 2017). Along with other LMS tools, such as course analytics, and formative and summative feedback, the educator is able to support students' progression and completion of the course (Iwata et al., 2017).

Metacognition can also function to aid cognitive presence within a Community of Inquiry, by fore fronting strategies that regulate learning. Cognitive presence is established by providing tasks that require scaffolding, and include higher order thinking, analyses and synthesis of information. The badge is a digital graphic linked to metadata, representing the knowledge gained towards the course learning outcomes (Facey-Shaw, 2018; Hensiek et al., 2017).

Benefits of Digital Badges for the Educator

There are articles pertaining to the benefits of digital badges for students, yet there is little published that evaluate the effect of the digital badge for the educator (Monty, Hope, Adams, 2018). Digital badges provide an additional tool when designing the interaction strategies of a course, an option which is responsive to individual student progress. Data generated within the LMS indicates task completion reports across the student cohort, along with a digital representation of the badges visible on the student profile, which represents useful feedback to the educator. With this feedback, educators are able to make decisions about the teaching within the course.

Engagement and Retention

Digital badges are an example of a way to increase the awareness of teacher presence in the online environment, which can connect students to the course. As the task is completed and the badges awarded, students are notified with a motivational or instructional message from the educator. This feedback is timely, seemingly personalized to the student (Jabbar, Gasser, & Lodge, 2016) and the educator is aware of their progress.

Students respond well to badges. Iwata et al. (2017) surveyed 102 students with a 100% response rate. 67% of students found them useful for achievement and 63.5% found useful for motivation. The digital representation of accomplishment serves to "recognise a learner's work (and) encourage learning (Finkelstein, Knight, & Manning, 2013, p. 2). Students who are engaged and motivated within higher education have been found to have greater course satisfaction (Abramovich, 2016) and higher retention within the course (Dowling-Hetherington & Glowatz, Ibabe, & Jauregizar, 2010; Pampline & Poirier, 2012; Xiong & Suen, 2018).

Achievement

Digital badges can be used to credential what students know or have completed (McDaniel & Fanfarelli, 2016) but also act as an additional piece of formative assessment to gain information as to how the student is progressing through the course. In many courses, highest weighted assessment tends to be towards the end of the course (Abramovich, 2016).

Educators can use tracking tools and course analytics (Tak & Wong, 2017) through the LMS to monitor students' progression through the course. Monitoring course participation, continual evaluation of assessment activities and using student feedback to improve pedagogy will increase perception of teaching presence within the course (Ford et al., 2015; Xiong & Suen, 2018). By having this data about the students, educators are able to communicate at timely intervals to support student engagement, thus increasing teaching presence.

Digital Efficiency

LMS systems have activity completion and digital badges embedded in the digital packages and free badges available online (http://moodlebadges.com/) which means badges are affordable (Facey-Shaw et al., 2018; La Magna, 2017), accessible (Cheng, Watson, & Newby, 2018) and instantaneous upon completion. Digital badges can reduce educator's workload and reduce human error, as they automatically send notifications to students upon completion of the task. This provides instantaneous feedback regardless of time of day, numbers of students, or physical location of academic staff.

LIMITITATIONS

To be successful in regulating learning, formative feedback provided through the digital badge must be of value to the students (Abramovich, 2016). Various studies have found that the design of the badge and type of learner impacted the value of the badge (Abramovich, 2016; Dowling-Hetherington & Glowatz, 2017). Badges that show a symbolic representation of the content or attained skill had a higher value by the student (Dowling-Hetherington & Glowatz, 2017).

Finkelstein, Knight & Manning (2013) suggest that students engaging with this pedagogy all must find meaning and value from the badge to gain significant benefit. Although most students have received badges through real life and gamification situations, Jones, et al. (2018) found that some students who have not experienced badges, lacked familiarity and were uncertain of their use within the higher education course. Dyjur and Lindstrom (2017) reported some students' first impressions of badges was that they were juvenile in design and appearance. Quantitative and qualitative data showed students are more likely to appreciate badges if they are familiar with their use (Dyjur & Lindstrom, 2017). Courses that implemented badges without educating students of the purpose and use of the badge were less likely to implement digital badges successfully.

In order for badges to be valued in gamification, a meaning needs to be assigned that underpins the work needed to complete the task. That is, students need to appreciate the representation of the badge as commensurate with the effort expended (Carey & Stefaniak, 2018). Dowling-Hetherington & Glowatz (2017) found that courses where digital badges had an implied or explicit value in the course, higher completion rates existed. Abramovich, (2016) reasoned that students felt the badge represented the parts of the course that identified the knowledge required to complete the course and therefore were motivated to complete these sections not just to earn a badge, but because the digital badge alluded to the value of the content.

A challenge similar to gaming applies to a classroom context. That is, badges are exciting while they are new technology and sparingly used, but as they increase in popularity students may lose interest (Zichermann, 2011). Furthermore, students could become so intent on earning the next badge, that they briefly engage in the task and complete it with minimum effort, to reach the next badge and activity (Cheng, Watson, & Newby, 2018). The intent of the digital badges is, alongside other methods of engagement and feedback, to assist with self-motivation, and self-regulation for the students.

Finally, with emerging technology, there is always the possibility that it or the processes attached to it may not work in the desired way (Carey & Stefaniak, 2018; Jones, Hope, & Adams, 2018). Ensuring an understanding of how the digital badges work with course completion and analytics will assist educators in setting up badges to maximise their effectiveness for the context of their course. Establishing good relationships with ICT services within the university is important to assist with technical support (Carey & Stefaniak, 2018; Facey-Shaw, 2018).

SOLUTIONS AND RECOMMENDATIONS

The use of badges and other forms of gamification in the higher education learning environment has potential for growth to engage, motivate and retain students (Iwata, et al., 2017). To ensure digital badges are an authentic feedback tool, it is important to design and implement badges according to course content and assessment. For badges to be most effective, students need to be aware of how to use badges and value their use within the learning environment. By examining the findings of early adopters of badges, the following recommendations are provided, to effectively implement digital badges within the higher education context.

Designing Badges for Higher Education

Identifying Learning Outcome

To ensure students find value in the badge, determining the reason for using badges is the first step in the implementation of digital badges. Badges may be required to credential tasks completed (Gamrat et al., 2014) and/or as tools to scaffold tasks (Ford, 2017). They may also be used to personalize learning and allow students to choose from a variety of learning tasks to cater for diversity of learning (Jabbar, Gasser, & Lodge, 2016) or provide flexible study options (Jabbar, Gasser, & Lodge, 2016). Badges may additionally, be used to motivate students and provide instant feedback on completion progress. To use the badge as an authentic assessment tool, educators must decide which learning outcomes are being assessed (Dyjur & Lindstrom, 2017) and how this will be evidenced, to receive the recognition of the badge (Carey & Stefaniak; LaMagna, 2017). Determining the purpose for assessment feedback will then guide the use of the badge as a pedagogical tool in the learning environment (Facey-Shaw, 2018).

Designing Digital Badges

The design plays a key role in students' initial connectedness of the badge (Abramovich, Schunn & Higashi, 2013; Dowling-Hetherington & Glowatz, 2017) and therefore careful consideration should go into the graphic design and metadata included within the badge (Carey & Stefaniak, 2018). Standard badges can be downloaded free of charge from the internet (eg http://moodlebadges.com/) or can be designed by learning designers or even by suitable JPG images targeted to the needs of the course (see Figure 1) designed with the unique cohort, course and community in mind and to match the teaching presence of the educator.

Course completion tools need to be embedded in the LMS to enable digital badges to be employed. Tasks need to be set up to show either demonstrate skill and mastery (Carey & Stefaniak, 2018) or completion of a series of scaffolded tasks (Jones, Hope, & Adams, 2018), which can be evidenced by receiving a badge. Badges can then be set up; including metadata which holds information about completion of tasks and how it met the learning objectives (Henseik et al., 2017), a message to students is added, which will then provide instantaneous feedback (Abramovich, 2016).

Communicating to Students

In order for the badges to work effectively, they should be used within a Community of Inquiry, where teacher presence, social presence and cognitive presence are forefront of the learning community. Facey-Shaw (2018) found that students wanted to know the reasons for receiving badges and how they met the learning outcomes. Educators must explain the badge (Henseik et al., 2017) and how it directly links to learning content, assessment and desired behaviors (Dyjur & Lindstrom, 2017). Time should be taken at the beginning of the course to model how the badges and activity completion can be used to aide their learning. Students should be encouraged to share their learning experiences, which may include tips for others on how to use the badges as a tool to motivate and track learning. This supports a social constructivist approach, fosters social collaboration, a sense of belonging and providing cognitive knowledge through peers within the learning environment (Facey-Shaw, 2018).

Evaluating Evidence Upon Submission

Once badges have been set up in the LMS, a digital badge will be set up to SMS instantaneously the student upon required task completion (Casilli & Hickey, 2016). This allows students to work flexibly through the course and the badge will be awarded as they complete the task. The timely feedback helps to establish a sense of personalization of the course and strong teacher presence. This information is automatically sent through the LMS and enables educators to use tracking tools and course analytics to provide formative assessment of how students are progressing within the course (Bowen & Thomas, 2014).

Monitoring and Evaluation

With the implementation of all new technology and pedagogical tools, it is important to monitor and evaluate the effectiveness within the specific course learning context (Facey-Shaw, 2018). The effect of digital badges will depend significantly upon individual behaviors of students in the course (Abramovich et al., 2013) and how the course has been set up to explain and implement the digital badges to the course participants (Dyjur & Lindstrom, 2017). Students should be encouraged to provide feedback about the effectiveness of the digital badges, through quantitative data such as usage and qualitative comments within forums, course opinion surveys and informal student comments. Use of course analytics and activity tracking can also provide excellent feedback to educators in regards to activity participation and effectiveness (Bowen & Thomas, 2014).

FUTURE RESEARCH DIRECTIONS

Further case studies on the effectiveness of digital badges in higher education is required across disciplines. Further use of gamification strategies within badges could be explored, such as 'completion logic' (McDaniel & Fanfarelli, 2016). This would provide even more personalized learning and greater cognitive engagement for the student receiving the badge (Carey & Stefaniak, 2018). Leaderboards (Gibson et al., 2013), the use of avatars (Falloon, 2010) and video as instantaneous and automated feedback may also be possible within online courses in the near future.

CONCLUSION

"Badges become a much more revolutionary way of being able to speak to the ways that people are educated in the modern world" (Carey & Stefaniak, 2018, p. 1223). Digital badges are an authentic assessment pedagogical tool that provide instantaneous feedback to engage students in learning, motivate, track progress and support a Community of Inquiry within higher education. Visually appealing badges provide seamless personalization to the learning environment. Students can track their progress and participation through the course, which may be useful in motivating them through their course. Continually monitoring and refining the use of digital badges can influence student engagement by acknowledging self-regulating behaviors necessary for future study and the workplace, whilst improving achievement and retention within higher education courses.

REFERENCES

Abramovich, S. (2016). Understanding digital badges in higher education through assessment. *On the Horizon*, *24*(1), 126–131. doi:10.1108/OTH-08-2015-0044

Abramovich, S., Schunn, C., & Higashi, R. (2013). Are badges useful in education? It depends upon the type of badge and expertise of learner. *Educational Technology Research and Development*, *61*(2), 217–232. doi:10.100711423-013-9289-2

Allen, I., & Seaman, J. (2014). *Grade change: Tracking online education in the United States*. Retrieved from https://www.onlinelearningsurvey.com/reports/gradechange.pdf

Bista, S., Nepal, S., Colineau, N., & Paris, C. (2012). Using gamification in an online community. *8th International Conference on Collaborative Computing: Networking, Applications and Worksharing*. Retrieved from https://www.infona.pl/resource/bwmeta1.element.ieee-pub-000006415460/tab/bContent

Bowen, K., & Thomas, A. (2014). Badges: A common currency for learning. *Change*, *46*(1), 21–25. doi:10.1080/00091383.2014.867206

Çakir, R., Korkmaz, Ö., Bacanak, A., & Arslan, Ö. (2016). An exploration of the relationship between students' preferences for formative feedback and self-regulated learning skills. *Malaysian Online Journal of Educational Sciences*, *4*(4), 14-30. Retrieved from http://ezproxy.usq.edu.au/login?url=http://search.ebscohost.com/login.aspx?direct=true&db=eric&AN=EJ1116318&site=ehost-live

Cano, F., Martin, A., Ginns, P., & Berbén, A. (2018). Students' self-worth protection and approaches to learning in higher education: Predictors and consequences. *Higher Education*, *76*(1), 163–181. doi:10.100710734-017-0215-0

Carey, K., & Stefaniak, J. (2018). An exploration of the utility of digital badging in higher education settings. *Educational Technology Research and Development*, *66*(5), 1211–1229. doi:10.100711423-018-9602-1

Casilli, C., & Hickey, D. (2016). Transcending conventional credentialing and assessment paradigms with information-rich digital badges. *The Information Society*, *32*(2), 117–129. doi:10.1080/01972243.2016.1130500

Cassidy, S. (2011). Self-regulating learning in higher education: Identifying key component process. *Studies in Higher Education*, *36*(8), 989–1000. doi:10.1080/03075079.2010.503269

Cheng, Z., Watson, S., & Newby, T. (2018). Goal Setting and Open Digital Badges in Higher Education. *TechTrends*, *62*(2), 190–196. doi:10.100711528-018-0249-x

Chiasson, K., Terras, K., & Smart, K. (2015). Faculty perceptions of moving a face-to-face course to online instruction. *Journal of College Teaching and Learning*, *12*(3), 231–240. doi:10.19030/tlc.v12i3.9315

Cramp, A. (2011). Developing first year engagement with written feedback. *Active Learning in Higher Education*, *12*(2), 112–124. doi:10.1177/1469787411402484

Dossel, S. (2016). Maths anxiety. *Australian Mathematics Teacher*, *72*(3). Retrieved from https://search.informit.com.au/documentSummary;dn=351118154615118;res=IELHSS

Dowden, T., Pittaway, S., Yost, H., & McCarthy, R. (2013). Students' perceptions of written feedback in teacher education: Ideally feedback is a continuing two-way communication that encourages progress. *Assessment & Evaluation in Higher Education*, *38*(3), 349–362. doi:10.1080/02602938.2011.632676

Dowling-Hetherington, L., & Glowatz, M. (2017). The usefulness of digital badges in higher education: Exploring the students' perspectives. *Irish Journal of Academic Practice*, *6*(1). Retrieved from https://arrow.dit.ie/cgi/viewcontent.cgi?article=1062&context=ijap

Dyjur, P., & Lindstrom, G. (2017). Perceptions and uses of digital badges for professional learning development in higher education. *TechTrends*, *61*(4), 386–392. doi:10.100711528-017-0168-2

Evans, C. (2013). Making sense of assessment feedback in higher education. *Review of Educational Research*, *83*(1), 70–120. doi:10.3102/0034654312474350

Facey-Shaw, L., Specht, M., Bartley-Bryan, J., & van Rosmalen, P. (2018). Technological and implementation issues in moodle-based digital badge system. *Proceedings of the European Conference on Games Based Learning*, 82–90. Retrieved from http://ezproxy.usq.edu.au/login?url=http://search.ebscohost.com/login.aspx?direct=true&db=ehh&AN=132746038&site=ehost-live

Falloon, G. (2010). Using avatars and virtual environments in learning: What do they have to offer? *British Journal of Educational Technology*, *41*(1), 108–122. doi:10.1111/j.1467-8535.2009.00991.x

Fanfarelli, J. R., & McDaniel, R. (2017). Exploring digital badges in university courses: Relationships between quantity, engagement, and performance. *Online Learning, 21*(2). Retrieved from http://ezproxy.usq.edu.au/login?url=http://search.ebscohost.com/login.aspx?direct=true&db=eric&AN=EJ1149350&site=ehost-live

Ferguson, P. (2011). Student's perceptions of quality feedback in teacher education. *Assessment & Evaluation in Higher Education*, *36*(1), 51–62. doi:10.1080/02602930903197883

Finkelstein, J., Knight, E., & Manning, S. (2013). The potential and value of using digital badges for adult learners final report. American Institutes for Research, 16.

Ford, E. (2017). To badge or not to badge? *College & Research Libraries News*, *78*(1), 20–21. doi:10.5860/crln.78.1.9602

Ford, E., Izumi, B., Lottes, J., & Richardson, D. (2015). Badge it!: A collaborative learning outcomes based approach to integrating information literacy badges within disciplinary curriculum. *RSR. Reference Services Review*, *43*(1), 31–44. doi:10.1108/RSR-07-2014-0026

Gamrat, C., Zimmerman, H., Dudek, J., & Peck, K. (2014). Personalised workplace learning: An exploratory study on digital badging within a teacher professional development program. *British Journal of Educational Technology*, *45*(6), 1136–1148. doi:10.1111/bjet.12200

Garrison, D., & Akyol, Z. (2013). *The community of inquiry theoretical framework*. Retrieved from https://www.researchgate.net/publication/284306348_The_Community_of_Inquiry_Theoretical_Framework

Garrison, D. R., Cleveland-Innes, M., & Fung, T. (2010). Exploring causal relationships among teaching, cognitive and social presence: Student perceptions of the community of inquiry framework. *The Internet and Higher Education, 13*(1), 31–36. doi:10.1016/j.iheduc.2009.10.002

Gibson, D., Ostashewski, N., Flintoff, K., Grant, S., & Knight, E. (2013). Digital badges in education. *Education and Information Technologies, 20*(2), 403–410. doi:10.100710639-013-9291-7

Gibson, J. J. (1977). The theory of affordances. In R. Shaw & J. Bransford (Eds.), *Perceiving, acting, and knowing: Toward an ecological psychology* (pp. 67–82). Hillsdale, NJ: Lawrence Erlbaum Associates.

Gikandi, J., Morrow, D., & Davis, N. (2011). Online formative assessment in higher education: A review of the literature. *Computers & Education, 57*(4), 2333–2351. doi:10.1016/j.compedu.2011.06.004

Goodyear, M., & Nathan-Roberts, D. (2017). Gamification and The Design of Badges in Relation to Educational Achievement. *Proceedings of the Human Factors and Ergonomics Society Annual Meeting, 61*(1), 1229–1233. doi:10.1177/1541931213601789

Hamari, J. (2017). Do badges increase user activity? A field experiment on the effects of gamification. *Computers in Human Behavior, 71*, 469–478. doi:10.1016/j.chb.2015.03.036

Hensiek, S., DeKorver, B., Harwood, C., Fish, J. O., Shea, K., & Towns, M. (2017). Digital badges in science: A novel approach to the assessment of student learning. *Journal of College Science Teaching, 46*(3), 28–33. doi:10.2505/4/jcst17_046_03_28

Hill, P., & Barber, M. (2014). *Preparing for a Renaissance in Assessment.* Retrieved from https://www.pearson.com/content/dam/one-dot-com/one-dot-com/uk/documents/educator/primary/preparing_for_a_renaissance_in_assessment_and_summary_text_december_2014.pdf

Hills, L., & Hughes, J. (2016). Assessment worlds colliding? Negotiating between discourses of assessment on an online open course. *Open Learning: The Journal of Open, Distance and e-Learning, 31*(2), 108-115. doi:10.1080/02680513.2016.1194747

Ibabe, I., & Jaureguzar, J. (2010). Online self-assessment with feedback and metacognitive knowledge. *Higher Education, 59*(2), 243–258. doi:10.100710734-009-9245-6

Ippoliti, C., & Baeza, V. (2017). Using digital badges to organize student learning opportunities. *Journal of Electronic Resources Librarianship, 29*(4), 221–235. doi:10.1080/1941126X.2017.1378541

Iwata, J., Telloyan, J., Murphy, L., Wang, S., & Clayton, J. (2017). The use of a digital badge as an indicator and a motivator. *International Association for Development of the Information Society*, 5. Retrieved from http://ezproxy.usq.edu.au/login?url=http://search.ebscohost.com/login.aspx?direct=true&db=eric&AN=ED579293&site=ehost-live

Jabbar, A., Gasser, R., & Lodge, J. (2016). Can new digital technologies support parasitology teaching and learning? *Trends in Parasitology, 32*(7), 522–530. doi:10.1016/j.pt.2016.04.004 PMID:27131629

Jirgensons, M., & Kapenieks, J. (2018). Blockchain and the future of digital learning credential assessment and management. *Journal of Teacher Education for Sustainability, 20*(1), 145–156. doi:10.2478/jtes-2018-0009

Jones, W., Hope, S., & Adams, B. (2018). Teachers' perceptions of digital badges as recognition of professional development. *British Journal of Educational Technology, 49*(3), 427–438. doi:10.1111/bjet.12557

LaMagna, M. (2017). Placing digital badges and micro-credentials in context. *Journal of Electronic Resources Librarianship, 29*(4), 206–210. doi:10.1080/1941126X.2017.1378538

Lehman, R., & Conceicao, S. (2014). *Motivating and retaining online students: Research based strategies that work*. San Francisco, CA: Jossey-Bass Press.

McDaniel, R., & Fanfarelli, J. (2016). Building Better Digital Badges: Pairing Completion Logic With Psychological Factors. *Simulation & Gaming, 47*(1), 73–102. doi:10.1177/1046878115627138

Ministerial Council on Education, Employment, Training and Youth Affairs (MCEETYA). (2008). *Melbourne declaration on educational goals for your Australians*. Retrieved from http://www.curriculum.edu.au/verve/_resources/National_Declaration_on_the_Educational_Goals_for_Young_Australians.pdf

Nelson, M., & Schunn, C. (2009). The nature of feedback: How different types of peer feedback affect writing performance. *Instructional Science, 37*(4), 375–401. doi:10.100711251-008-9053-x

Nicol, D., & MacFarlane-Dick, D. (2006). Formative assessment and self-regulated learning: A model and seven principles of good feedback practice. *Studies in Higher Education, 31*(2), 199–218. doi:10.1080/03075070600572090

Pamplin, M., & Poirier, M. (2012). A Learner-centred induction to moodle. *Proceedings of the European Conference on E-Learning*, 641-644. Retrieved from http://ezproxy.usq.edu.au/login?url=http://search.ebscohost.com/login.aspx?direct=true&db=ehh&AN=101746542&site=ehost-live

Roth, A., Ogrin, S., & Schmitz, B. (2016). Assessing self-regulated learning in higher education: A systematic literature review of self-report instruments. *Educational Assessment, Evaluation and Accountability, 28*(3), 225–250. doi:10.100711092-015-9229-2

Saville, B., Lawrence, N., & Jakobsen, K. (2012). Creating learning communities in the classroom. *New Directions for Teaching and Learning, 2012*(132), 57–69. doi:10.1002/tl.20036

Starkey, L. (2010). Teachers' pedagogical reasoning and action in the digital Age. *Teachers and Teaching, 16*(2), 233–244. doi:10.1080/13540600903478433

Thomas, L. (2012). *Building student engagement and belonging in Higher Education at a time of change; Final report from What Works? Student retension and success programme*. Retrieved from https://www.heacademy.ac.uk/system/files/what_works_final_report.pdf

Thompson, P., Vogler, J., & Xiu, Y. (2017). Strategic tooling: Technology for constructing a community of inquiry. *Journal of Educators Online, 14*(2), 117–124. doi:10.9743/jeo.2017.14.2.10

Weidlich, J., & Bastiaens, T. (2017). Explaining social presence and the quality of online learning with the SIPS model. *Computers in Human Behavior, 72*, 479–487. doi:10.1016/j.chb.2017.03.016

Wilson, G., & Stacey, E. (2003). Online interaction impacts on learning: Teaching the teachers to teach online. In G. Crisp, D. Theile, I. Scholten, S. Barker & J. Baron (Eds.), *Interact, Integrate, Impact: Proceedings of the 20th Annual Conference of the Australasian Society for Computers in Learning in Tertiary Education* (pp. 541-551). Retrieved from http://dro.deakin.edu.au/eserv/DU:30009593/stacey-onlineinteractionimpactson-2003.pdf

Xiong, Y., & Suen, H. (2018). Assessment approaches in massive open online courses: Possibilities, challenges and future directions. *International Review of Education, 64*(2), 241–263. doi:10.100711159-018-9710-5

Zhu, J., & Mok, M. (2018). Predicting primary students' self-regulated learning by their prior achievement, interest, personal best goal orientation and teacher feedback. *Educational Psychology, 38*(9), 1106–1128. doi:10.1080/01443410.2018.1497775

Zichermann, G. (2011). *Gamification has issues, but they aren't the ones everyone focuses on.* Retrieved from http://radar.oreilly.com/2011/06/gamification-criticism-overjustification-ownership-addiction.html

Zimmerman, N. (1990). Self-Regulated Learning and Academic Achievement: An Overview. *Educational Psychologist, 25*(1), 3–17. doi:10.120715326985ep2501_2

ADDITIONAL READING

Casilli, C., & Knight, E. (2012). *7 things you should know about badges.* Retrieved from https://library.educause.edu/resources/2012/6/7-things-you-should-know-about-badges

de-Marcos, L., Garcia-Lopez, E., & Garcia-Cabot, A. (2016). On the effectiveness of game-like and social approaches in learning: Comparing educational gaming, gamification & social networking. *Computers & Education, 95*, 99–113. doi:10.1016/j.compedu.2015.12.008

Dicheva, D., Dichev, C., Agre, G., & Angelova, G. (2015). Gamification in education: A systematic mapping study. *Journal of Educational Technology & Society, 18*(3), 75–88.

Hew, K., Huang, B., Chu, K., & Chiu, D. (2016). Engaging Asian students through game mechanics: Findings from two experimental studies. *Computers & Education, 92*, 221–236. doi:10.1016/j.compedu.2015.10.010

Kyewski, E., & Kramer, N. (2018). To gamify or not to gamify? An experimental field study of the influenve of badges on motivation, activity and performance in an online learning course. *Computers & Education, 118*, 25–37. doi:10.1016/j.compedu.2017.11.006

Mah, D. (2016). Learning Analytics and Digital Badges: Potential Impact on Student Retention in Higher Education. *Technology, Knowledge and Learning, 21*(3), 285–305. Retrieved from http://ezproxy.usq.edu.au/login?url=http://search.ebscohost.com/login.aspx?direct=true&db=eric&AN=EJ1113984&site=ehost-live

Roland, N., Frenay, M., & Boudrenghien, G. (2016). Understanding academic persistence through the theory of planned behavior: Normative factors under investigation. *Journal of College Student Retention, 20*(2), 215–235. doi:10.1177/1521025116656632

Shields, R., & Chugh, R. (2017). Digital badges - rewards for learning? *Education and Information Technologies, 22*(4), 1817–1824. doi:10.100710639-016-9521-x

Sullivan, A. (2018). *Open badges and student motivation: A study of their relationship to student assessment scores*. Retrieved from http://gateway.proquest.com.ezproxy.usq.edu.au/openurl?url_ver=Z39.88-2004&rft_val_fmt=info:ofi/fmt:kev:mtx:dissertation&res_dat=xri:pqm&rft_dat=xri:pqdiss:10748817

KEY TERMS AND DEFINITIONS

Community of Inquiry Framework: Recognizes social presence, teaching presence, and cognitive presence as elements of the online environment, which afford learning.

Digital Badge: An online representation of an achievement or award.

Feedback: The communication exchange regarding student progress.

Gamification: The application of game design elements on non-game settings, in order to influence motivation and participation.

Learning Management System: A platform for online learning.

Self-Regulated Learning: Refers to the knowledge and skills students have about their learning; the forethought to plan and set goals, employ task-related academic skills, and reflect upon and monitor performance.

Chapter 4
Utilizing Internet Resources in TESOL:
The Design of English Language Learning and Formative Assessment Practices

Vineetha Hewagodage
University of Southern Queensland, Australia

ABSTRACT

Advancements in digital communication technology (DCT) have significant potential to impact on language teaching and learning by providing new pedagogical opportunities accessible via the internet. Resources that support the design of English language learning and formative assessment have been seen as a fundamental component of the English language learning environments in English language intensive courses for overseas students (ELICOS) and the like. While its importance is universally recognized, the facilitation of a move from the traditional approach to languages learning to a more contemporary approach in keeping with social constructivist theory remains a challenge for both teachers and learners. Part of this challenge involves the need for teachers to reconceptualize the way that they teach to move away from the teacher as instructor to that of facilitator whose pedagogical approach ensures students have opportunities to use English for meaningful purposes. Since this demands an understanding of students' different proficiency levels and the ability to monitor improvement to ensure learning experiences are at the optimum level to facilitate growth, effective formative assessment practices are a vital monitoring tool. Thus, this chapter examines how internet resources can move pedagogy towards a social constructivist approach to improving English language learning bearing in mind the need for a framework to support formative assessment in the monitoring of improvements in proficiency.

DOI: 10.4018/978-1-7998-0426-0.ch004

INTRODUCTION

Today, it is evident that the Internet provides access to the most usable digital communication technologies (DCTs) for supporting the teaching, learning and assessment process. With the evolution of various software applications many people across the globe, and particularly in the Western world and Asia, in particular, (O'Neill, 2019) are well connected such that digital communication technology and its easy access through mobile devices has a significant impacted on their lives (Tabatabaei & Gui, 2011). Thienthong and Lian (2014) note that:

This raises questions for educators and researchers on how online resources and applications can be used to enhance students' learning and meet their needs. An increasing number of language teachers are showing an interest in making use of resources on the Internet for language pedagogy (p. 108).

This is reinforced by Stockwell and Hubbard (2013), Tochon (2012), Towns and Loo (2012) and O'Neill (2015) who show how the availability of the CLOUD has created a new globally accessible virtual learning space where a wealth of data can be automatically recorded to support monitoring of learning, assessment and evaluation. While the CLOUD is applicable no matter which discipline this chapter explores its ability to transform languages learning in the case of English as an additional language (EAL) (including those learning English as a second language (ESL) and English as a foreign language (EFL). Learning English therefore, continues to be a major activity for adults and children around the world, since it is the medium that connects people for many different purposes e.g. social/family connections, employment, international business, and education where teaching and assessing English for work and entry to higher education constitutes an international business. Of note is the fact that when developing countries adopt the use of DCTs for learning and training they can expect an increase in their GDP (Van Tol, 2016) and increasing proficiency in English the language to enhance international communication is a large part of this success.

Thus, in the context of international education, and the area of teaching English to speakers of other languages (TESOL) DCTs are increasingly being applied to facilitate and increase the effectiveness of pedagogy and assessment practices. New digital tools such as Google, YouTube, Facebook, Viber, Whatsapp, WeChat, Skype and Zoom are some examples that apply to enhancing pedagogy and assessment of learning (Martin, et al., 2011). At the same time their use in education coincides with the everyday social and business needs of all who are connected to the Internet. Moreover, the ability of these tools or Apps to facilitate authentic communication in English (or other learners' target language) provide the learning opportunity that has been lacking in most languages learning programs.

Traditionally, languages teachers have needed to create communicative interactions for students to use the target language but these have been seen as contrived because of the absence of a real-life purpose and involvement of first language speakers of the target language. This has been in spite of changes to approaches (Hismanoglu, 2012; O'Neill, 2012a) and strategies to integrate the macro skills. Traditionally, opportunities for authentic communication in language programs has been addressed through short visits to the target language country. However, such approaches have been shown to have limitations for authentic language use (Magnan & Back, 2012), and shorter visits have potential to reinforce stereotypes rather than foster intercultural literacy (Peterson & Helms, 2013). But in the case of the provision of English Language Intensive Courses for Overseas Students (ELICOS), international EFL students participate in English language courses in English speaking countries that are especially designed to prepare them to meet the English language proficiency benchmarks required for entry into that country's degree programs in higher education. Such courses, in their provision of a full-time focus of intensive English language learning in an English-speaking country, are generally very effective in progressing students from beginners' level, general English through elementary, pre-intermediate, intermediate, upper intermediate and advanced, leading to mastery of English for academic purposes. This is because this hierarchy of proficiency levels are well described, researched and resourced in terms of learning materials, and assessment practices. Explicit indicators of evidence of students' progression, including assessment of students' entry levels gauged through a placement test are provided. In this respect the formative assessment of students' improvement in English language proficiency can be both formatively and summatively assessed, according to the proficiency goal.

Policy to guide the identification of the associated language proficiency levels is commonly determined internationally as advised through United Nations Educational, Scientific and Cultural Organization (UNESCO) in terms of the Common European Framework of Reference for languages (CEFR). In addition, the tests of English most commonly accepted in higher education in English speaking countries, include the International English Language Test System (IELTS) and Test of English as a Foreign Language (TOEFL), that includes an online version. However, in being able to provide indicators to support formative assessment practices, the less well known International Second Language Proficiency Scales (ISLPR) uses detailed, descriptive scales of English language proficiency for Listening, Speaking, Reading and Writing based on the American Council on the Teaching of Foreign Languages (ACTFEL). It is a test that is adaptive in terms of assessing English language proficiency for vocational purposes as well as academic.

With the advent of the Internet, resources that inform and support the teaching, learning and assessment practices involved in providing English language education are prolific. It has created a 'virtual world', in Moje, Ciechanowski, Kramer, Ellis, Carrillo, and Collazo's (2004) terms of 'a third space', where through the CLOUD, learning in general but language learning in particular, is on the cusp of a transformation (Lamb, Hatoss & O'Neill, in press). It is with this in mind, and the recognition that learning needs to be assessed formatively as learners progress and need feedback, this chapter sets out to (1) explore how Internet resources can contribute to the design of authentic English language learning experiences and (2) identify the implications for building formative assessment practices into pedagogy for TESOL in higher education. It considers personalized learning environments, and collaborative approaches to knowledge building and the use of mobile technology accessed via mobile phones, iPads and Tablets, and the use of video conferencing tools like Skype and Zoom and the potential of video games.

The role of formative assessment practices and how these can be enhanced through the use of DCT in these ways in teaching and learning is at the core of the discussion. Within this it addresses the issue of pedagogical change in contemporary times and considers the research that supports a shift towards a learner-centred approach in keeping with Tochon's (2014) concept of deep learning and cognitive apprenticeship.

BACKGROUND

People born after 1995 are categorized as 'generation Z'. They expect pedagogy to reflect that which they experienced in this time frame, making them the most sophisticated learners to date, since they have not known any other environment than the current, where the omnipresence of the Internet, social media and mobile technologies are central to their life style (Povah & Vaukins, 2018). As Hashemi, Azizinezhad, Najafi and Nesari (2011, p. 2477) explain, "the processing power of held-devices has grown exponentially while becoming more affordable and even ubiquitous due to the demand for games, business communications, and in general the connected lifestyle of the wireless society". For those reared in this era, access to the Internet and associated resources is taken for granted as are the mobile devices such as smart phones, iPads and Tablets and the many applications (Apps) that facilitate their use. Internet access has the capacity to change the way a foreign or a second language is taught or learnt because it can enable learners to communicate, interact with each other, experience new ideas and knowledge through virtual reality and simulated activities to create real-life authentic learning experiences previously unachievable through traditional methods of language learning that are typically driven by monologic text book exercises.

Alongside formal education, everyday opportunities to access learning resources through mobile devices have multiplied. Language learning Apps, such as Babbel, Duolingo and Hello Talk are prevalent and are motivating because of their ease of use and their structuring of learning activities according to capability. They have the ability to personalise learning through a built-in approach to formatively assess a student's starting point as well as monitor progress, giving feedback, on an ongoing individualised basis across the key skills. This amounts to providing opportunities for personalised learning, providing student agency and a degree of independence in learning. These DCTs also have the capacity to facilitate language learners connecting and communicating with each other through both speaking and writing, and support experiential approaches that can facilitate deep language learning that may be project- or problem- based (Tochon, 2014).

The application of DCTs to language learning is also in keeping with Krashen's (2003) theory of second language acquisition in his emphasis on the need to provide comprehensible input for the learner, in that he sees this as vital to development. From this perspective, the many choices of resources and information in multimedia and multimodal forms can provide the virtual reality in building funds of knowledge to develop proficiency and intercultural literacy. There are already studies that provide evidence that technology plays a positive role in second language acquisition (Gee, 2003; Young 2013) and as Larsen-Freeman and Anderson (2011) point out, technology provides unlimited educational resources for both teachers and students. Integration of technology into the ESL or EFL learning environment therefore can assist teachers to better motivate and engage learners to improve their own learning process (Gilakjani, 2017; Tomlinson, 2009). Today's language learners have also been called digital natives (Thompson, 2013), since as generation z they have not experienced life without the Internet. They are considered adept in their use of DCTs and typically have unlimited access to a huge amounts of information, via the Internet, that in traditional teacher-centred classrooms, teachers were unable to provide (Gilakjani, 2017). However, DCT as a pedagogical tool is not necessarily adopted by teachers in spite of its usefulness since they may not have the skills to use it for that purpose. Li and Walsh (2011) and Mndzebele (2013) note that difficulties remain not only in countries where EFL is core curriculum but also in ESL contexts in Western settings. The adoption of DCTs for teaching and learning English also depends on teachers and students being able to access the Internet through smart phones, iPads and Tablets that provide ease of mobility and timeliness of communication. These tools, among others, have made it possible to form a virtual classroom where participants can communicate and work collaboratively in real time as well as asynchronously, regardless of the location. In other words, the emergence of DCTs has removed the barriers of distance making it possible and easy for all to connect to each other.

Nevertheless, to use DCTs as a key teaching strategy requires adequate resourcing, although learners could be encouraged to use their own equipment, although teachers today may still feel inadequate and even fearful when it comes to understanding how to teach EFL/ESL in this way (Blake, 2013). Working with Apps, websites, blogs, wikis, video games and Vimeo resources and the like, require students to take charge of their learning to a large extent and teachers to facilitate that learning (Boyle, 2013). This contrasts with the traditional approach that would rely on discrete textbook exercises carried out with limited opportunity for collaborative authentic meaning making (Albadry, 2017; Taylor, 1994). Farooq and Soomro (2018) link the lack of pedagogical change to "the lack of time for out-of-school training in combination with the natural difficulty in incorporating new working schemata within their own classrooms" (p. 11). Thus, the challenge is for teachers to be able to confidently use DCTs and the Internet as major resource for English language learning and to understand how it can help maximize results i.e. improve proficiency levels (Gilakjani, 2017; Larsen-Freeman & Anderson, 2011). To achieve this, teachers need to understand the teaching, learning and assessment cycle and levels of proficiency. They need to be able to design and build in formative assessment practices that link to markers that signify students' growth in proficiency (Ingram & Wylie, 1979/1999). Assessment literacy for teachers is a relatively new concept but from the TESOL perspective and the common use of tests in programs like ELICOS, the need for teachers' professional development to better understand how their practices can be enhanced through DCTs in terms of internet resources is well recognized in the literature.

REVIEW OF THE LITERATURE

The High Stakes Context of English Language Requirements in Higher Education

At the beginning of the 21st century more attention than ever before is being paid to issues such as the efficiency and quality of personnel, the assessment of performance, cost-benefit analysis of human resources, and accountability in all walks of life. As a consequence, "high-stakes assessment has been extended to an increasing number of contexts worldwide" (Coniam & Falvey, 2015, p. 443). For example, testing knowledge of English language becomes high stakes when the outcomes are used to make decisions about admission requirements to tertiary institutions, assessment of professional competence and certification, membership to a professional body, promotion, graduation, salaries increment, citizenship acquisition and even to terminate an employment (Green, 2006) High-stakes testing is also often associated with public reporting of testing outcomes as a way to bring attention to the assessment results (Jones & Ennes, 2018). High-stakes tests include IELTS and TOEFL as mentioned earlier and can include the ISLPR and also the Pearson Test of English (PTE Academic). IELTS is comprehensive in applicability since besides being a test designed for educational and vocational purposes, it is used to measure the English language proficiency of people who want to study, work and migrate to countries where proficiency in English is a prerequisite. It is administered on a very large scale worldwide. As reported by the British Council (2017) more than three million IELTS tests were taken in 2016, which was seen as reflecting its growing importance as the world's leading test of English for international higher education and migration. It is the most widely used test of English for migration to Australia, Canada, New Zealand and the UK and is recognized by more than 10,000 universities, schools, employers and immigration bodies. TOEFL is similarly accepted worldwide. Meanwhile the Australian ISLPR is commonly used for the eligible of non-English speaking background teachers for their registration requirements, entry requirement for non-English background speakers to many Technical and Further Education (TAFE) college competency-based courses in Australia, New Zealand, United Kingdom, United States of America, Canada and Republic of Ireland (International Second Language Proficiency Ratings, 2019). The Pearson Test of English Academic (PTE Academic) as a computer-based academic English test also forms a significant part of Australia's international education industry aimed at non-English background speakers wanting to study abroad. Most of these tests are already taking advantage of delivery online, although as high stakes issues of security need to be managed carefully. However, testing English for such high stakes purposes is summative in providing proficiency levels to inform whether the required benchmark is met or not. Thus, what is more relevant to learners in a course of English language learning is the provision of information at various points during a course so as to feed forward to help progress their learning towards achieving the high stakes summative goal. In other words, there is a need for effective formative assessment practices if teachers are to maximize students' learning. As Thienthong and Lian (2014) point out:

This raises questions for educators and researchers on how online resources and applications can be used to enhance students' learning and meet their needs. An increasing number of language teachers are showing an interest in making use of resources on the Internet for language pedagogy [and assessment] (p. 108).

The Importance of Using Technology in Language Pedagogy and Assessment

Traditionally, pencil and paper assessment processes were the established practice but now, as noted above, computer-based assessment is at the fore, although some countries are slower to adopt it than others (Schaffhauser, 2012). As Thienthong and Lian (2014, p.108) explain there are many kinds of online resources and applications that language teachers can utilize to support English language teaching regardless of their personal pedagogical style. A multitude of search engines are available (e.g., Google, Ask, Bing), as well as free synchronous and asynchronous communicative tools (e.g., Skype, Gmail). In addition, social networking provides many opportunities for authentic meaning-making, where students might participate in Facebook, Instagram and Second Life, or via microblogging in terms of Twitter, Blogger, and WordPress). Media sharing and creating personal or group project resources may also be facilitated via YouTube, Flickr, and Pininterest and the like. There are also a range of authoring and publishing tools through Wiki and Wikipedia such that all-in-all English language learners today are able to engage with their peers and tutors in real-life/real-world learning experiences that extend beyond the walls of the classroom.

This use of DCTs has considerable potential to not only stimulate learning but it has the ability to support teachers formative assessment of studets' skills and knowledge during the teaching and learning process. Providing immediate feedback during the formative assessment process, with the aim of improving students' performance, is crucial in learning and teaching (Elmahdi, Al-Hattami & Fawzi, 2018, p. 182). Chapelle and Voss (2016) emphasize that assessment of learners' language ability is a vital part of language education, which can be assisted through the use of computer technology. They note that it is also important in diagnosing students' learning needs, which relates to the formative assessment process turn. However, language teachers require specific knowledge and skills to create formative assessment tasks that link to established benchmarks of English proficiency across the macro skills. This involves assessment design that may relate to a variety of formats. The process may also be seen as burdensome and time consuming when feedback is demanded by a certain time. However, the use of technology in the formative assessment of language learning, as Litzler and García Laborda (2007) advises:

can be faster, more efficient, and less costly than traditional paper-and-pencil testing. Additionally, multimedia prompts can help make the test feel more 'real'. Adaptive tests can facilitate the difficult task of rapid diagnosis, and self-correcting tests can accelerate the process of correction, feedback, and reporting (p. 8).

In other words, there are several advantages of using modern technology for language assessment. It can motivate students, as most assessments are designed on interactive formats and feedback is frequent and built in to the design. Assessment responses can also be corrected regardless of students' location usually automatically and learners have easy access via their own mobile devices connecting to the Internet. Thus, the use of DCTs are becoming the norm and offer a less expensive delivery and more environmentally friendly than paper-based formats (Bulut & Kan, 2012; Fu, Fu, & Beasley, 2012).

The integration of technology into language learning and assessment experiences therefore is becoming a necessity not only to respond to generation z's existing skill set but to facilitate a communicative, social constructivist approach, where formative assessment is recognized as necessary to inform students' learning. This is reinforced by Elmahdi et al. (2018, p.182) who point out that today, "for effective learning, where that improves learning . . . the road to motivate and encourage students to learn is paved with their passion for technology and digital tools". With the increasing advancements in DCTs, educators are using them in different formats to increase efficiency in their courses and to educate individuals who need to adapt their practice (Gürışık & Demirkan, 2019). In addition, since learners have access to so many educational resources via the Internet, they have the advantage of being able to focus on their interests, and teachers are able to foster studnets' autonomous learning by engaging them in problem- or project- based learning experiences (Tochon, 2014) where they need to communicate for more authentic purposes. In this way the teacher becomes more of a facilitator of learning as students grow in proficiency. Through formative assessment teachers are able to identify proficiency markers along the way to monitor learner success and ownership, bearing in mind this requires intentional accountability systems to avoid learners wasting their time because of lack of explicit goals (Brown 2010; Furtak et al., 2016). The ISLPR, as a set of scales that describe language performance in detail for each macro skill, provides a highly detailed basis to assess formatively (Ingram & Wylie, 1979/1999). Elmahdi et al. (2018) also argue for online classroom response systems that have the capacity to enhance students' learning by incorporating formative assessment as opposed to only summative. According to Black and Wiliam (1998) formative assessment is "all those activities undertaken by teachers, and/or their students to modify teaching and learning activities in which they [the students] are engaged [to progress learning]" (p. 8). As Bloom (1969, p. 48) also explains, formative assessment practices are invoked "to provide feedback and correctives at each stage in the teaching-learning process". As opposed to being summative and monologic, formative assessment is dialogic (Dann, Dann, & O'Neill, 2018) such that it involves critical reflection and two-way conversations based around performance markers that support a student's progress. Rather than being thought of as a test instrument in itself, Black and Wiliam (2009) explain:

practice in a classroom is formative to the extent that evidence about student achievement is elicited, interpreted, and used by teachers, learners, or their peers, to make decisions about the next steps in instruction that are likely to be better, or better founded, than the decisions they would have taken in the absence of the evidence that was elicited (p. 9).

Thus, as Furtak et al. (2016) advise, formative assessment is intended to have a positive effect on students' performance, but at the same time as Bennett (2011) warns that research shows that where there is a wide variety in implementation of formative assessment the effects would be expected to vary widely between student cohorts. Thus, as noted early formative assessment practices applicable to developing students' English language proficiency need to be underpinned by sound curriculum and explicit benchmarked measures (ACTFEL; Ingram & Wylie, 1979/1999); O'Neill, 2012b).

How Technology Is Providing Simulated Second Language Learning Experiences

DCTs are reasonably well established as supporting language teachers to design and implement educational programs that provide high quality teaching and learning experiences both on and off line. There is a wealth of opportunities that can enrich and increase authenticity in learning design and in turn enhance formative assessment practices. For example, as highlighted by Tabatabaei and Gui (2011, p. 514), "3D virtual worlds like Second Life are now being used by language teachers to provide simulation of real life environments and allow students to discover more creative and more realistic ways to improve their language skills". Similarly, Aydin and Yildiz (2014, pp. 160–161) note DCTs "offer new ways of teaching by allowing authoring, information sharing, knowledge building and easier collaboration opportunities among learners". The availability of tools for authoring together with collaborative learning experiences are increasingly being recognized as core to providing authenticity in language learning and increasing students' interest and motivation. In particular, access to 3D virtual worlds such as Second Life and video games (Soyoof, 2018a) can engage students in deep learning. Such learning environments that involve virtual residents (avatars), and where learners take on a specific role in a game, where their survival may be involved, they become emotionally engrossed in their character and the use language relative to the experience, within imaginative virtual worlds. In addition, as with Second Life learners are able to construct their own learning activities as Ata (2016, p. 11) specifies, "an innovative culture of teaching and learning within Virtual World . . . evoke[s] paradigms such as learner-centred, self-directed, power and responsibility, immersiveness, embodied presence, social and peer-to-peer, networked, inquiry-based, collaborative, and co-learning". Larsen-Freeman and Anderson (2011) support the fact that technology-based activities provide learners rapid information and motivate them to use the language of the experience in meaningful ways. While video games can build in levels of challenge that may link to language learning objectives that take account of gradually increasing proficiency levels there remains the opportunity for designs that are able to both address this need as well as provide the appropriate stimulating experience (Soyoof, 2018b).

Thus, although the development of virtual learning experiences, as mentioned above, have been described as beneficial for both teachers and students, through their interactivity, efficiency, flexibilty and program design features, Richardson et al. (2011) raise concerns about financial costs and the potential of being seen to promote what might be potentially a negative learning experience. These concerns relate to learning how to use Second Life in that it can be perceived as difficult to understand regarding navigation, simulation, time pressures, and logistical issues such as the impact of slow Internet connection/poor reception (Chang et al., 2009; Inman, Wright, & Hartman, 2010). Similarly, without close analysis or specific design for language learning purposes video games may prove to condone violence and cause stress from the impact of deep involvement in the role adopted, and increasing challenges and requirements to play and survive in character (DeHaan, 2005;). While this has serious implications for ensuring appropriate learning experience design and the need to schrutize exiting products' potential, as emphasized by Tabatabaei and Gui (2011):

this new medium using technology to teach a language offers many opportunities and calls for a new way of teaching language. Therefore, one should not try to exactly duplicate the classroom activities to teach the language online and second life as a medium. The second life itself should be viewed as a classroom providing new set of activities to practice in teaching the language" (p. 515).

IMPLICATIONS OF FORMATIVE ASSESSMENT PRACTICES FOR LANGUAGE TEACHERS' PEDAGOGY IN HIGHER EDUCATION SETTINGS

It goes without saying that the role of assessment in English as a Foreign Language (EFL) and English as a Second Language (ESL) must closely connect to the pedagogy involved (Torres, 2019, p.1). It is also becoming increasingly recognized that English language learners from other language back grounds may have diverse needs as do first language learners (Shore, Wolf, & Heritage, 2016).Thus, the need for descriptive feedback on learning progression tied to explicit benchmarks in formative assessment is essential if English language programs are to be effective and meaningful for diverse groups. The definition of formative assessment in the United Kingdom (UK), where it is termed as 'assessment for learning' (Assessment Reform Group, 2002) is: "the process of seeking and interpreting evidence for use by learners and their teachers to decide where the learners are in their learning, where they need to go and how best to get there'' (p. 2). Ten principles are applied with respect to how the role of assessment in classroom practice relates to pedagogy, which involve rely upon the provision of timely and effective feedback that can build learner's active engagement and learning capability. In keeping with the notion of teachers as facilitators of learning their ability to manage and adjust both content and assessment procedures is seen as vital to progress students' growth in learning. Thus, as opposed to the traditional view of learning as information transmission, in keeping with Tochon's (2014) deep approach, learners are seen as having an active role. As Brown (2010) notes part of formative assessment may include teachers' qualitative feedback in the form of modelling the task at hand. Moreover, important for students' ongoing motivation to learn is the need to ensure feedback is constructive and not couched in the negative (Mellati & Khademi, 2018; Torres, 2019).

The shift from the traditional ways of teaching, learning and assessing students' English language improvement towards a more student-centred model (that recognizes students' agency in how they learn and the need for real-life, authentic communicative tasks), formative assessment practices emerge as central to the change. Hamp-Lyons (2007) finds the complexity in assessment practices between a traditional testing culture and that which is classroom-based involves a variation in principles and purpose. This is because the traditional testing culture's is on accuracy and proficiency, typically echoing the high stakes testing of English proficiency compared with classroom feedback to individuals mis course. Dann et al. (2018) also highlight the shift in practice as teachers needing to move from monologic – one-way communication of feedback to students to two-way conversations that are therefore dialogic and so allow for a range of information able to feed forward for students to understand how they can improve their learning and performance. This differentiation is further explained by Proyor and Crossouard (2005) citing the work of Torrance and Pryor (1998) where teachers are seen as moving between between convergent and divergent practices. As outlined in Table 1, Torrance and Pryor (1998) see 'convergent' practices as characteristic of a behaviorist approach, compared with 'divergent' practices as being in keeping with social constructivism, that supports the use of 'helpful questions' to scaffold learning as opposed to 'testing questions' that make a summative judgment. However, they note that the two approaches are not necessarily mutually exclusive in practice, since teachers may need to switch between the two depending on the situation. In addition, point 8 in convergent formative assessment acknowledges its close relationship to summative assessment, but the emphasis in this chapter is on divergent assessment practices that identify the formative markers that are able to be linked to the student's growth in learning, since these have the ability to feed forward as exemplified in points 12 and 13, 'An intention to teach in the zone of proximal development' and 'Part of an on-going dialogue between and amongst learners and teachers where learners initiate as well as respond, ask questions as well reply'.

Both divergent and constructivist approaches to assessment practices involve the teacher in utilising interactions and assessment evidence to measure the level of a student's understanding of the content thus formative assessment positively impacts the teachers' performance in the classroom as well as on the development of language skills by learners. From the perspective of the teachers, it serves as a reflection state within the course so that weaknesses can be identified and remediated (Torres, 2019, p.7). However, past studies found that effective, formative assessment practices have been overlooked in many programs owing to the complex structure of assessment criteria, limited negotiation within the context of the formative assessment, and inadequate knowledge of the teacher in the assessment area thus pointing to the importance of teacher professional development (Campion, 2016; Mellati & Khademi, 2018).

Table 1. Implications of convergent and divergent assessment practices for pedagogical approaches

Formative Assessment	
Convergent	**Divergent**
Assessment which starts from the aim to discover if the learner knows, understands or can do a predetermined thing. *Reflects a traditional approach to teaching and learning*	Assessment which starts from the aim to discover what the learner knows, understands or can do. *Reflect a social constructivist approach to teaching and learning*
1. A view of assessment as accomplished mainly by the teacher. 2. Precise planning by the teacher and an intention to stick to it. 3. Recording via check lists and can-do statements; 4. Closed or pseudo-open teacher questioning and tasks. 5. A focus on contrasting errors with correct responses. 6. Authoritative, judgmental or quantitative feedback. 7. Feedback focused on performance and the successful completion of the task in hand. 8. Formative assessment focused on communicating criteria usually closely related to those used in summative assessment. 9. Involvement of learners as recipients of assessments. 10. An analysis of the interaction of learners and the curriculum from the point of view of the curriculum. 11. May conform to either a behaviorist or a constructivist view of education. 12. An intention to teach or assess the next predetermined thing in a linear progression. 13. An interaction usually embedded within an Initiation–Response-Feedback (IRF) sequence. 1.	2. Flexible planning or complex planning which incorporates alternatives. 3. Open forms of recording (narrative, quotations etc.). 4. Primarily open tasks with questioning by teachers and learners directed at 'helping' rather than testing. 5. A focus on miscues - aspects of learners' work which yield insights into their current understanding - and on prompting metacognition. 6. Exploratory, provisional or provocative descriptive feedback aimed at prompting further engagement from the learners. 7. Discussion prompting reflection on the task and its context with a view to constructing understanding of future situations in which new knowledge might be applied. 8. Formative assessment focused on a holistic view of criteria, the learners' understandings of them and how they fit into wider notions of knowledge and competence. 9. Involvement of the learners as initiators of assessments as well as recipients. 10. An analysis of the interaction of learners and the curriculum from the point of view of both learners and the curriculum. 11. Conforms to a socio-cultural view of education with an acknowledgement of the importance of the context for the assessment. 12. An intention to teach in the zone of proximal development. 13. Part of an on-going dialogue between and amongst learners and teachers where learners initiate as well as respond, ask questions as well reply. 14. A view of assessment as a collaboration between and amongst teachers and students.

Source: Adapted from Torrance and Pryor (1998, p. 93)

PEDAGOGICAL CHANGE THROUGH UTILISING INTERNET RESOURCES IN THE DESIGN OF ENGLISH LANGUAGE LEARNING AND FORMATIVE ASSESSMENT EXPERIENCES: STIMULATING 'ACTIVE PEDAGOGY'

There are many studies that describe how the utilisation of Internet resources in the design of English language learning formative assessment experiences is changing practice (Litzler & García Laborda, 2016; Yan & Cheng, 2015). However, what is clear is that the integration of DCTs in the ESL/EFL classroom offers new transformative pathways, as mentioned earlier, that can increase flexibility, introduce new learning spaces and more active involvement in learning. As teachers and learners become more involved in the application of DCTs through mobile devices such as smart-phones, iPads and Tablets, some key pedagogical changes have emerged that support the idea of 'activist languages education'. Lamb, Hatoss and O'Neill (in press), define activist languages education as being underpinned by the following six principles. It can be appreciated that the availability of DCTs is core to this changing practice as per principle 6 that emphasizes the importance of interactivity and collaboration:

1. Languages education policy, curriculum, pedagogy and practices that create social connectedness, and value home and heritage languages and cultures as key resources in inquiry-based contexts for learning.
2. Multidimensional lens, anti-racist education, socially inclusive policy, practices, and education, and active approach for socially just outcomes.
3. Valuing of participants' existing, knowledge, skills, needs and aspirations.
4. Participants' agency and voice in active learning design.
5. Mentoring, collaborative capacity building and mutual benefits.
6. Bottom-up, authentic participatory, dialogic, interactive learning spaces, supported by access to digital communication technologies.

With respect to this chapter's exploration of how Internet resources can be utilized in the design of English language learning and formative assessment practices in TESOL the following section considers the personalization of learning environments, opportunities for blended learning, developing a collaborative approach to knowledge building, open educational resources, use of video games, and mobile learning or e-learning.

Personal Learning Environments (PLEs)

Personalized learning, as never before, is seen as being a manageable strategy through the advent of DCTs. Through the use of artificial intelligence and machine learning, advances in online learning where students move through learning activities that are hierarchically organized and have built in formative assessment are possible. This reflects the design of video games and other interactive communicative learning initiatives (Deen, 2015; Osman & Bakar, 2013). Goff (2017) argues personalized learning responds to the need to develop skills that promote lifelong learning and pedagogical reform more broadly, besides the individual needs of students. The promise of personalized learning environments (PLEs) has become prevalent as the opportunity to support such learning through DCTs has occurred. In terms of language learning it is in keeping with Tochon's (2014) concept of cognitive apprenticeship in the learner having agency in the approach, engaging in personalized collaborations with peers.

Thus, PLEs can be understood "as complex knowledge systems helping students organize their learning freely and thus able to take ownership of it" (Tochon, Karaman & Ökten, 2013, p. 2). According to Buchem, Attwell and Torres (2011) "capturing the individual activity, or how the learner uses technology to support learning, lies at the heart of the PLE concept" (p. 1). Tochon et al. (2013) cite Entwistle's (2008, p. 25) characteristics of teachers and the teaching-learning environment in their argument for PLEs ability to enhance deep language learning. These characteristics reflect the variables that impact on learning, although not exhaustive, include students' background, knowledge and aspirations, teachers' design of the learning environment, enthusiasm, beliefs about pedagogy, and support for learning, as well as student peer group attitude and mutual support and congruence of the learning experience with the course objectives. Tochon et al. (2013) further note, "deep learning depends upon the characteristics of both learner and teacher and the quality and depth of learning is determined by the course aims, students' aspirations, the peer group and mutual support, and the selection, organization, presentation, and assessment of the course materials" (p. 6). This shows the inter-dependency from one context to the other and as Kravcik and Klamma (2012, p. 710) state PLEs comprise "tools, communities, and services that constitute the individual educational platforms that learners use to direct their own learning and pursue educational goals". Therefore, in the provision of English language learning teachers require the pedagogical skills to utilize DCTs in the design of PLEs and the assessment literacy to build-in formative assessment to ensure students are able to understand their growth in learning in relation to shirt term and long term goals.

Blended Learning

The concept of of blended learning in higher education originated with the quality framework created by Sloan Consortium (Sloan-C), which is a non-profit organization dedicated to integration of online education into the mainstream of higher education (Sloan-C website). This framework aims to raise the quality of learning effectiveness, student satisfaction, faculty satisfaction, cost effectiveness, and access. Tang (2013, p. 80) stated, "blended learning is an integration of online learning and regular class learning. Until now, there's no consensus on how it is defined and different terms, such as 'blended', 'mixed-mode' and 'hybrid', are used". It represents an innovative pedagogical approach that highlights the way learning environments can incorporate both diverse teaching styles and delivery modes, that may include face-to-face, and online synchronous and asynchronous teaching and learning, different media formats or a combination (La Banca et al., 2013). Blended learning opens up a multitude of ideas that are applicable to supporting Tochon's deep approach to language learning and through the use of online platforms can provide formative assessment data. This is reinforced by Gibson (2013) who states: "this rapid growth [in blended learning] requires a focus on the types of instructional strategies that will best serve as effective tools to draw students into online discussions that are constructivist, collaborative approaches to creating meaning (p. 60).

Nevertheless, the effectiveness of blended learning still depends on teachers being able to meet the challenge of using the learning management system tools, designing learning experiences to foster learner communicative interactions, provide face-to-face support, and ensure the quality of the DCTs. Given the variation in learner backgrounds, the way students might engage with the DCTs to use and improve their English language proficiency, and their need to understand the changed pedagogy also is an important consideration. For instance, student from developing countries may not be aware of such innovative teaching methods, and users who experience difficulties with technology may drop out unless there is opportunity to discuss (Hofmann, 2014). As Tang (2013, p. 81) put it "ESL teachers should guide students to integrate the online learning with the class learning to conduct blended learning, which can provide the ESL learner a better learning environment, and encourage them to conduct self-regulated learning, collaborative learning". Blended learning, therefore does have the capacity to enable English language use for authentic purposes across macro skills, connecting students in collaborative tasks that link to formative assessment benchmarks.

Collaborative Approach to Knowledge Building

According to Vygotsky (1978) collaborative learning is necessary for assisting students to move through their Zone of Proximal Development (ZPD). Vygotsky (1978) defined the ZPD as "the distance between the actual development level as determined by independent problem solving and the level of potential development as determined through problem solving under adult guidance or in collaboration with more capable peer" (p. 86). Applied to Tochon's (2014) deep approach this occurs during students' work in collaboration with peers in project- or problem- based learning. In this way the ESL/EFL student has opportunity to internalize new concepts, and restructure skills as they need to respond to the demands of a task. Humphries (2010) explains that collaborative activities provide students with authority and responsibility in their learning. Carrio-Pastor and Skorczynska (2015) emphasize, "this learning process entails concepts such as collective learning, information acquisition, technological skills and processes of meaning negotiation. This contributes to the reason why collaborative activities should take advantage of DCTs, as they facilitate collaborative activities (p. 33). The pedagogy of collaborative knowledge building requires new learning assessments that can measure both individual and collective knowledge and provide feedback on students' inquiry and collaboration during the process (Chen & Zhang, 2016). As Gibson (2013, p. 61) notes, "the asynchronous online discussion environment offers unique opportunities for students and instructors to construct knowledge through formative assessment". Discussions may be designed to engage learners in writing and thinking skills, or verbal discussions via video conferencing for example. As dialogic activity it lends itself to a collaborative and reflective approach where learners become active agents in their own learning (Chen & Zhang, 2016; Tochon, 2014). Carrio-Pastor and Skorczynska (2015, p. 33) identify the following five aspects to support the design of collaborative learning activities: (1) to group students, taking into consideration language level, (2) to give clear and simple instructions to the groups, (3). to encourage interaction among the members of the collaborative activity, (4) to restrict the information to be retrieved by students, (5) to encourage work in close cooperation (to discuss, to plan, and to assume responsibilities). In this way the learning environment is controlled sufficiently to maximize opportunities for learning.

Open Educational Resources (OER) and Use of Video Games

There are numerous online educational resources freely available for both teachers and learners. These include digital media, multimodal texts, such as YouTube videos, including TED talks, the Khan Academy, Open Textbook Library, academic earth, MIT's OpenCourseWare, Apple's iTunes, and the UK Open University's OpenLearn, documentaries, short movies, lectures, animations, simulations, virtual worlds and the like that enable teachers and students to access and apply knowledge in a wide variety of ways. However, barriers to selection and integration of appropriate OER by teachers and students has been raised as an issue by Bonk, Lee, Kou, Xu and Sheu (2015).

One of the important parts of learning EFL is the acquisition of vocabulary. Using educational games, instructional video games for teaching English vocabulary have been very popular for several decades (Hamari, Shernoff, Rowe, Coller, Asbell-Clarke, & Edwards, 2016; Soyoof, 2018). In Chen and Yang's (2013) reported that students who played video games improved their English vocabulary acquisition significantly. 'The fun factor' was identified as the reason that students engaged in playing games, while vocabulary learning ocurred incidentally. Overall, students were found to hold positive attitudes toward the use of video games for language learning (Chen & Yang, 2013; Kafai & Burke, 2015). This was reinforced by Cabraja's (2016) who also found Swedish L2 learners increased their English vocabulary through playing of video games compared with those who did not play. On the other hand, Rankin, Gold, and Gooch's (2006) research into university ESL students who played 'EverQuest 2' video game also found students' learned incidentally, but they also increased in confidence in their English language skills use. This video game experience also proved beneficial as the results were found to be similar to outcomes where students had opportunity to interact with native speakers, thus reinforcing the importance of ESL students participating in opportunities to interact in the English language.

Similarly, Thorne, Black and Sykes' (2009) investigation of ESL students' online interactions using video games, e.g. *The World of Warcraft*, found they were motivated to interact multilingually, which developed opportunities for intercultural communication, and promoted the idea of learning languages other than English. Participants were found to be mutually cooperative in their efforts to make meaning, and supported each other making corrections. Video games that are suitable for educational purposes have been found to contain substantial dialogues, which students as players interact with and communicate, and ultimately use and practice. This gaming experience was found to help the students learn the language freely besides acquire problem solving skills. However, other research has not been as convincing in terms of fostering language learning through the use of video games. For example, Fujii's (2010) research into the the effects of an adventure puzzle video game on Japanese EFL university students during a one month period found only slight improvement in English proficiency. In the case of deHaan, Reed, and Kuwada's (2010) study that compared the English language improvement of students who played the game versus those who watched the game found the 'watchers' learnt more. deHaan et al. (2010) suggested that the cognitive demands required for playing the game may have detracted from their ability to focus on the language needed to play the game per se. Thus, the cognitive demand of playing video games was suggested as a possible interference in the English language learning process. The design of video games from the perspective of providing an opportunity to use the English language for real-life purposes needs to be carefully considered (Gee, 2007). Such games need to incorporate principles of effective learning and teaching besides being geared to understanding how second languages are acquired.

Video game design from the perspective of language use demands and vocabulary needs to be linked to benchmark levels of proficiency e.g. the ISLPR or (ACTFEL). It is not sufficient to know that students enjoy playing a particular, video game, rather in keeping with Tochon's (2014) deep approach, they need to be engaged in using the language at their optimal level with the challenge to develop further through appropriate formative assessment that is built in to the game. In this way through strategic game design students would gradually move up the proficiency scale. Needing English to play video games provides a strong incentive for many students to acquire the language through need to communicate. Tochon's (2014) advocacy for students to engage in project- or problem- based learning similarly provides a context to stimulate the use the language for real-life purposes. However, feedback to students on their ability to progress their English language proficiency is essential to maintain motivation and this needs to be ongoing and therefore formative. A strength of video games, is the ability to design activities at different levels, expanding vocabulary, increasing sentence complexity yet supported by the story and structure (Howanitz, 2013; Kafai & Burke, 2015; Parsayi & Soyoof, 2018; Shokri & Abdolmanafi-Rokni, 2014). As well, OER are also being used to provide opportunities for ESL/EFL students to verbally communicate in English through the use mobile phone technology with automated speaking assessment to measure oral proficiency (Blake et al., 2008). This kind of system was able to collect evidence of improvement to students' oral proficiency as part of ongoing formative assessment, which was seen as comparable to the results of more conventional learning experiences.

There are also OER that involve software applications that provide Classroom Response Systems (CRS) that are useful to support teachers' formative assessment. These include, but are not limited to, Plickers, Clickers, Socrative, Kahoot, Plickers and RecaP (Elmahdi et al., 2018). Plickers, for example, is a free tool that allows teachers to create assessment activities accessible online. Plickers and the like have broad application in:

monitoring of students' work that allows for adaptivity [and] also creates the opportunity to gather an additional type of data that can be relevant to assessing students' abilities: the amount of time they spend on each item can be recorded and used as an indicator of the automaticity or fluency of test takers' performance (Chapelle & Voss, 2016, p.118).

Teachers are able to access guidelines for use and can keep a record of the assessment items they develop thus building a questions bank (Elmahdi et al., 2018; Gürışık & Demirkan, 2019; Winkle, 2017).

When looking at the role of technology in language learning instructions, assessments and their effect in the whole educational process, a wide body of research has provided evidence of the significant contribution that technology makes in improving the interactive learning environment, teaching methodology, subject knowledge and skills (Elmahdi et al., 2018; Gürışık & Demirkan, 2019; Irving, 2015; Thienthong & Lian, 2014). Elmahdi et al. (2018) found Plickers was very useful as a formative assessment tool in the learning process because it improved students' learning and saved time. It was also found to provide more equal participation opportunities and created a more dynamic learning environment that students enjoyed. Similarly, Gürışık and Demirkan (2019) found it valuable in working with high school students where for the purposes of formative assessment students were able to receive immediate feedback on their incorrect answers, which increased their motivation to learn. They preferred Plickers to conventional pen and paper tests.

Other studies have considered readily available DCTs as learning resources showing them to be advantageous in the learning and assessment space. Alsaleem's (2014) study on the use of WhatsApp, where uses communicate using a variety of media, found students who used it to compile English dialogue journals improved writing, vocabulary, word choice, and speaking ability. Whereas, Chen and Cheng (2008) investigate MyAccess! software which also supports communicative interactions that involve authoring tools online. They found that as a pedagogical tool and a recording device it provided diagnostic feedback that formed an important part of teachers' formative assessment practices. Examination of Wiki technology by Lin and Yang (2011) also found that this technology was able to improve students' writing skills. Students were engaged in write passages on a Wiki page that was shared with among peers. The strategy entailed students reading their peers' work and answering questions about it. This practice was found to improve students' vocabulary, spelling, and sentence structure. The fact that the activity was recorded and easily accessible by all participants adds power to its usefulness, since it equates to being able to build-in the collection of data to facilitate monitoring of students' performance and formative assessment. This is also borne out by Peregoy and Boyle's (2012) research that found students' access to Internet resources provided encouraged their use of reading and writing skills.

The richness of the learning environment was seen as assisting students to communicate in English for more authentic purposes and to learn more effectively and faster. With regard to these skills the technology enhanced learning and assessment environment can also promote the metacognitive strategies involved (Walsh, 2006; Wang, 2013). Teo (2012), for instance, in the teaching of EFL in Taiwan, found that the web-based Computerized Dynamic Assessment (C-DA) program helped improve "the learners' metacognitive strategies, especially with the intention of training them to be strategic and reflective readers" (Teo, 2012, p. 12). Similarly, Wei, Chen & Adawu (2014) found that the use of multimedia graphic-organizer software to teacher ESL beginners metacognitive writing strategies through a Cognitive Academic Language Learning Approach (CALLA) was very successful in teaching planning and organising strategies. These are only some of the many applications and strategies that can enhance learning through DCTs. As the Internet becomes more available worldwide these OER have the capacity to change traditional practices to allow learners to learn more effectively by using the language to communicate in more authentic ways.

The use of computers and DCTs in language learning have been found to be able to better tailor learning to students' individual abilities and needs, (NAEYC & Fred Rogers Center for Early Learning and Children's Media, 2012). The monitoring and assessment of students' work and learning that allows for adaptivity also facilitates the gathering of additional data that is relevant to learning. This might be recording the amount of time spend on different items of tasks, the generation of automaticity or fluency of performance and students' preferences for learning. When assessments are adaptive, which technological designs are able to create the students are able to work more at their optimum level and be challenged without fear of failure (Chapelle & Voss, 2016). However, as Lazar, (2015) points out teachers may lack professional development to apply pedagogically new educational technologies. This aligns with Campion (2016) view that teachers of ESL/EFL may require professional development to teach the critical reflective skills as well as the intercultural literacy skills (Liddicoat, Papademetre, Scarino, & Kohler, 2003) to be effective in the current information and data rich learning environment. Given that many teachers remain in education sectors, internationally, where policy, curriculum and assessment/testing may still require a traditional view of learning the fact that educational technology and its operation require a set of skills based on pedagogy, psychology, cybernetics didactics, computer sciences,and informatics (Neupokoeva, Chapaev, Akimova, Shcherbin & Borovikov, 2016; Stošić, 2015) there needs to be a major shift to maximize the uptake of change (Bindé, 2005; O'Neill, 2019).

Mobile Learning or e-Learning

Mobile learning (m-learning) or e-learning has emerged way of learning through personal mobile devices, such as smartphones, iPads and Tablets, as well as MP3/ MP4 players and handheld gaming devices like Sony PSP, Nintendo DS, Ultramobile PCs (UMPCs), mini notebooks, handheld GPS or voting devices, and specialist portable technologies applicable to science and engineering. These technologies are also supported by video conferencing capability through Skype, Zoom and Facetime videotelephony. The opportunities to support pedagogy, learning and assessment through the uptake of such mobile technologies possibilities appear endless. The fact that they can be used anywhere, anytime and are mostly in the public domain (Bonk, et al., 2015; UNESCO, 2012), those with access have the potential to maximize teaching and learning. As Bindé (2005) points out access to the Internet or not is the new divide across the world with regards to education. Education in the higher education sector is largely accessible online across the world and many courses to upskill are offered through open networks, universities and many private providers. Five different ways of e-learning include: (1) learning that is learner-led, (2) learning that is teacher led, (3) learning that is embedded or (4) facilitated, and learning that is (5) telementored (Gulbahar, 2009), all of which can be mediated through mobile devices and used to support language learning and formative assessment. Large scale education operations such as universities provide their online education through specially designed learning management systems such as Modular Object-Oriented Dynamic Learning Environment (Moodle) (Suppasetseree & Dennis, 2010; Tan, 2013). Moodle provides an online learning environment that has flexibility on the way educators may design. However, it has the capacity to support asynchronous and synchronous learning, provide links to Internet resources and support video conferencing technology. It supports the building in of assessment tasks and other

Figure 1. Teaching design of the blended learning of ESL on a Moodle platform
Source: Tang (2013, p. 32).

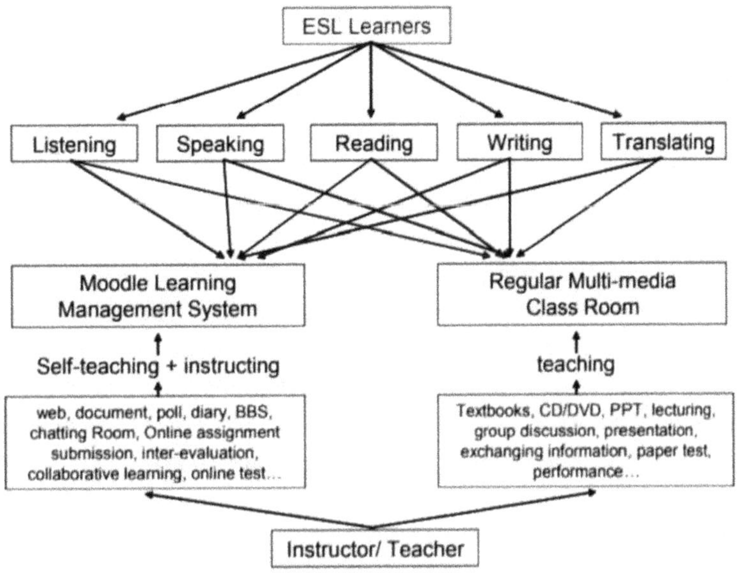

feature in keeping with the principles of social constructivist learning. allowing learners to contribute to their learning in different ways. For example, posting comments or discuss a topic on online discussion board (Tang, 2013). Figure 1 illustrates the teaching design of blended learning in ESL on a Moodle platform as developed by Tang (2013, p, 32).

Figure 1 outlines how the various aspects of this context for ESL learning relate to each other and provide both direct instruction and opportunities for students to be independent in their learning. Thorough the Moodle platform teachers design and provide resources and tasks, and particularly communicative interactive activities where the English language needs to be used. This may include meet-ups, forums, online learning modules and resources to work with, group work, and various assessments, including self-assessment, formative assessment and summative assessment instructions.

Additionally, the learning system would support multimedia resources such as graphics, video or audio (e.g., MP3 files), PowerPoint, or Flash-based applications and Java applets text-based or Html formatted documents. Learning systems such as Moodle support interactivity and collaboration, the grouping of students for online tutorials. They foster flexibility and are distance friendly in facilitating learning designs that are "personally meaningful to students in their achievement of educationally worthwhile outcomes" (Anderson, Rourke, Garrison, & Archer, 2001, p. 5). Moreover, the Moodle system is seen as being able to address both small and large scale education provider needs (Shweta & Shri Ram, 2009) and allow teachers to create learning environments that "support ESL/EFL students' communicative competency" (Soliman, 2014, p. 755) and their active learning (Baran & Correia, 2009).

However, as noted above access to this new world of learning and assessment depends on access to a stable Internet connection, which is not universal and there remains some learners who prefer learning face-to-face in reality.

FUTURE RESEARCH DIRECTIONS

When considering technology enhanced formative assessment in the area of TESOL there are advantages in terms of its design. Since learning programs are typically designed around the theory of second language acquisition and are able to draw upon well researched development scales and a range of other resources especially designed to identify students' starting points and milestones in ESL/EFL proficiency development, this makes the design of formative assessment somewhat easier. The challenge lies in the pedagogical transformation that is needed for teachers to be able to use the Internet resources discussed here to enhance students language learning. These resources have the capacity to bridge the gap in current language teaching, learning and assessment practices that lack a consistent approach to providing students with systematic 'real-life' opportunities to communicate in English in authentic ways. The second challenge involves the need for research to investigate teachers good practice in using these DCTs to enhance both teaching and assessment, and also what this means for practice in relation to the added need to work with the wide range and wealth of resources available in terms of teaching metacognitive skills and critical thinking. Finally, as more exemplars of students' and teachers' success with using DCTs to achieve greater success with improving English proficiency there is a need to share and investigate how formative assessment was designed, and how a more social constructivist approach was implemented.

CONCLUSION

This chapter addressed how Internet resources feature in the design of English language learning and enhancement of formative assessment practices in TESOL. The review of the literature highlighted how the advent of the Internet has facilitated access to a myriad of readily available resources that can be used to improve the TESOL. However, it needs to be noted that while the development of English language proficiency is well documented and resources are available to guide formative assessment design there are significant issues that currently may influence the success of using DCTs to improve practice and outcomes as would be expected. These relate to the professional learning of language teachers in their ability to adopt and use DCTs in their teaching and modify their approach to move away from traditional pedagogies to those that support students having agency in relation to their own learning and in terms of Tochon's deep approach acknowledge the potential of experiential learning afforded by students being involved in projects and problem-based learning. This signifies a conceptual change for many teachers and students alike yet without this there is less scope for enhancing pedagogy and learning through technology enhanced formative assessment. While the literature review is convincing in identifying the advantages of DCTs in improving language learners' learning and indeed formative assessment that feeds forward, clearly, it is important for teachers to participate in continuous professional development and training to have the knowledge and skills to select and use the variety of technological devices and software applications available.

REFERENCES

Alsaleem, B. I. A. (2014). The effect of "WhatsApp" electronic dialogue journaling on improving writing vocabulary word choice and voice of EFL undergraduate Saudi students. *21st Century Academic Forum Conference Proceedings*. Retrieved from http://www.readwritethink.org/lesson_images/lesson782/Rubric.pdf

Anderson, T., Rourke, L., Garrison, D. R., & Archer, W. (2001). Assessing teaching presence in a computer conference environment. *Journal of Asynchronous Learning Networks*, *5*(2), 1–1.

Assessment Reform Group. (2002). *Assessment for learning: 10 principles*. Nuffield Foundation. Retrieved from http://www.assessment-reform-group.org/CIE3.pdf

Ata, R. (2016). An exploration of higher education teaching in Second Life in the context of blended learning. *Turkish Online Journal of Educational Technology*, *15*(3), 9–26.

Baran, E., & Correia, A. P. (2009). Student-led facilitation strategies in online discussions. *Distance Education*, *30*(3), 339–361. doi:10.1080/01587910903236510

Bennett, R. E. (2011). Formative assessment: A critical review. *Assessment in Education: Principles, Policy & Practice*, *18*(1), 5–25. doi:10.1080/0969594X.2010.513678

Bindé, J. (Coord.) (2005). *Towards knowledge societies: UNESCO world report. UNESCO reference works series*. Paris: UNESCO. Retrieved from http://unesdoc.unesco.org/images/0014/001418/141843e.pdf

Black, P., & Wiliam, D. (1998). *Inside the black box: Raising standards through classroom assessment*. London: King's College School of Education.

Black, P., & Wiliam, D. (2009). Developing the theory of formative assessment. *Educational Assessment, Evaluation and Accountability*, *21*(1), 5–31. doi:10.100711092-008-9068-5

Blake, R. (2013). Brave new digital classroom: Technology and foreign language (2nd. ed.). Georgetown University Press.

Blake, R., Wilson, N. L., Cetto, M., & Pardo-Ballester, C. (2008). Measuring oral proficiency in distance, face-to-face, and blended classrooms. *Language Learning & Technology*, *12*(3), 114–127. Retrieved from http://llt.msu.edu/vol12num3/blakeetal.pdf

Bloom, B. S. (1969). Some theoretical issues relating to educational evaluation. In R. W. Tyler (Ed.), *Educational evaluation: New roles, new means. The 63rd yearbook of the National Society for the Study of Education, Part 2* (Vol. 69, pp. 26–50). Chicago, IL: University of Chicago Press.

Bonk, C. J., Lee, M. M., Kou, X., Xu, S., & Sheu, F.-R. (2015). Understanding the self-directed online learning preferences, goals, achievements, and challenges of MIT OpenCourseWare Subscribers. *Journal of Educational Technology & Society*, *18*(2), 349–368.

Boyle, J. (2013). *Great online resources for ESL/EFL teachers. A selection of Internet tools that can be highly useful to teachers of English as a second (or foreign) language*. Retrieved from http://www.emergingedtech.com/2013/07/8-great-online-resources-for-esl-efl-teachers/

British Council. (2017). *IELTS numbers rise to three million a year*. Retrieved from https://www.britishcouncil.org/organisation/press/ielts-numbers-rise-three-million-year

Brown, H. D. (2010). Language assessment, principles and classroom practices (2nd ed.). New York: Pearson Longman.

Buchem, I., Attwell, G., & Torres, R. (2011), Understanding personal learning environments: Literature review and synthesis through the activity theory lens. *PLE Conference*. Retrieved from http://journal.webscience.org/658/1/PLE_SOU_Paper_ Buchem_Attwell_Torress.doc

Bulut, O., & Kan, A. (2012). Application of computerized adaptive testing to entrance examination for graduate studies in Turkey. *Eurasian Journal of Educational Research*, *49*, 61–80.

Cabraja, A. (2016). *The effects of video games on the receptive vocabulary proficiency of Swedish ESL students*. Retrieved from http://www.diva-portal.org/smash/get/diva2:936329/FULLTEXT01.pdf

Campion, G. C. (2016). The learning never ends: Exploring teachers' views on the transition from General English to EAP. *Journal of English for Academic Purposes*, *3*, 59–70. doi:10.1016/j.jeap.2016.06.003

Carrio-Pastor, M., & Skorczynska, H. (2015). Collaborative learning and communication technologies in teaching business English. *Procedia: Social and Behavioral Sciences*, *178*, 32–37. doi:10.1016/j.sbspro.2015.03.142

Chang, V., Gütl, C., Kopeinik, S., & Williams, R. (2009). Evaluation of collaborative learning settings in 3D virtual worlds. *International Journal of Emerging Technologies in Learning*, *4*(3), 6–17. doi:10.3991.ijet.v453.1112

Chapelle, C. A., & Voss, E. (2016). 20 years of technology and language assessment in *Language Learning & Technology, 20*(2), 116–128. Retrieved from http://llt.msu.edu/issues/june2016/chapellevoss.pdf

Chen, C.-F. E., & Cheng, W.-Y. E. (2008). Beyond the design of automated writing evaluation: Pedagogical practices and perceived learning effectiveness in EFL writing classes. *Language Learning & Technology*, *12*(2), 94–112. Retrieved from http://llt.msu.edu/vol12num2/chencheng.pdf

Chen, J., & Zhang, J. (2016). Design collaborative formative assessment for sustained knowledge building using idea thread mapper. In *Proceedings of the International Conference of the Learning Sciences* (ICLS 2016). Singapore: International Society of the Learning Sciences.

Common European Framework of Reference (CEFR). (n.d.). *Common European framework of reference for languages: Learning, teaching and assessment*. Strasbourg: Cambridge University Press. Retrieve from https://rm.coe.int/1680459f97

Dann, B., Dann, C., & O'Neill, S. (2017). Formative assessment via video feedback on practicum: Implications for higher education and professional teacher accreditation bodies. In T. Richardson, B. Dann, C. Dann, & S. O'Neill (Eds.), *Formative assessment practices for pre-service teacher practicum feedback: Emerging research and opportunities* (pp. 158–183). Hershey, PA: IGI Global.

Deen, M. (2015). *G.A.M.E., Games autonomy motivation and education: How autonomy-supportive game design may improve motivation to learn*. Eindhoven: Technische Universiteit Eindhoven.

DeHaan, J. (2005). Learning language through video games: A theoretical framework, an evaluation of game genres and questions for future research. In S. P. Schaffer & M. L. Price (Eds.), *Interactive convergence: Critical issues in multimedia* (pp. 229-239). London: Inter-disciplinary Press. Retrieved from http://interdisciplinary.net/publishing/id-press/ebooks/interactive-convergence-criticalissues-inmultimedia

DeHaan, J., Reed, W. M., & Kuwada, K. (2010). The effect of interactivity with a music video game on second language vocabulary recall. *Language Learning & Technology, 14*(2), 74–94.

Elmahdi, I., Abdulghani, A.-H., & Fawzi, H. (2018). Using technology for formative assessment to improve students' learning. *TOJET: Turkish Online Journal of Educational Technology, 17*(2), 182–188.

Farooq, M. U., & Soomro, A. F. (2018). Teachers and technology: Trends in English language teaching in Saudi Arabia. *International Journal of English Linguistics, 8*(5), 10–19. doi:10.5539/ijel.v8n5p10

Fu, I., Chuang, Y., Fu, S., & Beasley, R. (2012). Implementing Cloud technology of NETPAW at I-Shou University. *The APAMALL Higher Education Journal of Language Learning Technologies, 2*(1), 91–119.

Furtak, E. M., Kiemer, K., Circi, R. K., Swanson, R., de Leon, V., Morrison, D., & Heredia, S. C. (2016). Teachers' formative assessment abilities and their relationship to student learning: Findings from a four-year intervention study. *Instructional Science, 44*(3), 267–291. doi:10.100711251-016-9371-3

Gee, J. P. (2003). *What video games have to teach us about learning and literacy.* Basingstoke, UK: Palgrave Macmillan. doi:10.1145/950566.950595

Gee, J. P. (2007). *Good video games and good learning: Collected essays on video games, learning and literacy.* New York: Peter Lang Publishers. doi:10.3726/978-1-4539-1162-4

Gibson, K. M. (2013). Fostering collaboration and learning in asynchronous online environments. *Journal of Teaching and Learning with Technology, 2*(2), 60–78.

Gilakjani, A. P. (2017). A review of the literature on the integration of technology into the learning and teaching of English language skills. *International Journal of English Linguistics, 7*(5), 95–106. doi:10.5539/ijel.v7n5p95

Green, A. (2006). Washback to the learner: Learner and teacher perspectives on IELTS preparation course expectations and outcomes. *Assessing Writing, 11*(2), 113–134. doi:10.1016/j.asw.2006.07.002

Groff, J. (2017). *Personalized learning: The state of the field & future directions.* Center for Curriculum Redesign.

Gürışık, A., & Demirkan, Ö. (2019). Opinion of high school students about plickers. *International Journal of Scientific Research and Innovative Technology, 1*(6), 12–25.

Hamari, J., Shernoff, D. J., Rowe, E., Coller, B., Asbell-Clarke, J., & Edwards, T. (2016). Challenging games help students learn: An empirical study on engagement, flow and immersion in game-based learning. *Computers in Behavior, 54*, 170–179. doi:10.1016/j.chb.2015.07.045

Hamp-Lyons, L. (2007). The impact of testing practices on teaching. In E. Hinkel (Ed.), *The handbook of research in second language teaching and learning* (pp. 487–504). London: Lawrence Erlbaum Associates. doi:10.1007/978-0-387-46301-8_35

Hashemi, M., Azizinezhad, M., Najafi, V., & Nesari, A. J. (2011). What is mobile learning? Challenges and capabilities. *Procedia: Social and Behavioral Sciences, 30*, 2477–2481. doi:10.1016/j.sbspro.2011.10.483

Hismanoglu, M. (2012). Prospective EFL teachers' perceptions of ICT integration: A study of distance higher education in Turkey. *Journal of Educational Technology & Society, 15*(1), 185–196.

Hofmann, J. (2014). *Solutions to the top 10 challenges of blended learning. Top 10 challenges of blended learning*. Available on cedma-europe.org

Howanitz, G. (2013). 'Metro 2033' – More than a cinegame? *Digital Icons: Studies in Russian, Eurasian and Central European New Media, 8*, 97-116. Retrieved from http://www.digitalicons.org/issue08/gernot-howanitz/

Ingram, D. E., & Wylie, E. (1979/1999). *The International Second Language Proficiency Ratings (ISLPR)*. Brisbane: Centre for Applied Linguistics and Languages, Griffith University.

Inman, C., Wright, V. H., & Hartman, J. A. (2010). Use of Second Life in K-12 and higher education: A review of research. *Journal of Interactive Online Learning, 9*, 44–63.

International Language Testing System (ELTS). (2017). *IELTS*. London: British Council. Available at https://www.ielts.org/

Jones, M. G., & Ennes, M. (2018). High-stakes testing. In A. Hynds (Ed.), *Oxford Bibliographies in Education* (pp. 1–12). New York: Oxford University Press. doi:10.1093/OBO_dataset_home

Kafai, Y. B., & Burke, Q. (2015). Constructionist gaming: Understanding the benefits of making games for learning. *Educational Psychologist, 50*(4), 313–334. doi:10.1080/00461520.2015.1124022 PMID:27019536

Krashen, S. (2003). *Explorations in language acquisition and use: The Taipei lectures*. Portsmouth, NH: Heinemann.

Kravcik, M., & Klamma, R. (2012), Supporting self-regulation by personal learning environments. In I. Aedo, R.-M. Bottino, N.-S. Chen, C. Giovannella, D. G. Sampson, & Kinshuk (Eds.), *Proceedings of 12th IEEE International Conference on Advanced Learning Technologies* (pp. 710-711). Rome: IEEE Computer Society. 10.1109/ICALT.2012.192

LaBanca, F., Worwood, M., Schauss, S., LaSala, J., & Donn, J. (2013). *Blended instruction: Exploring student centered pedagogical strategies to promote a technology-enhanced learning environment*. Litchfield, CT: Education Connection.

Lamb, T. E., Hatoss, A., & O'Neill, S. (in press). Challenging social justice in superdiverse contexts through 'activist' languages education. In R. Papa (Ed.), *Handbook on Promoting Social Justice in Education*. Springer.

Larsen-Freeman, D., & Anderson, M. (2011). *Techniques and principles in language teaching*. Oxford, UK: Oxford University Press.

Liddicoat, A. J., Papademetre, L., Scarino, A., & Kohler, M. (2003). *Report on intercultural language learning*. Canberra: Department of Education, Science and Training.

Lin, W., & Yang, S. (2011). Exploring students' perceptions of integrating Wiki technology and peer feedback into English writing courses. *English Teaching, 10*(2), 88–103.

Litzler, M. F., & García Laborda, J. (2016). *Students' opinions about ubiquitous delivery of standardized English exams*. Retrieved from https://files.eric.ed.gov/fulltext/ED577152.pdf

Magnan, S. S., & Back, M. (2007). Social interaction and linguistic gain during study abroad. *Foreign Language Annals, 40*(1), 43–61. doi:10.1111/j.1944-9720.2007.tb02853.x

Mellati, M., Khademi, M., & Shirzad, A. (2015). The relationships among sources of teacher pedagogical beliefs, teaching experiences, and student outcomes. *International Journal of Applied Linguistics and English Literature, 4*(2), 177–184. doi:10.7575/aiac.ijalel.v.4n.2p.177

Mndzebele, N. (2013). Teachers' readiness in using ICT in the classroom: The case of a developing country. *International Journal of Information and Education Technology (IJIET), 3*(4), 409–412. doi:10.7763/IJIET.2013.V3.309

Moje, E. B., Ciechanowski, K. M., Kramer, K., Ellis, L., Carrillo, R., & Collazo, T. (2004). Working toward third space in content area literacy: An examination of everyday funds of knowledge and discourse. *Reading Research Quarterly, 39*(1), 38–70. doi:10.1598/RRQ.39.1.4

NAEYC & Fred Rogers Center for Early Learning and Children's Media. (2012). *Technology and interactive media as tools in early childhood programs serving children from birth through age 8*. Joint position statement. Washington, DC: NAEYC; Latrobe, PA: Fred Rogers Center at St. Vincent College. Retrieved from www.naeyc.org/content/technology-and-young-children

Neupokoeva, E. E., Chapaev, N. K., Akimova, O. B., Shcherbin, M. D., & Borovikov, E. A. (2016). Open access using educational technology in applications as element of teaching for special discipines. *International Journal of Environmental and Science Education, 11*(16), 8858–8872.

O'Neill, S. (2012a). Implementing NETPAW's test of English proficiency in Australia: A case study. [Winter.]. *Language Learning Technologies: An International Online Journal, 2*(1), 121–144.

O'Neill, S. (2012b). Teaching and assessment of persuasive writing: Juggling the language features and grasping the metalanguage. *International Journal of Pedagogies and Learning, 7*(1), 84–98. doi:10.5172/ijpl.2012.7.1.84

O'Neill, S. (2015). EFL teachers' professional learning needs: Working with multimedia and the CLOUD. In L. T. Wong & A. Dubey-Jhaveri (Eds.), *English language education in a global world: Practices, issues and challenges* (pp. 47–60). New York: Nova Science Publishers.

O'Neill, S. (2018). Building students' capacity to write English for academic purposes: Pedagogy and the demands of writing persuasively. In L. T. Wong & W. L. Wong (Eds.), *Teaching and learning English for Academic Purposes: Current research and practices* (pp. 69–96). New York: Nova Science Publishers.

O'Neill, S. (2019). New opportunities for languages learning through 21st Century knowledge building communities. In R. Arber, M. Weinmann, & J. Blackmore (Eds.), *Rethinking languages education: Directions, challenges and innovations*. New York: Routledge.

Osman, K., & Bakar, N. A. (2013). Teachers and students as game designers: Designing games for new pedagogical approaches in game enhanced learning. *Curriculum Integration*, 102-113.

Parsayi, F., & Soyoof, A. (2018). Video games: The interface between language learning and storytelling. *International Journal of Pedagogies and Learning*, *13*(2), 103–118.

Peregoy, S., & Boyle, O. (2012). *Reading, writing and learning in ESL: A resource book for teachers.* New York: Allyn & Bacon.

Peterson, P. M., & Helms, R. M. (2013). *Challenges and opportunities for the global engagement of higher education.* Occasional Paper produced for CIGE: American Council on Education. Washington, DC: American Council of Education.

Povah, C., & Vaukins, S. (2018). *Generation Z is starting university – but is higher education ready?* Retrieved from https://www.theguardian.com/higher-educationnetwork/2017/jul/10/generation-z-starting-university-highereducation-ready

Pryor, J., & Crossouard, B. (2008). A socio-cultural theorisation of formative assessment. *Oxford Review of Education*, *34*(1), 1–20. doi:10.1080/03054980701476386

Richardson, A., Hazzard, M., Challman, S. D., Morgenstein, A. M., & Brueckner, J. K. (2011). A "second life" for gross anatomy: Applications for multiuser virtual environments in teaching the anatomical sciences. *Anatomical Sciences Education*, *4*(1), 39–43. doi:10.1002/ase.195 PMID:21265036

Schaffhauser, D. (2012). As high-stakes online testing approaches, Will your iPads work? *Assessment, Digital Edition.* Retrieved from http://thejournal.com/Articles/2012/10/11/You-May-Now-Open-Your-Test-Tablets.aspx

Shokri, H., & Abdolmanafi-Rokni, S. J. (2014). The impact of computer games on EFL learners' spelling: A qualitative study. *Studies in English Language Teaching*, *2*(3), 266–274. doi:10.22158elt.v2n3p266

Shore, J. R., Wolf, M.-K., & Heritage, M. (2016). A case study of formative assessment to support teaching of reading comprehension for English learners. *Journal of Educational Research and Innovation*, *5*(2), 4. Retrieved from http://digscholarship.unco.edu/jeri/vol5/iss2/4

Shweta, P., & Shri Ram, P. (2009). *Developing a more effective and flexible learning management system (LMS) for the academic institutions using moodle*. ICAL.

Sloan Consortium. (2019). *Online learning consortium.* Retrieved from http://sloanconsortium.org/

Soliman, N. (2014). Using e-Learning to develop EFL students' language skills and activate their independent learning. *Creative Education.*, *5*(10), 752–757. doi:10.4236/ce.2014.510088

Soyoof, A. (2018). Video games: A springboard for enhancing students' L2C. *International Journal of Pedagogies and Learning*, *13*(2), 137–151.

Stockwell, G., & Hubbard, P. (2013). *Some emerging principles for mobile-assisted language learning.* Monterey, CA: The International Research Foundation for English Language Education; Retrieved from http://www.tirfonline.org/english-in-the-workforce/mobile-assisted-language-learning

Stošić, L. (2015). The importance of educational technology in teaching. *International Journal of Cognitive Research in Science Engineering and Education, 3*(1), 111–114.

Suppasetseree, S., & Dennis, N. K. (2010). *The use of Moodle for teaching and learning English at tertiary level in Thailand. The International Journal of Humanities.*

Tabatabaei, M., & Gui, Y. (2011). The impact of technology on teaching and learning languages. In A. Méndez-Vilas (Ed.), *Education in a technological world: Communicating current and emerging research and technological efforts* (pp. 513–517). Retrieved from http://www.formatex.info/ict/book/isbn-contents.pdf

Tang, J. (2013). The research on blended learning of ESL based on Moodle platform. *Studies in Literature and Language, 6*(2), 30–34. doi:10.3968/j.sll.1923156320130602.3612

Tang, J. (2013). The research on blended learning of ESL based on a Moodle platform. *Studies in Literature and Language, 6*(2), 30–34. doi:10.3698/j.sll.1923156320130602.3612

Taylor, D. 1994). Inauthentic authenticity or authentic inauthenticity? *TESL-EJ, 1*(2). Retrieved from http://www.tesl-ej.org/wordpress/issues/volume1/ej02/ej02a1/

Teo, A. (2012). Promoting EFL students' inferential reading skills through computerized dynamic assessment. *Language Learning & Technology, 16*(3), 10–20. Retrieved from http://llt.msu.edu/issues/october2012/action.pdf

Thienthong, A., & Lian, A. (2014). The use of Internet resources and applications for Language instruction. *Voices in Asia., 1*, 107–116.

Thompson, P. (2013). The digital natives as learners: Technology use patterns and approaches to learning. *Computers & Education, 65*, 12–33. doi:10.1016/j.compedu.2012.12.022

Thornc, S. L., Black, R. W., & Sykes, J. M. (2009). Second language use, socialization, and learning in internet interest communities and online gaming. *Modern Language Journal, 93*, 802–821. doi:10.1111/j.1540-4781.2009.00974.x

Tochon, T., Karaman Cendel, A., & Ökten, C. (2014). Online instructional personal environment for deep language learning. *International Online Journal of Education and Teaching, 1*(2), 147–173.

Tomlinson, B. (2009). *Materials development in language teaching*. Cambridge, UK: Cambridge University Press.

Torrance, H., & Pryor, J. (1998). *Investigating formative assessment. Teaching, learning and assessment in the classroom*. Buckingham, UK: Open University Press.

Torres, J. O. (2019). Positive impact of utilizing more formative assessment over summative assessment in the EFL/ESL classroom. *Open Journal of Modern Linguistics, 9*(01), 1–11. doi:10.4236/ojml.2019.91001

Towns, S. G., & Loo, D. B. (2012). Current issues in the development ofMultimedia language learning software for mobile devices: Platforms to reach Thai university students. *AsiaCALL 2011 Proceedings and Presentations*, 1-8.

United Nations Educational, Scientific and Cultural Organization (UNESCO). (2012). *Mobile learning for teachers in Europe: exploring the potential of mobile technologies to support teachers and improve practice.* Author.

Van Tol, C. (2016). The global English language job market is changing: what this means for ESL teachers. *University of Toronto TEFL Online*. Retrieved from https://medium.com/accelerated/the-global-english-language-job-market-is-changing-what-this-means-for-esl-teachers-92c8d87cf8eb

Walsh, M. (2006). The "textual shift": Examining the reading process with visual and multimodal texts. *Australian Journal of Language and Literacy*, *29*(1), 24–37.

Wang, K. (2013). Showing or telling a story: A comparative study of public education texts in multimodality and monomodality. *English Language Teaching, 6*(2), 136. doi:10.5539/elt.v6n2p136

Wei, J., Chen, J. C., & Adawu, A. (2014). Teaching ESL beginners metacognitive writing strategies through multimedia software. *The CATSOL Journal*, *26*(1), 60–75.

Winkle, A. (2017). The power of Plickers. *Texas Adult Education & Literacy Quarterly*, *21*(2), 8.

Young, Y. F. (2013). Exploring students' language awareness through intercultural communication in computer-supported collaborative learning. *Journal of Educational Technology & Society*, *16*(2), 325–342.

Section 2
Video Feedback and Feedforward

Chapter 5
Feedback to "Feedforward": Promoting Student Reflection and Learning Through Teacher Evaluation

Carolyn L. Berenato
Cabrini University, USA

ABSTRACT

In recognizing that effective feedback and pedagogical approaches that support social interaction in learning are vital to students' achievement, this chapter focuses on students' need to be actively involved in their learning and assessment. It reports research into how undergraduate education students utilized assessment feedback. The students provided their perceptions of the feedback strategies employed by completing a survey. This included an investigation into whether the students used the feedback to "feed forward," that is, to answer the question as to whether it was used to enhance their responses in their next assessments. The results revealed a contrast between the students' approach to their learning compared with that of the pedagogy underpinning the course, which sought to empower them in their learning through teaching and assessment practices that engaged them in critical thinking. The students appeared to lack assessment literacy since they treated the feedback as corrective of the assignment being marked and not relevant to them taking-action to improve their future work. It is recommended that this disconnect be further explored since without students' understanding of the purposes of assessment and feedback, in keeping with their educators' intent they are limited in their ability to learn.

INTRODUCTION

One of the foremost responsibilities for a professor in higher education is teaching, formative assessment, and feedback. It is known from meta-analyses of research across all levels of education that the factors which have the greatest positive effect of student's achievement are feedback and "whole-class interactive teaching" (Hattie & Timperley, 2007; Hendry, White, & Herbert, 2016; Petty, 2009). Formative assessment can generally be defined as an "assessment which provides the learner with information that allows them to improve their learning and performance" (Gedye, 2010, p. 40). Hattie and Timperley

DOI: 10.4018/978-1-7998-0426-0.ch005

(2007) argue that the "main purpose of feedback is to reduce the gap between current understandings and performance and a goal" (p. 86). Feedback will demonstrate appropriate ways of enhancing the performance on the task, and offer strategies that invite more learner responsibility to improve. Too often professors tend to focus on correctional feedback rather than instructional aspects of feedback (Hattie & Timperley, 2007). In order for feedback to be useful, it should be constructive and help students to develop skills to evaluate their own performance, as well as provide them with opportunities to close gaps between current and desired performance (Nicole & Macfarlance-Dick, 2006). Feedback is an important part of the learning process, students frequently express disappointment and frustration in relation to the feedback process (Edeilken-Cooperman & Berenato, 2014). Students need to be actively involved in learning, knowing the criteria, understanding the goals and purposes of feedback. Feedback needs to be explicitly linked to the assessment criteria with students involved in the feedback process. If the conversation around assessment and feedback includes students' participation through active learning, then feedback will be useful to the students' learning. Both the professor and student should use the phase "feed-forward" to focus on future learning.

The aim of this study was to determine how undergraduate students, Elementary and/or Special Education majors, use professor feedback on their performance on graded assignments or formative assessments. Formative assessment and feedback should empower students and assist them in the development of critical thinking as well as to accelerate learning in a "feed forward" process. A survey was developed that questioned students' perceptions of feedback strategies as well as to discover if students use feedback to "feed forward" and to make subsequent improvement on their assessments.

FORMATIVE ASSESSMENT OVERVIEW

Research on Elementary and/or Special Education Students: Implications for the Understanding University Students Views in the Area of Formative Assessment

Research in undergraduate Elementary and/or Special Education contributed the following summary of the students' views of:

1. Assessment as a formative measure
2. Assessment in the form of personal verses electronic feedback and how useful is the personal and/or electronic feedback
3. Feedback as "feed forward" for students' next assessments

The views of the students' are discussed below.

View 1: Student Assessment as a Formative Measure

It is understood that assessment shapes learning. Assessment defines what students regard as important, how they spend their time, and how they come to see themselves as students and graduates (Brown & Knight, 1994, p. 12). Formative assessment encompasses learning; it attends to the learning process and focuses on lifelong learning. Feedback on assignments/assessments encompasses a variety of assessment purposes, e.g., correction, reinforcement, forensic diagnosis, benchmarks, longitudinal development, and to "feed forward" future assignments. Before assignments are completed, the criteria for grading or marking must be made clear and concise in advance to the students. It is essential that assignment/assessment grading or marks must be graded fairly and that no bias is present. Teacher or professor feedback on students' work must be prompt. The feedback should be prompt so students are able to use the feedback while completing future assignments. The feedback should be in the form of detailed comments that are given to the students. Feedback on the students' work should help to clarify things that were not understood and confirm the validity of correct assumptions to improve the students overall academic performance. Brown (2004) purports the following:

To ensure that assessment is part of the learning process, it should be learner-centered assessment and should reflect a learner-centered curriculum. Assessment methods and approaches need to be focused on evidence of achievement rather than the ability to regurgitate information (p. 82).

Formative assessment encompasses learning; it attends to the learning process and focuses on lifelong learning. Assessment is a key part of the learning process and needs to focus on areas of achievement. The quality and effectiveness of this feedback is essential to promote the learning process and indicate whether the student has adequately met the professor's goals and expectations of the assignment/assessment (Cooperman & Berenato, 2014, p. 79).

View 2: Student Assessment in The Form of Personal Verses Electronic Feedback and How Useful is The Personal and/or Electronic Feedback

Effective feedback, in the electronic or handwritten form, is a valued component of curriculum and instruction in higher education. Feedback, as defined in this article, is information that is provided to a student by a professor or instructor that indicates to that student their level of performance on an assignment/assessment. When feedback is combined with correctional review, the feedback and the instruction should become intertwined and this process should take on the form of new instruction (Cooperman & Berenato, 2014; Hattie & Timperley, 2007). Shute (2008) emphasizes that effective feedback can improve the learning processes and outcomes, but this feedback must be properly processed in order to accomplish this result. Brown (2004) states, "if assessment is to be integral to learning, feedback must be the heart of the process" (p. 84). Generally, feedback is delivered to students in the form of written comments about an assignment/assessment, but today the delivery of these comments can be presented personally or electronically. Minimal research focusing on a students' point of view has been conducted (Brown, 2004), significant research has been completed concerning feedback but it lacks the critical component of the student interpretation of this feedback. (Budge, 2011; Budge & Gopal, 2009; Cooperman & Berenato, 2014; Higgins, Harley, & Skelton, 2002; Murtagh & Baker, 2009; Rowe & Wood, 2008; Weaver, 2006). Feedback constitutes a central aspect of learning and surprisingly current educational research has not focused on feedback as a primary issue, particularly from the students' perspective (Cooperman & Berenato, 2014; Rowe and Wood, 2008).

View 3: Feedback as ""Feed Forward"" for Students' Next Assessments

Biggs and Yang (2011) would argue, in higher education, formative assessment and feedback should be used to empower students as self-regulated learners. Where the assessment feedback model would argue the engagement of the students is minimal (Murtagh et al., 2009). Learners who in theory should use the provided feedback to help develop "deep learning." Feedback strategies should cause the student to "feed forward" and seek opportunities to develop new academic strengths. "Feed forward" is a process accomplished by the student once information has been "communicated to the learner that is intended to modify his or her thinking or behavior for the purpose of improving learning" (Shute, 2008, p. 154). The process of reviewing feedback should link formative assessment with effective pedagogy: teachers sharing learning intentions with students, students engaging in self-evaluation, feedback leading students to next steps, and teachers confident that all students can improve (Murtagh & Baker, 2009). Students should not just be concerned with grades, they should "feed forward" their learning through feedback. The student will modify his or her thinking or behavior for the purpose of improving learning through "feed forward" for the student once information has been communicated to the learner by the professor or instruction in the form of feedback (Shute, 2008). Students should connect with feedback strategies and should "feed forward" to inform and improve future work.

Questions to Guide Analysis of Elementary and/or Special Education Students' Experience of Formative Assessment

1. What do the narratives of the university students participating in the study reflect about the delivery of formative assessments?
2. If these university students' responses do reflect using professor/instructor feedback, how is this manifested?
3. To what extent do participating students use professor's feedback in order to improve future assignments? Do students keep track of feedback given from professor/instructors and use the data to inform future assignments? How can I use what I learned in the feedback process to shape and inform future assignments?

Framing the Study: Undergraduate Student's Experiences within Feedback to Inform Future Assignments

This analysis considers what university students gain from involvement with formative assessment results. This study tends to the cognitive and, more specifically, the reflective: the knowledge, understanding, ideas and insights that university students develop as they use feedback from their formative assessments to "feed forward" in future assignments. We refer to this professor/ instructor feedback as developmental gains contributing to the university students' understanding of their work. In line with this view, we frame feedback as:

1. Information communicated to the university student in the form of written comments (handwritten or electronic);
2. Intended to improve learning by offering constructive criticism, comments, and opportunities for revision;
3. Allowing ways to improve future assignments, "feed forward".

Formative assessment (Black and William, 1998) found that effective assessment can play a powerful role in the learning experience. Students' tasks need to be aligned, or on target, with learning goals, and students receive meaningful timely feedback on their performance, as well as target follow-up work. To regulate the students' learning effectively, students need to understand three things: the measure on which they will be judged, where they stand on these measures, and how they can improve (p. 14). Providing effective feedback to students on their work is an important part of teaching, and its significance is widely recognized in education (Ramsden, 1992). Formative assessments are designed with the intention of providing feedback on students' work (Knight, 1995). The effectiveness of feedback can be judged by the extent to which students use the information to alter future work. We offer that university students develop their understanding of feedback by reflecting on their work as they completed it and after it. With their reflectiveness, university students in effect learn about their learning on the spot and initiating processes of learning in the future.

STUDY DESIGN AND METHOD

This study derives from 136 undergraduate students from the Education Unit at an East Coast, private, coeducational, Roman Catholic Jesuit University to take part in an investigation of the students' perceptions pertaining to formative assessment and feedback on completed assessments/assignments. These students were majoring in the areas of Elementary and/or Special Education. Chang, Watson, Bakerson, Williams, McGoron, and Spitzer, (2012) found using a mixed methodology research design is a sound way to obtain different data on student perceptions pertaining to formative assessment and feedback. They note "the mixed methodology approach combines the differing strengths and weaknesses of the quantitative methods with those of qualitative methods" (p. 5). A mixed methodology approach was used in this study. Fifty students responded to the provided survey, making the return rate 37%. Table 1, shows the majority of the participants (36) were sophomores and juniors in each case, thus, 72% of the sample. Freshman and seniors comprised the remaining 14 participants (28% of the sample). It displays the students' demographic data; 98% identified as females, while 2% identified as males. All were ranging from 18 to 22 years of age and Table 2 shows fifty-four percent of the participants indicated that their GPA was in the 3.5-4.0 range. Forty-two percent of the participants reported a GPA range of 3.0-3.4 and four percent of reported a 2.5-2.9 GPA range. Sixty-eight percent of the students were double majors of elementary and special education, while 32% were elementary majors, pre-kindergarten to grade 4, and middle school majors.

Two months after the spring semester started, all undergraduates admitted into the Teacher Education and Special Education preparation programs were invited to participate in the study via their university email. The survey questions were developed by the two researchers and member checked by a faculty member with expertise in instructional technology. The University's registrar's office sent the potential participants to a prompt on an online site with a consent letter that informed them of the purpose of the study, ensured confidentiality, and that participation in the study was voluntary. If the potential respon-

Table 1. Student demographic background

Variables	Types	% (n)
Gender	Male	2% (1)
	Female	98% (49)
Academic Class	Freshman	12% (6)
	Sophomore	36% (18)
	Junior	36% (18)
	Senior	16% (8)

Table 2. Student academic background

Variables	Types	% (n)
Major	Elementary (PreK-4)	26% (13)
	Elementary/Middle (4-8)	6% (3)
	Spec Ed/Elementary	68% (34)
	Secondary	0% (0)
	Undeclared	0% (0)
GPA Range	3.5-4.0	54% (27)
	3.0-3.4	42% (21)
	2.5-2.9	4% (2)
	Below 2.5	0% (0)

dents agreed to participate, they continued on to complete the survey. The survey was available directly in the student's email account. The participants had to respond to all questions and the data was collected anonymously. An online application survey was used for data collection. Responses to the survey were recorded within Survey Monkey® and then nonparametric tests were utilized along with an analysis of qualitative responses.

The researchers read and analyzed the respondents' responses for emergent themes. Four themes were identified from the provided data parameters. These were: 1) accessibility to the professor, 2) timeliness of the graded assessments, 3) legibility of the professor comments and quality of the feedback, and 4) the degree of personal attention from the professor. Once the coders further defined the themes, the researchers then checked the codes for reliability. For the purpose of coding and analysis, as explained earlier, "feed forward" is defined as a process accomplished by the student once information has been "communicated to the learner that is intended to modify his or her thinking or behavior for the purpose of improving learning" (Shute, 2008, p. 154).

Analysis of Elementary and/or Special Education Students Experiences with the narratives of the university students participating in the study reflect about the delivery of formative assessment

In response to the open-ended question, "Please define electronic feedback below and then provide one example to help us understand it better," students were in agreement that this referred to receiving feedback either on Blackboard9™, which is the learning management strategy currently in use at this University, or by email. Examples referenced included receiving online comments, grades, insight, or completed rubrics attached to the assignment being graded. Further analysis showed that with regards to preferred feedback timeliness this was highly regarded. In responding to the question students noted, "This type of feedback allows me to get prompt feedback on academic performance", but only 2% reported they received prompt feedback 100% of the time, while 12% received prompt feedback 90-99% of the time, 46% received prompt feedback 80-89% of the time, 32% received prompt feedback 70-79% of the time, and 8% did not feel they ever received prompt feedback. Students strongly agreed that they kept track of the feedback that professors/instructors gave and used the data to construct future assignments. However, students perceived it did not matter whether the feedback was electronic or handwritten in response to "this type of feedback allows me to get information easily."

Table 3 shows a summary of responses to the survey closed questions. The percentage of the students that preferred electronic feedback was 36%, compared with handwritten feedback, which was preferred by 64%. It should be noted that this two-thirds majority preference of handwritten comments is an area that needs to be explored for providing effective feedback to students. The students overwhelmingly agreed that it is easier to get information with both handwritten and electronic feedback (66%), as seen in Table 3, as opposed to handwritten feedback (16%), and electronic feedback (18%). In addition, Table 3 indicates that students felt as though they were encouraged to read feedback when it was handwritten (34%), rather than the 10% that were encouraged by electronic feedback and both handwritten and electronic (56%). Students believed that when they received both handwritten and electronic feedback the professors cared about them and the other students (66%), and that they appreciated the professor's time and attention with handwritten feedback as well (60%), also exemplified in Table 3.

Table 3. Responses to closed-ended questions

Question	Electronic	Handwritten	Both	Neither
What kind do you prefer?	36% (18)	64% (32)		
This feedback enables me to read the feedback. (Legibility)	78% (39)	4% (2)	18% (9)	
This feedback allows me to get information easily.	28% (14)	14% (7)	54% (27)	4% (2)
This feedback allows me to establish a rapport with my professor.	18% (9)	26% (13)	56% (28)	
This feedback allows me to receive and send information conveniently.	78% (39)	8% (4)	14% (7)	
This feedback allows me to receive feedback fast.	80% (40)	0% (0)	20% (10)	

Analysis of Elementary and/or Special Education Students Experiences with Responses to Reflect Using Professor/ Instructor Feedback: How it is Manifested

Table 4 reports students responses to closed ended questions and related comments. In response to the question, "I use professor's feedback in order to improve future assignments," students were in strong agreement that they used comments and feedback to improve future assignments. Students commented, "a professor submits a comment on my grade or complete electronic rubrics. They use the information constructively and are actively aware of this commentary when submitting future papers." But when it came to hearing the term "feed forward", the idea of feeding forward feedback from a professor was an unfamiliar term. The students described using "feed forward" when they asked themselves, "how can I use what I learned in the feedback process to inform future assignments". Yet the students did not refer to it as "feed forward". It is critical to discover whether or not this term "feed forward" resonates with students in an effective manner.

Table 4. Responses to closed-ended questions with comments

Closed Question	%(n) Response
Have you ever heard of the term "feed forward"?	98% (49) No
I use professor's feedback to improve future assignments?	96% (48) Yes
Do you ask yourself "how can I use what I learned in the feedback process" to inform future assignments?	94% (47) Yes

Analysis of Elementary and/or Special Education Students' Experiences with Professors' Feedback to Improve Future Assignments

This section provides the analysis of the education students' experiences with their professor's feedback to improve future assignments and the tracking of feedback given from professors/instructors using the data to inform future assignments, along with learning in the feedback process to shape and inform future assignments. The study informed the researchers that although 96% of the students who responded to the survey use what they have learned in the feedback process to inform them for future assignments, 98% of them had never heard of the term "feed forward". In addition, it was determined that 94% of the students store the information received or keep track of the feedback from the professor to inform them of expectations on future assignments. A majority of the students, 64%, preferred handwritten feedback rather than electronic feedback, but 68% of the students believed that either handwritten or electronic feedback allowed them to make revisions and improve on future assignments. Lastly, 32% of the students reported that electronic feedback enabled them to understand the expectations of the professor as opposed to 8% who report that handwritten feedback accomplishes this task, but overall 60% of the students reported that either handwritten or electronic feedback informed them of the professor's expectations and goals of the assignment/assessment. The opened ended questions all resulted in similar themes. The overall theme that was expressed in the open-ended questions was that feedback should be returned in a timely manner in order to encourage students to do well in their coursework, and to give them an opportunity to improve or correct mistakes on the following assignment/assessment.

CONCLUSION AND IMPLICATIONS: DEEPENING VIEWS OF "FEED FORWARD" IN STUDENTS' WORK

As indicated in Sadler's (1989) research, three conditions must be present for students to benefit from the process of feedback:

1. What good performance is (i.e. the student must possess a concept of the goal or standard being aimed for).
2. How current performance relates to good performance (for this, the student must be able to compare current and good performance).
3. How to act to close the gap between current and good performance (Nicol & Macfarlane-Dick, 2006, p. 204).

Students' perceptions feed forward reported here provided a surprising discovery that many of these students did not use "feed forward" productively. The analysis of the data suggests that students appear to be driven solely by individual graded assignments and not the overall improvement of their critical thinking. A disconnect was revealed between professor and the students' interpretation of that feedback. It appears that students view individual assignment feedback as corrective of only that particular assignment as opposed to an indication of areas to improve on all future work. With the data provided, this project should be further researched taking into consideration the prior discovers researchers have revealed that "good quality feedback must be accurate, timely, comprehensive, and appropriate" (Quinton & Smallbone, 2010, p. 128; Gibbs & Simpson, 2004; Sadler, 1998; Weaver, 2006). Students should use this feedback "to monitor the strengths and weaknesses of their performances, so that aspects associated with success or high quality can be recognized and reinforced, and unsatisfactory aspects modified or improved" (Sadler, 1998, p. 20). In contrast, it has been suggested by some researchers that university students are primarily concerned with a mark or grade on an assessment and display minimal interest in written or oral comments that have been offered by the professor (Duncan, 2007; Wotjas, 1998).

Future research needs to further examine this disconnect between educators and their students to understand why "feed forward" is not being utilized to its full potential. Currently, the effort placed into grading assignments with constructive comments by educators appears to be a waste of time and energy when the majority of students do not take the time to constructively interpret the effort placed by the professors to better the students' overall performance. The data provide by the set of 50 current college students was extremely useful in regards to constructive thought as well as interpretation of students in consideration of whether an assignment was graded with digital feedback versus paper feedback. Further research should be conducted to discover why a disconnect exists between young millennials and the way they interpret education feedback whether it is handwritten or digital from their educators. A possible idea for further research is to conduct a survey with students of what they expect from their educators with graded assignments, which can then lead to discovering where the disconnect exists.

REFERENCES

Biggs, J. B., & Tang, C. (2011). *Teaching for quality learning at university* (4th ed.). New York, NY: McGraw-Hill/Society for Research in Higher Education/ Open University.

Black, P., & William, D. (1998). *Inside the black box: Raising standards through classroom assessment.* London: King's College.

Brown, S. (2004-05). The changing nature of assessment. *Learning and Teaching in Higher Education, 1*(1), 81–89.

Brown, S., & Knight, P. (1994). *Assessing learners in higher education.* London: Routledge, Taylor & Francis Group.

Budge, K. (2011). A desire for the personal: Students perceptions of electronic feedback. *International Journal on Teaching and Learning in Higher Education, 23*(3), 342–349.

Budge, K., & Gopal, S. (2009). Feedback: Working from the student perspective. *Proceedings of the Assessment in Different Dimensions, AT Assessment Conference.*

Chang, N., Watson, A. B., Bakerson, M., Williams, E. E., McGoron, F. X., & Spitzer, B. (2012). Electronic feedback or handwritten feedback: What do undergraduate students prefer and why? *Journal of Teaching and Learning with Technology, 1*(1), 1–13.

Cooperman, N., & Berenato, C. L. (2014). Students' perceptions of electronic feedback as an alternative to handwritten feedback: One University's inquiry. *Journal of Education and Learning*, *3*(1), 79–85.

Duncan, N. (2007). "Feed forward": Improving use of tutors' comments. *Assessment & Evaluation in Higher Education*, *32*(3), 271–283. doi:10.1080/02602930600896498

Gedye, S. (2010). Formative assessment and feedback: A review. *Planet*, *23*(1), 40–45. doi:10.11120/plan.2010.00230040

Gibbs, G., & Simpson, C. (2004). Conditions under which assessment supports students' learning. *Learning and Teaching in Higher Education*, *1*(1), 3–31.

Hattie, J., & Timperley, H. (2007). The power of feedback. *Review of Educational Research*, *77*(1), 81–112. doi:10.3102/003465430298487

Hendry, G. D., White, P., & Herbert, C. (2016). Providing exemplar-based 'feedforward' before an assessment: The role of teacher explanation. *Active Learning in Higher Education*, *17*(2), 99–109. doi:10.1177/1469787416637479

Higgins, R., Hartley, P., & Skelton, A. (2001). Getting the message across: The problem of communicating assessment feedback. *Teaching in Higher Education*, *6*(2), 269–274. doi:10.1080/13562510120045230

Knight, P. (1995). *Assessment for learning in higher education*. London: Kogan Page.

Leibold, N., & Schwarz, L. M. (2015). The art of giving online feedback. *The Journal of Effective Teaching*, *15*(1), 34–46.

Murtagh, L., & Baker, N. (2009). Feedback to "feed forward": Student response to tutors' written comments on assignments. *Practitioner Research in Higher Education*, *3*(1), 20–28.

Nicol, D. J., & Macfarlane-Dick, D. (2006). Formative assessment and self-regulated learning: A model and seven principles of good feedback practice. *Studies in Higher Education*, *31*(2), 199–218. doi:10.1080/03075070600572090

Petty, G. (2009). *Evidence-based teaching: A practical approach*. Cheltenham, UK: Nelson Thornes.

Price, M., Handley, K., Millar, J., & O'Donovan, B. (2010). Feedback: All that effort, but what is the effect? *Assessment & Evaluation in Higher Education*, *35*(3), 277–289. doi:10.1080/02602930903541007

Quinton, S., & Smallbone, T. (2010). Feeding forward: Using feedback to promote student reflection and learning-a teaching model. *Innovations in Education and Teaching International*, *47*(1), 125–135. doi:10.1080/14703290903525911

Ramsden, P. (1992). *Learning to teach in higher education*. New York, NY: Routledge Press.

Rowe, A. D., & Wood, L. N. (2008). Student perceptions and preferences for feedback. *Asian Social Science*, *4*(3), 78–88.

Sadler, D. R. (1998). Formative assessment: Revisiting the territory. *Assessment in Education: Principles, Policy & Practice*, *5*(1), 77–85. doi:10.1080/0969595980050104

Sadler, R. (1989). Formative assessment and the design of instructional systems. *Instructional Science*, *18*(2), 119–144. doi:10.1007/BF00117714

Shute, V. J. (2008). Focus on formative feedback. *Review of Educational Research*, *78*(1), 153–189. doi:10.3102/0034654307313795

Weaver, M. (2006). Do students value feedback? Student perceptions of tutors' written responses. *Assessment & Evaluation in Higher Education*, *31*(3), 379–394. doi:10.1080/02602930500353061

Wotjas, O. (1998, September). Feedback? No, just give us the answers. *Times Higher Education*, 25.

Chapter 6
A Theory-Practice Research Framework for Video-Enhanced Learning, Assessment, and Feedback

James McDowell
Lancaster University, UK

ABSTRACT

This chapter discusses a two-year project that explored the impact of video-enhanced learning, assessment, and feedback on undergraduate first-year students in higher education. Underpinned by a pragmatist epistemology, and arguing the case for a design-based methodological approach within a theoretical framework embracing the cognitive theory of multimedia learning, the community of inquiry, and the conversational framework, the chapter explores contemporary research into assessment and feedback, uses of technology-enhanced learning to promote inclusivity, and educational applications of asynchronous video.

INTRODUCTION

Background and Context

Following the introduction of the National Student Survey (NSS) in 2005, undergraduate students in the UK consistently signalled their desire for improvement around their experience of assessment and feedback (Higher Education Funding Council for England, HEFCE, 2011). Statistical data from the NSS has confirmed research suggesting that students feel they receive feedback which is either incomprehensible or lacking in detail, and that they receive this feedback too late for it to have any impact on their academic performance (Hounsell, 2007; Hounsell, McCune, Hounsell, & Litjens, 2008). Echoing these sentiments, the National Student Forum called for greater use of technology in the assessment and feedback process (National Student Forum, 2009).

DOI: 10.4018/978-1-7998-0426-0.ch006

An increase in the number of students entering university in recent decades has been accompanied by ever-larger numbers of students declaring a disability on entry to these institutions, with the number of students undergoing diagnostic assessment for specific learning difficulties during their time in higher education also increasing significantly over the last twenty years (Richardson & Wydell, 2003). Concurrently, legislation has been enacted which affords equality of educational opportunity for students affected by a learning difficulty such as dyslexia or an autistic spectrum condition, with a requirement placed on universities to ensure that any such students are not disadvantaged by the assessment and feedback process (Her Majesty's Stationery Office, HMSO 2001; 2010).

While research has focused on the use of audio in the assessment and feedback process (e.g. Nortcliffe & Middleton, 2007; Rotherham, 2009; Stewart, 2009), video has been relatively overlooked. The process of creating audio has traditionally been simpler and less expensive than that required in the production of video, and the respective file-sizes of audio and video assets make audio an easier option for the purposes of distribution. In recent years, however, there have been significant improvements in both compression techniques and in the availability of bandwidth required for the distribution of video over the internet, resulting in research into the influence of video on the assessment and feedback process being both practical and timely.

Against this background, this chapter develops a theory-practice research framework which forms the backdrop to a qualitative investigation into the influence of asynchronous video on the learner experience of assessment and feedback conducted at a university in the north of England. The second of these three chapters (Chapter 7) discusses the design, development, refinement and integration of a series of bespoke video-based interventions to form an integrated model of video-enhanced learning, assessment and feedback. The third of these three chapters (Chapter 2) evaluates the student experience of video-enhanced learning, assessment and feedback, presenting illustrative cases which highlight the impact of the interventions on the experience of six students, including two affected by dyslexia and one with Asperger's Syndrome (AS).

Epistemological and Ontological Backdrop

The study emerged against an underlying framework that draws on research into both cognitive and social theories of learning, and is underpinned by a pragmatist epistemological position which follows philosopher and educator John Dewey's twin assertions that "[the] educational process has two sides, one psychological and one sociological, and that neither can be subordinated to the other or neglected" (Dewey, 1929, p. 291). To this end, the investigation took place in a broadly constructivist space, drawing on and situated between social and cognitive constructivism.

LITERATURE REVIEW

Overview of the Literature

In the domain of theory, the investigation was broadly framed by social constructivist theorisations around learning that include the *Community of Inquiry Framework* (Garrison, Anderson & Archer, 1999) and the *Conversational Framework* (Laurillard, 2002), while also drawing on the body of theoretical work from the field of cognitive psychology that forms the *Cognitive Theory of Multimedia Learning* (Mayer, 2001).

In the domain of practice, the investigation has taken place in a space which overlaps a range of areas explored by contemporary research into higher education, including assessment and feedback (Bloxham & Boyd, 2007), threshold concepts (Meyer & Land, 2005), blended learning communities (Akyol, Garrison & Ozden, 2009), applications of technology-enhanced learning to support inclusivity (e.g. Woodfine, Nunes & Wright, 2005), and video-enhanced learning interventions (e.g. Griffiths & Graham, 2009).

Assessment and Feedback

The importance of assessment in the learner experience of higher education should not be underestimated. In an increasingly consumer-driven environment, time spent at university is viewed as a stepping stone into gainful employment, and assessment tasks as the hoops which must be jumped through in order to progress to certification (Rust, 2002; Gibbs & Simpson, 2004). With life as an undergraduate student now frequently predicated on the retention of a part-time job, academic tutors report that learner engagement with the subject area is often limited to a focus on the strategic completion of the necessary assessment tasks (Kneale, 1997), while Bloxham and Boyd note that many tutors "despair of trying to persuade students to undertake study which does not contribute in some way to their grades" (2007, p. 3).

Traditionally, university assessment schedules were dominated by summative assessment, which manifested itself as end-of-year examinations for which coursework was viewed as preparation. While this summative approach to assessment was widely regarded as appropriate within UK HEIs prior to the expansion of the sector which took place in 1992, a range of cost-efficiency pressures, including larger class sizes, modularisation, and widening participation initiatives all contributed to a situation where the traditional model became unsustainable (Bloxham & Boyd, 2007). Another change in the demographic of students entering UK HEIs can be seen in the numbers of students declaring unseen disabilities at the point of registration (Higher Education Statistics Agency, HESA 2013), and the next sub-section turns to consider the place of inclusivity in the design of assessment and feedback practices.

Inclusivity in Assessment and Feedback

In the UK, successive legislative instruments including the *Disabilities Discrimination Act* (HMSO, 2005), the *Special Educational Needs and Disability Act* (HMSO, 2001) and the *Equality Act* (HMSO, 2010) have placed a duty of care on UKHEIs to ensure that "disabled students are not placed at a substantial disadvantage in comparison with students who are not disabled" (HMSO, 2001, p. 25). Where a student is unable to participate in an assessment task on an equal basis with peers, UKHEIs must allow students affected by learning difficulties to undertake assessment and receive feedback in alternative forms, and two conditions relevant to this investigation are explored here: Asperger's Syndrome and dyslexia.

Asperger's Syndrome

Asperger's Syndrome (AS) is recognised as a pervasive developmental disorder, and is a condition located on the autistic spectrum characterised by a triad of impairments (Wing & Gould, 1979), comprising problems with social interaction, communication and imagination (Attwood, 2000). This latter characteristic manifests itself as an inability to empath, or being "unable to attribute mental states to others" (Benford, 2008, p. 32). Often demonstrating above normal results in intelligence quotient tests (Klin & Volkmar, 2000), students with AS are aware of their difficulties in communicating (Attwood, 2000; Baron-Cohen, 2008; Benford, 2008). Notably, within a university context, learners affected by AS will often find direct face-to-face communication particularly stressful, and are likely to avoid eye contact while offering only short, or even one-word answers in response to questions (Attwood, 2000).

Unusually high levels of sensory sensitivity experienced by students with AS, particularly to audio stimuli (Attwood, 2000), can make learning difficult in lecture theatres and studio-environments where noise levels are frequently above normal, and in cases of high cognitive load the audio channel may become overloaded (Paivio, 1990).

Studies exploring uses of video with this group have found that students with AS respond very positively to the visual medium (Charlop & Milstein, 1989; Shipley-Benamou, Lutzker & Taubman, 2002; Rayner, Denham, & Sigafoos, 2009), and that video can promote participation in dialogic processes (Lang, Shogren, Machalicek, Rispoli, O'Reilly, Baker, & Regester, 2009). This highlights the possibility that the opportunity for delayed response characteristic of asynchronous text-based communications explains the preference for text over audio, and there is a strong argument for further research to investigate whether and how asynchronous video might be employed as an effective learning, assessment and feedback medium for students with AS.

Dyslexia

According to the British Dyslexia Association Management Board: "Dyslexia is a specific learning difficulty that mainly affects the development of literacy and language related skills . . . characterised by difficulties with phonological processing, rapid naming, working memory, [and] processing speed" (British Dyslexia Association, 2007, Para. 5). Learners with dyslexia can find themselves placed at a significant disadvantage by approaches which rely on asynchronous text-based definitions of blended learning (Woodfine, Nunes & Wright, 2005; Woodfine, Nunes & Wright, 2008; McDowell, 2011).

While the impact of dyslexia varies across a broad spectrum encompassing spelling, writing, verbal memory, and sensory processing (Hari, Renvall, & Tanskanen, 2001), learners affected by the condition typically experience discontinuity in the visual processing of textual information (e.g. Asbjørnsen, Obrzut, Eikeland, & Manger, 2010). Observable effects in a student's ability to engage with text-based forms of assessment and feedback therefore range from minor misspelling of words to substantial deficits in the ability to process or produce textual forms of information (Woodfine, Nunes & Wright, 2008).

The opportunity to present information in visual and/or audio form represents a significant distinction between media-rich learning environments and their text-based counterparts. A distinction between visual and audio found early expression in *Dual Coding Theory*, which posits that humans use separate channels for processing visual and verbal information (Paivio, 1990), and this was later adopted as the dual-channel assumption that formed a cornerstone of the *Cognitive Theory of Multimedia Learning* (Mayer, 2001), as discussed later in this chapter.

Technology-Enhanced Approaches to Learning, Assessment and Feedback

The use of technology to enhance the learning, assessment and feedback process has received substantial attention in the literature since Laurillard's (1993) seminal work addressed issues in this field, however the majority of this work has, until recently, continued to focus on technology-enhanced approaches which rely on asynchronous text (e.g. Garrison & Kanuka, 2004). An exploration of the use of non-textual technologies in learning, assessment and feedback opens up two significant areas of educational research underpinned by radically different epistemologies: (i) conceptualisations of learning based upon a social constructivist pedagogy; and (ii) theorisations of learning broadly termed as cognitive constructivist.

At a theoretical level, research into technology-enhanced approaches to assessment and feedback emerging from the social constructivist camp has invited consideration of the notions of cognitive presence (CP), teaching presence (TP) and social presence (SP) as espoused by Garrison, Anderson and Archer (1999) in the context of their *Community of Inquiry* (CoI) framework. On the cognitive constructivist side, earlier research in the field of cognitive and educational psychology includes work such as *Dual-Coding Theory* (Paivio, 1990), *Cognitive Load Theory* (Chandler & Sweller, 1991), and the *Cognitive Theory of Multimedia Learning* (Mayer, 2001), all of which inform the design of assessment tasks requiring the production of multimedia artefacts to demonstrate learning, and the presentation of visually-oriented feedback on such assessment tasks within a dialogic framework.

The next two sections of this chapter will therefore address these areas, starting with an examination of the social considerations above, and then through an exploration of the cognitive considerations which should be addressed when employing diverse media in the assessment and feedback process.

Learning in a Community of Inquiry

Originally developed to facilitate the study of asynchronous text-based learning in online environments, the *Community of Inquiry* (CoI) framework (Garrison, Anderson & Archer, 1999), is comprised of three elements: cognitive presence (CP); teaching presence (TP); and social presence (SP). Since its inception in 1999, the CoI framework has evolved in response to the emergence of blended learning approaches, leading Akyol et al. (2009) to argue that the identification of strong indicators of each of the three interdependent elements is key to the formation of a successful learning community.

Cognitive Presence and the Practical Inquiry Model

Early in the evolution of the CoI framework, cognitive presence was believed to require "sustained reflection and discourse" (Garrison, Anderson, & Archer, 2001, p. 11), whereas in an investigation examining cognitive presence within a blended community of inquiry through exploration of deep learning approaches and outcomes, Akyol and Garrison (2011) describe cognitive presence in terms of "the progressive phases of practical inquiry leading to resolution of a problem or dilemma" (p. 235). Rooted in pragmatist thinking which recognises both a private world experienced by each individual and the social world shared between individuals (Dewey, 1944), the practical inquiry model (PIM) incorporates four phases through which learners are said to progress in the process of constructing understanding, developing knowledge and refining skills within their discipline: a triggering event; exploration; integration; and resolution (Garrison et al., 2001).

Garrison, Anderson and Archer (2001) note that having achieved resolution to a triggering event, learners may appear to shortcut the cycle of inquiry, describing this process as an intuitive leap. An analysis of PIM based upon cognitive theories of multimedia learning suggest that the apparent leap may be explained by reference to the prior knowledge component of the *Individual Differences* principle, explored later in this chapter, in which prior knowledge contributes to the process of challenging or consolidating the construction of those mental models formed where an incomplete picture is presented (Mayer, 2002).

Teaching Presence, Learner-Tutor Interactions, and the Role of the Tutor

Anderson, Rourke, Garrison and Archer (2001) define teaching presence as the "design, facilitation, and direction of cognitive and social processes for the purpose of realizing personally meaningful and educationally worthwhile learning outcomes" (p. 5), which suggests that strong TP is developed through actively supporting the development of strong CP and SP within a community, effectively requiring a switch in the role of the tutor from sage to guide (King, 1993).

McConnell (2006) notes potential pitfalls associated with the changing role of the tutor in the transition from traditional approaches to learning towards the development of e-learning communities, pointing out that when greater scaffolding is provided, tutors may "become slightly more distant and less participatory" (p. 194). The establishment of strong learner-learner and learner-tutor ties based on a combination of trust and mutual understanding is widely recognised as vital in the formation and maintenance of online learning communities (McConnell, 2006; Smith, 2008; Hemmi, Bayne, & Land, 2009), and learner-tutor interactions form the basis for identification of teaching presence in this study.

Social Presence, Learner-Learner Interactions, and Ownership and Control

While learner-learner interactions play a key role in the formation of online learning communities (McConnell, 2006; Smith, 2008; Hemmi et al., 2009), "social presence reflects the development of climate and interpersonal relationships" (Akyol & Garrison, 2011, p. 235) within those communities. The extent to which learners feel a sense of ownership and control has been highlighted as another key factor influencing the successful development of those learning communities (e.g. McConnell, 2006; Hemmi et al., 2009).

Evidence of the emergence of SP through the development of interpersonal relationships within an *online* learning community can be found in learner-learner interactions which take place within a virtual learning environment or e-portfolio system. It must be acknowledged however, that the nature of a *blended* learning context creates opportunities for substantial development of those interpersonal relationships outside institutionally supported learning spaces, through both face-to-face and virtual interactions between learners.

Challenging the Primacy of Text-Based Learning

A common feature of the Oxford tutorial system, the Open University's distance learning programmes, and many non-practice-based taught courses offered in UKHEIs has been the tradition for learning and assessment through two text-based forms of activity: reading and writing.

The problems experienced by learners affected by dyslexia caused by an insistence on text-based activities are well documented (e.g. Carey, 1997; Woodfine, Nunes & Wright, 2005; Woodfine, Nunes & Wright, 2008; McDowell, 2011). Despite both the findings of this research and requirements of the legislation cited above which is designed to ensure that all learners are afforded equality of opportunity within education, the recent proliferation of online distance learning (ODL) programmes based on the asynchronous text-based (ATB) model and the pedagogical research upon which these courses have been founded appears to consolidate the primacy of the ATB model.

Indeed, many educational researchers and practitioners have reached a broad conclusion regarding the primacy of ATB learning in promoting reflexive learning (e.g. Salmon, 2000; Garrison & Kanuka, 2004; McConnell, 2006), with some suggesting that alternative media should be used in only limited and judicious ways (e.g. Palloff & Pratt, 2005). It is notable however that many of the studies have been conducted primarily with learners engaged in discursive subject areas such as education and history where ATB learning naturally lends itself to assessment based on text-based outputs such as essays and dissertations. By contrast, work conducted with learners in prescriptive, numerate and creative fields has found the introduction of multimedia and video tutorials in disciplines such as computer science (Schaffer, Doube & Tuovinen, 2003), and subject areas such as statistical analysis (DeVaney, 2009), can lead to improved engagement and understanding for learners.

Asynchronous Text and Technology-Enhanced Approaches in Assessment and Feedback

The technological context in which the ATB model has apparently secured a sustainable position within higher education ODL programmes has, however, been accompanied by the emergence of visual alternatives to the textual model, such as multimedia learning (Mayer, 2001). While proponents of the ATB model insist that ODL should retain the text-based model above the visual model (e.g. Garrison & Kanuka, 2004; Humfrey, 2010), a growing body of work has found that the use of instructional tutorial videos has been led to increased levels of understanding in learners (e.g. Schaffer, Doube & Tuovinen, 2003; Hardin & Ellington, 2005; DeVaney, 2009), suggesting that strong CP can be established through the introduction of alternatives to text-based learning materials.

Research has found that text-based communication can be impersonal, leading to a greater sense of isolation (Ice, Curtis, Phillips, & Wells, 2007; Croft, Dalton, & Grant, 2010), and that the absence of those social cues and non-verbal indicators encountered in face-to-face settings can lead to reduced levels of social cohesion within online learning communities (McConnell, 2006). By contrast, stronger cohesion has been shown to exist in face-to-face and blended learning communities where learners engage in direct synchronous communication (Doolan & Simpson, 2010), and efforts to overcome isolation and improve cohesion in asynchronous online communities would seem to demand further exploration of the potential offered by audio-visual technologies.

Asynchronicity, Reflexivity and Alternatives to Text

Historically, text-based communications were necessarily asynchronous, however the development of the web from command-line interface to audio-visually rich virtual environment has seen a blurring of the boundaries between asynchronous text and synchronous video as forms of expression, facilitating the emergence of both synchronous text and asynchronous video as viable modes of communication. Indeed, the facility to pause and rewind live television signals, for example, gives rise to a new temporal space, in which social presence and presenter immediacy are maintained while allowing individuals to reflect immediately on issues arising in news streams or other real-time televised events. The implication is that the opportunity for reflexivity afforded by ATB approaches can be argued as emerging not from the use of text, but from the pause for reflection inherent within asynchronous communication.

Research into the use of asynchronous audio in learner-tutor and learner-learner communications has found learners expressing exceptionally high levels of satisfaction where feedback has been provided in audio form, and that higher levels of SP and TP can be achieved through the adoption of strategies which employ voice (Belfer & Morgan, 2005; Ice et al., 2007); strong levels of engagement have also been directly attributed to the introduction of audio feedback (Doolan & Simpson, 2010). Audio-supported strategies have also been found to promote greater inclusivity for learners with dyslexia and other learning difficulties associated with text-based learning (Woodfine, Nunes, & Wright, 2005), however research into the potential of video to improve inclusivity and participation by promoting SP and TP remained in its early stages at the outset of the study (e.g. Griffiths & Graham, 2009).

The opportunity to present information in visual and/or audio form represents a significant distinction between media-rich learning environments and their ATB counterparts. A distinction between visual and audio found early expression in *Dual Coding Theory*, which posits that humans use separate channels for processing visual and verbal information (Paivio, 1990), and this was later adopted as the dual-channel assumption that formed a cornerstone of the *Cognitive Theory of Multimedia Learning* (Mayer, 2001).

Against the background of this literature, the next sub-section turns to examine previous research which has investigated how multimedia and video have been used in higher education, focusing first on the theoretical case for its use developed by authors in the field of cognitive psychology, and then on a range of contemporary, practice-based examples of how different forms of multimedia and video have been used in teaching, learning, assessment and feedback.

The Cognitive Theory of Multimedia Learning

Starting from the premise that "a multimedia instructional message is a communication using words and pictures that is intended to promote learning" (Mayer, 2001, p. 21), the *Cognitive Theory of Multimedia Learning* is underpinned by evidence from a broad range of experiments in the fields of cognitive psychology. Mayer posits that meaningful learning is more likely to occur successfully as a result of a multimedia message when the design of the intervention has considered relevant aspects of both cognitive and educational psychology, and is founded on three key assumptions:

1. **The Dual Channel Assumption:** That independent channels process visual and auditory information.
2. **The Limited Capacity Assumption:** That the auditory and visual channel have finite capacity for information processing.
3. **The Active Learning Assumption:** That learning is dependent upon completion of a structured series of cognitive processes.

According to Mayer, visual information from sources such as pictures is mainly processed by the visual channel, and auditory information such as speech is mainly processed by the auditory channel, however printed text is believed to be processed differently; initially, it is claimed, printed text is processed visually, however secondary processing employs the auditory channel. On the basis of these three assumptions, Mayer expands on the *Cognitive Theory of Multimedia Learning*, drawing support from both theoretical and empirical sources to derive seven principles, six of which are relevant here: multimedia; temporal contiguity; coherence; modality; redundancy; and individual differences.

The Multimedia Principle

Stating that "students learn better from words and pictures than from words alone" (Mayer, 2001, p. 63), the *Multimedia Principle* forms a cornerstone of the case for the use of asynchronous video in teaching, learning, assessment and feedback in this investigation. At a theoretical level, Mayer argues that the use of both printed text and pictures affords learners opportunities to construct multiple mental models, both visual and aural, first by processing both the text and the pictures as visual information via the visual channel, and subsequently by processing the printed textual information through the auditory channel as this passes through the phonological loop. Importantly, the construction of these mental models is complemented by the identification of connections or disconnections between these models, thereby further informing the construction of appropriate models and consolidating learning. In contrast, Mayer suggests, the presentation of words unaccompanied by pictorial information, whether printed or spoken, offers an opportunity to construct a verbal mental model based upon the processing of textual information, but significantly reduces the likelihood of the construction of mental models derived from information processed by the visual channel, thereby limiting opportunities to consolidate learning afforded by formation of connections between the two types of mental models.

The Temporal Contiguity Principle

Considerations of the temporal proximity of text in relation to images or animations are covered by Mayer's third principle, which states that "students learn better when corresponding words and pictures are presented simultaneously rather than successively" (p. 96). The audio-visual nature of video as a combination of narration and animation made this principle significantly more relevant to this study than the *Spatial Contiguity Principle*. Mayer argues that the simultaneous presentation of related elements of narration and animation allows concurrent processing of these elements by both auditory and visual channels, enabling the construction of multiple mental models which are maintained simultaneously in working memory, thereby amplifying the construction and consolidation of connections between these mental constructs. Similarly, the converse proposition is held to be true, however Mayer also acknowledges grey areas between these two poles; he suggests that where an auditory message and visual representation are processed within a short space of time, effective connections between the auditory and visual mental models may still be constructed, whereas the likelihood of forming strong connections between textual and pictorial information is inversely proportional both to increased time delay and increased duration of either the narrative or animated elements.

The Coherence Principle

In the fourth principle of the *Cognitive Theory of Multimedia Learning*, Mayer turns his attention to the notion of the coherence of the content of multimedia messages, suggesting that "students learn better when extraneous material is excluded rather than included" (p. 113). The relevance of content within both the textual and visual elements of a message is examined, while the effectiveness of multimedia messages from which extraneous material has been removed is also explored. The theoretical rationale for the *Coherence Principle* borrows from one of the central tenets of the *Spatial Contiguity Principle*, i.e. that the spatial dis-contiguity of printed text and pictorial elements of a multimedia message place additional demands on the processing capacity of their respective channels. By forcing learners to scan a screen for content, extraneous content absorbs cognitive resources that would otherwise be available to process relevant information, to construct appropriate mental models, to hold these in working memory, and to establish and consolidate connections between them.

The Modality Principle

The *Modality Principle* is another cornerstone in the case for the use of asynchronous video; Mayer states that "students learn better from animation and narration than from animation and on-screen text; that is, students learn better when words in a multimedia message are presented as spoken text rather than printed text" (p. 134). The *Modality Principle* is therefore consistent with work examining the adoption of assessment and feedback strategies employing voice, which found the use of asynchronous audio in learner-tutor and learner-learner communications within online distance learning communities led to higher levels of social presence and teaching presence (Belfer & Morgan, 2005; Ice et al., 2007). The presentation of both pictorial and textual elements of a multimedia message in visual format carries the potential for overloading the visual channel, while leaving the auditory channel unused or underused.

The theoretical rationale behind the *Modality Principle* suggests that a multimedia message is more effectively conveyed when textual information is presented orally, as information presented through the medium of the spoken word is processed by the auditory channel, thereby reducing the load on the visual channel. Taken in conjunction with the *Multimedia Principle*, it is to be expected that, by balancing the load and sharing the information processing across both visual and auditory channels, multiple mental models can be constructed, enabling the learner to forge connections between them to facilitate effective and consolidated learning.

The Redundancy Principle

Arguably the simplest of the seven principles, and connected to the *Modality Principle* by the potential for overload of the visual channel when both pictorial and textual information are presented visually, the *Redundancy Principle* holds that "students learn better from animation and narration than from animation, narration, and text" (Mayer, 2001, p. 147). Mayer found stronger results for tests of retention in cases where a multimedia message presented a combination of animation and narration than in cases which also included the visual presentation of text, suggesting that the high demand for visual information processing led to an overload of the visual channel.

The Individual Differences Principle

In the seventh and final principle of his *Cognitive Theory of Multimedia Learning*, which he describes as the *Individual Differences Principle*, Mayer contends that "design effects are stronger for low-knowledge learners than for high-knowledge learners, and for high-spatial learners rather than for low-spatial learners" (p. 147). This principle invokes two distinct cognitive concepts with the potential to impact on the learner's experience of a multimedia message: prior knowledge, and visuo-spatial awareness. In connection with the first of these, Mayer posits a hypothetical position wherein a multimedia message fails to provide adequate guidance, arguing that prior knowledge affords learners opportunities to construct alternative mental models through which to challenge or confirm both the models constructed in response to the multimedia message, and the connections forged between them. In tests of retention and transfer, Mayer found that learners with a low level of prior knowledge experienced more significant benefits from the introduction of well-designed multimedia messages than did learners with a higher level of prior knowledge. This is consistent with the findings of research into the use of instructional tutorial videos (ITVs), which found that ITVs were more effective for learners approaching a subject at an introductory or intermediate level than for learners who had progressed to more advanced stages (DeVaney, 2009).

With regard to visuo-spatial awareness, a higher capacity to construct appropriate mental models and to forge connections between them was observed in learners who scored highly in visuo-spatial awareness tests than in those achieving lower scores. The theoretical position supposes that members of this latter grouping experience a high level of demand on the cognitive resources available to them, limiting the capacity of working memory to construct appropriate mental models and forge connections between them.

Reducing Cognitive Load in Multimedia Learning

Building on the dual-channel assumption and responding to the problem of cognitive overload identified in the *Cognitive Theory of Multimedia Learning* (Mayer, 2001), Mayer and Moreno (2003) highlight five scenarios in which cognitive capacity can become overloaded, and introduce nine load-reduction strategies to address this problem in the design of multimedia learning. Five of these strategies directly inform the design of instructional tutorial videos to support the development of skills in the use of software development environments: *off-loading*, *segmenting*, *pretraining*, *weeding*, and *synchronizing* (Mayer & Moreno, 2003).

Drawing on the *Modality Principle*, the first strategy, *off-loading*, is a method which reassigns "some essential processing from visual channel to auditory channel" (Mayer & Moreno, 2003, p. 46) by presenting words as narration rather than on-screen text. In a scenario where "both channels are overloaded with essential processing demands" (p. 47), the second strategy, *segmenting*, recommends designing multimedia learning in bite-size units so that learners have time to process information before moving on to subsequent materials, while closely related to this, the third strategy, *pretraining*, proposes an incremental approach to the design of learning materials such that the introduction of components allows the learner to build successively on previous mental representations. In response to a scenario where "one or both channels are overloaded by the combination of essential and incidental processing demands" (p. 48), *weeding* draws on the *Coherence Principle*, recommending that extraneous material should be excluded in order to ensure effective information processing. Emerging from consideration of the *Temporal Contiguity Principle*, *synchronizing* addresses the problem where "one or both channels are overloaded by the combination of essential processing and representational holding" (p. 49) by proposing that narration and animation are presented simultaneously in order to ensure that a requirement to hold multiple mental representations in working memory is minimised.

Teaching and Learning: Instructional Tutorial Videos

Of the three types of asynchronous video examined here, a diverse range of uses in an instructional-tutorial context can be found in the literature (e.g. Rumble, 2001; Anderson, Dickey & Perkins, 2001; Shephard, 2002), however it is important to differentiate between video captured using a camera, and that captured directly from a computer's graphics processing unit and displayed on the screen. This latter is known as 'screencasting', and as a relatively recent approach, the application of screencast videos to scaffold the development of software skills has only been documented over the last decade (e.g. Veronikas & Maushak, 2005; Hardin & Ellington, 2005; Lee, Pradhan, & Dalgarno, 2008; DeVaney, 2009). Notably, where the intention is to scaffold the development of software skills, step-by-step demonstrations of key techniques have been found to assist learners achieve competency at both introductory and intermediate levels (Hardin & Ellington, 2005; DeVaney, 2009).

Lee et al. (2008) describe a study that examined the effectiveness of screencasts as learning objects to scaffold the development of object-oriented programming skills. Using a visual programming tool as the environment for the implementation of the coding concepts, the authors developed a series of structured exercises that were demonstrated in screencast videos. Unusually, and despite acknowledging the existence of "a vast body of literature available to help guide instructional design decisions when combining spoken and/or visual text with animation" (Lee et al., 2008, p. 75), these screencasts did not include any narrative element, and subsequently the study found "no significant effect of the provision of screencasts during learning" (p. 75). The authors suggest that this finding ran contrary to the theoretical predictions, however specific predictions are not made explicit, and it should be noted that the production of screencasts without the inclusion of any narration ignores the dual channel assumptions found in both *Dual Coding Theory* (Paivio, 1990; Clark & Paivio, 1991) and the *Model of Working Memory* (Baddeley & Hitch, 1974; Baddeley, 1992). Further, the non-inclusion of an audio narrative runs directly contrary to the recommendations of both the *Cognitive Theory of Multimedia Learning*'s 'Multimedia Principle', that "students learn better from words and pictures than from words alone" (Mayer, 2001, p. 63), and the *Modality Principle*, that "students learn better from animation *and* narration" (p. 134, *my emphasis*). As such, Lee et al. (2008) exclusion of an audio-narrative led to findings which appear to support, rather than contradict these particular theoretical predictions, and while the authors suggest that finding no significant effect may have been due to factors including sample size and learners' abilities, the *Cognitive Theory of Multimedia Learning* suggests that this finding can be attributed to the omission of audio from the screencasts.

Evaluating the introduction of video tutorials in a postgraduate module designed to develop skills with statistical analysis software in an ODL context, DeVaney (2009) had been seeking an effective alternative to the provision of 'user manuals' based on step-by-step instructions and screenshots. DeVaney's work draws on Mayer's *Cognitive Theory of Multimedia Learning* (2001) and both the *Dual-Coding Theory* (Paivio, 1990; Clark & Paivio, 1991) and *Model of Working Memory* (Baddeley & Hitch, 1974; Baddeley, 1992) underpinning it, and this theoretical framework informed the development of seventeen video tutorials introducing the *SPSS* statistical analysis package.

While DeVaney's study was unusual for having clearly identified the demographic of the participant group, there is a paucity of examples in the literature of work having explored how ITVs might influence the experience of students from specific learner groups. Further, while one study that emerged following the final round of data collection in this investigation makes a brief reference to ITVs in connection with dyslexic students (Sekovanić, Vukovac & Podbojec, 2012), a clear gap in the literature exists around the use of video tutorials as a tool to scaffold learning for students affected by specific learning difficulties (SpLDs) such as dyslexia and Asperger's Syndrome.

Assessment: Video-Diaries in Assessment

The use of video as a tool to promote self-reflection and to facilitate peer feedback is a theme which has been reported by Collins, Cook-Cottone, Robinson, and Sullivan (2005), O'Donoghue and Cochrane (2010), and more recently Jordan (2012), while Cheng and Chau's (2009) focus on how video can be leveraged to promote self-reflection within an e-portfolio context is also a study with particular relevance to this investigation.

Citing evidence from studies in teacher education and professional training which found video offering learners an alternative to the text-based route to self-reflection, Cheng and Chau (2009) highlight teaching and learning strategies in higher education which emphasise the use of e-portfolios to promote reflexivity, aiming to explore how video influences self-reflection within an eportfolio system. Against a backdrop of Fleming and Mills' (1992) model of four key learning styles characterised as *visual*, *aural*, *reader* and *kinaesthetic*, Cheng and Chau suggest that a broad range of learning materials "can be digitalized into web-enabled video formats to support learners with different learning preferences" (2009, p. 338).

Drawing on the dual channel assumption and the work of Baddeley (1992), Paivio (1990), and Moreno and Mayer (2000) to justify the cognitive value of using video, Cheng and Chau identify that video can be "an effective adapt-and-adopt medium for a variety of learning styles" (2009, p. 338). Adopting Keller's (1983) *ARCS* model as the analytical framework, the authors found evidence to support the suggestion that video promotes reflexivity within an eportfolio context, and call for the design of eportfolio systems to incorporate functionality which would better accommodate sharing of video content.

The learner-generated content ethos underpinning these examples is consistent with a move away from passive consumption towards active production of video, or "participatory video" (Thornhill, Asensio & Young, 2002, p. 17). The feature which is common to all these examples is that the video presents the learner directly using a web-cam or similar, while examples of learners using screencasting technologies to demonstrate skills in areas of the Computing discipline such as software engineering or computer games design remain elusive. The study discussed in the next chapter addresses that gap in the literature, with students afforded the opportunity to document the development of a body of practical work through a series of regular, screencast, video-diary entries.

Feedback: Video-Feedback and Learner-Tutor Dialogue

Video-feedback is a relatively new phenomenon, and practice-based reports on the use of video for feedback on students' work were slow to emerge following Stannard's (2007) use of screen capture tools to produce video-feedback on written work for students studying English as a foreign language, however the number and range of examples has grown since this investigation first began. Kerr and McLaughlin (2008) employed a screen capture approach to produce video-feedback on essays written by students in a biosciences course. Using a system which employed a combination of a bespoke marking tool integrated with *MS-Word* and the *Camtasia* screen capture software, 90 students from a cohort of 484 received video-feedback on a piece of written summative assessment. The mixed methods evaluation, which examined students' perceptions of the quality, quantity and usability of the feedback, found that a higher proportion of students who received video-feedback regarded their feedback as high quality than those who received written feedback. Further, none of the students in the video-feedback group expressed concerns that the feedback was of a lower quality than they had expected, in comparison to the 10% of students who did so from the written feedback group.

Regarding perceptions of quantity of feedback, the study found a higher proportion of the written feedback group reported that the quantity of feedback was greater than expected than of the video-feedback group, although a smaller proportion of the video-feedback group felt that the quantity had been too little than of the written feedback group. The evaluation of the usability of the feedback saw a significantly higher proportion of students from the video-feedback group regard the feedback comments more generically applicable to their writing skills than those from the written feedback group. Notably, the authors hypothesise that this "*could be due to the students engaging much more with the video commentary as it allows them a greater connection with the marker . . . than is possible through written comments alone which are open to differing interpretations*" (Kerr & McLaughlin, 2008, p. 162), citing a range of qualitative data in support of this conjecture, suggesting that both teaching presence and student engagement with feedback can be enhanced through the use of video-feedback.

Another case study from the field of biosciences showcases the use of *Jing* screen capture software to record video-feedback on students' written work (Hope, 2010). Viewing 2000-word essays through the *Grademark* system within the institutional VLE, Hope primarily used video-feedback as a correction tool, with areas highlighted for improvement accompanied by a real-time commentary, to support more general textual feedback on both strengths and weaknesses provided in a general feedback form. In their evaluation of this approach, students were unanimously in favour of video-feedback, reporting that the feedback was both more detailed and more personalised than that received in the written form, while one student affected by dyslexia indicated that video-feedback afforded improved accessibility over the text equivalent. Outside the United Kingdom, further studies examining the use of video to enhance feedback on written work report broadly similar findings, with work from the United States (Mathieson, 2012; Thompson & Lee, 2012), Norway (Mathisen, 2012) and Canada (Séror, 2013) also adopting the use of *Jing* to produce video-feedback on written assessments.

The JISC-funded *ASSET* project at the University of Reading sought to explore the potential of video technology to address issues of timeliness, quality, efficiency and student engagement around the provision of feedback. This institution-wide project saw practitioners from a broad range of disciplines equipped with both handheld video cameras and PC-based screen capture software; these academics were given training in their use before making a free choice on the application of the technology to produce generic, 'module-level' video-feedback in their subject area. A pre-study questionnaire found that students were least likely to have prior experience of audio-visual feedback, or to express a preference for feedback through this medium, and only 22.9% of students had previously engaged in dialogue with tutors around feedback on their work. The post-study questionnaire found that although academics perceived video-feedback to have had no impact on student engagement, 80% of students reported a preference for academics to continue using generic feedback at a module level, while 31.4% of students expressed an interest in a broadening of the approach to encompass the provision of personalised, one-to-one video-feedback (Crook, Mauchline, Maw, Lawson, Drinkwater, Lundqvist, Orsmond, Gomez, & Park, 2012).

In the examples outlined thus far, (i) the feedback has been generic (with the exception of Mathieson (2012)), (ii) the feedback has responded to summative assessment, and, as a corollary, (iii) there has been no attempt to engage students in dialogue around the feedback using video by encouraging a shift towards participatory video (Thornhill, Asensio & Young, 2002). Further, while Hope concludes that video-feedback "may be a particularly useful form of feedback for dyslexic students" (2010, p. 4), there is no indication of any subsequent formal evaluation of that suggestion.

All the examples above focus on the production of video-feedback on written work using face-to-camera techniques. While screencasting had been used to assist the development of students' skills in Computing and Statistics, at the outset of the investigation there were no examples in the academic literature of using screencast forms of video-feedback for students in primarily visual, creative and numerate disciplines, and this study aimed to address that gap. The investigation therefore explored the application of video-feedback in non-written formatively assessed forms of academic work, aiming to engage students in an asynchronous video dialogue around that feedback, with a view to the development of a holistic model for the integration of asynchronous video across learning, assessment and feedback.

The Emerging Research Questions

Thus far, this chapter has reviewed both theory-led and practitioner-driven literature relating to assessment and feedback, and multimedia learning. In so doing, gaps in the literature have been identified which the investigation discussed in the next chapter aims to address through one overarching primary research question (PRQ) and two inclusivity-related secondary research questions (SRQs):

PRQ: *How might asynchronous video influence the learner experience of assessment and feedback?*
SRQ1: *How might asynchronous video influence the experience of learners diagnostically assessed as affected by dyslexia?*
SRQ2: *How might asynchronous video influence the experience of learners clinically diagnosed with Asperger's Syndrome?*

The next section of this chapter therefore examines the methodological approach and research methods used to address these questions.

METHODOLOGICAL APPROACH

Research Design, Methods and Methodology

The research participants in this investigation were students for whom the interventions were designed, and whose academic success they were intended to enhance. This group was comprised of undergraduate first-year students, all of whom were male in the 18-25 age range, while their course of study led to degree-level certification subject to successful completion of a mix of mandatory (or 'core') and optional modules. As a researcher-practitioner, the author had regular, direct contact with all participants throughout the study, with sole responsibility for the delivery of a double-weighted core module which entailed substantial practical development work, and which all learners were required to pass in order to proceed from Year 1 to Year 2. A blended learning approach was employed in the delivery of this core module, with learners afforded direct contact in a studio environment for a single two-hour session once per week, and support outside that session available through asynchronous online contact. In an effort to mitigate a potential power imbalance within the learner-tutor relationship which might impact on the study, the author employed a pedagogical approach based in a collapsed teacher-student hierarchy (McDowell, 2010).

The investigation adopted design-based research as a methodological approach that remains consistent both with practice-based educational research (Brown, 1992; Collins, 1992), and with the adoption of multiple, mixed methods for data collection and analysis (Design Based Research Collective, 2003; Cobb, diSessa, Lehrer & Schauble, 2003).Two academics at the institution acted as expert witnesses (Yin, 2003), to whom preliminary and final findings of each research cycle were presented for challenge, verification or alternative interpretation: (i) the Course Leader for the degree programme offered comment on the impact of the interventions on the engagement of students as a group; (ii) the Disabilities Support Coordinator for the School focused on the individual experience of those students recognised by the institution as affected by a specific learning difficulty.

Table 1. The data collection methods and analytical approaches employed in the investigation

Method	Raw Data	Approach	Analysis Mode or Purpose
Questionnaires using multiple choice, and open-ended questions	Individual online responses saved on server	Quantitative	Numeric
		Qualitative	Categorisation using coding and inference
Interviews	Transcripts of audio recording of participants	Qualitative	Categorisation using coding and inference
Interrogation of online e-portfolio system	Eportfolio statistics of student usage	Quantitative	Numeric
	Learner-generated video-based content	Qualitative	Categorisation using coding and inference
Observations (direct and indirect)	Transcripts of field notes	Qualitative	Categorisation using coding and inference
Expert Witness referral	Transcripts of audio recording of colleagues	Qualitative	Methodological triangulation of findings
Interrogation of course results	Individual participants' module marks	Quantitative	Numeric

Consistent with the design-based approach, the outputs of the research included design exemplars for video-enhanced learning, assessment and feedback within an integrated model of video-enhanced learning, which as noted earlier is discussed in the next chapter, while the impact on the student experience is discussed in the third of these three chapters, through a series of illustrative cases that adopt the individual as the unit of analysis. One case highlights the experience of students assessed as affected by dyslexia, and another illustrates the case of a student with a clinical diagnosis of Asperger's Syndrome, while the remaining cases describe the experience of two neurotypical peers.

Design-Based Research

The origins of design-based research (DBR) can be found jointly in Brown's (1992) seminal paper on the area, in which she characters her work as an "attempt to engineer innovative educational environments and simultaneously conduct experimental studies of those innovations" (p. 141), and in both Brown's and Collins' discussions of *design experiments* (Brown, 1992; Collins, 1992) in the context of educational technology; this emphasis on the development of innovative learning environments continued as DBR matured.

As a contemporary educational research methodology which rejects the methodological polarisation of the paradigm wars (Gage, 1989), which seeks a balance between positivist and interpretivist approaches, and which embraces mixed methods, DBR is consistent with the pragmatist epistemology as outlined earlier. By forging a close relationship between theoretical discussion and practical application, DBR also responds to claims of a disconnect between theory and practice in educational research (Cobb et al., 2003).

Consideration of Case-Study as the Research Methodology

Case study methodology is underpinned by an epistemological basis that facilitates the collection of data using various methods and from a variety of sources, and as a methodology that recommends the referral of evidence and its interpretations to expert witnesses (Yin, 2003) to ensure validity and authenticity of the findings, the adoption of case study offered clear benefits. Despite this, case study methodology remains rooted in the tradition wherein a researcher reports on the impact of an intervention developed by a third party, rather than actively seeking direct involvement in the design, implementation, amendment, refinement and evaluation of that intervention as might be expected in an action research or design-based research study.

While rejecting case study methodology as the primary approach to this study, the investigation sought to address a research question and pursue research aims that it was anticipated would be both of interest and of practical relevance to practitioners in the field, following Reeves' (2000) approach to design-based research. Consistent with a pragmatist philosophical outlook, this approach recognises the benefits of presenting findings in forms that are appropriate for practitioners, and illustrative case studies provided that vehicle.

Taking the individual as the unit of analysis, the purpose of developing these case studies was to provide practitioners with the key points of lessons learned, and a rich description of the relevant contexts within which the research has taken place. As the investigation moved through the series of design-based research cycles, opportunities for purposive sampling arose which, in conjunction with progressive focusing, allowed the identification of participants whose experience highlighted interesting or unusual cases, following Stake (1995); these are presented as a series of illustrative cases that can be found in the third of these three chapters.

Data Collection and Analysis Methods

Data collection methods employed over the course of this study included paper questionnaires, online questionnaires, direct observation, indirect observation, blog entry capture, semi-structured interviews, expert witness referrals (Yin, 2003), video-diary entries, dialogic interviewing (Knight & Saunders, 1999; Denzin, 2001), and interrogation of course results data.

Table 1 presents the various data collection methods used over the course of the investigation and the analytical frameworks employed for their interpretation.

Standardisation of the data collected from a range of mixed-media sources using these methods required all sources to be reduced to the lowest common denominator; as text-based questionnaires and blog entry capture were employed this entailed the reduction of the audio from the semi-structured interviews and focus groups and the audio component of the video-diary entries to be transcribed into text files for analysis.

Reliability and Verification

The process of dialogic interviewing provided a strong mechanism that facilitated rapid verification of the data and confirmation of its validity. While the study was ongoing, the use of a dialogic interviewing approach to data collection in interactions with learner-participants saw a process of 'constant referral' employed to confirm the reliability and authenticity of any interpretations made. Raw data, researcher-interpretations and subsequent findings were also referred to expert witnesses (Yin, 2003) for verification and corroboration throughout the lifecycle of the study; expert witnesses included the Course Leader for the degree programme on which the participants were studying, academic colleagues teaching the participants in other modules of the degree programme, and the School's Disabilities Support Coordinator as someone with both a strong awareness of the impact of dyslexia and Asperger's Syndrome, and detailed knowledge of the circumstances surrounding individual participants affected by these learning difficulties.

Grounded Theory and the Emergent Research Design

In studies such as this where an intervention is developed and refined through a series of iterative cycles, the research design is necessarily emergent and dependent upon interaction between researcher and participants. As such, the design-based research (DBR) methodology employs strategies to address research questions that are divergent from those of the traditional scientific method. While this latter might typically see experiments designed to test the validity of a theory through the formulation and testing of a hypothesis, the nature of DBR lends itself more naturally to an inverted approach, whereby the researcher generates theory from analyses of the data (Glaser & Strauss, 1967).

The Data Analysis Process

Further to the distinction made earlier between the collection of data for internal and external purposes, the data analysis procedure described here focuses on those externally-facing data required for the purposes of addressing the study's research questions, rather than those internally-facing data collected for the purpose of addressing the development and refinement of the intervention.

Remaining consistent both with the methodological approach adopted and the dialogic nature of much of the data collection strategy employed, data analysis was conducted as an ongoing exercise in two phases. In the initial phase, the use of constant referral during the dialogic interviewing process saw data analysis take place hand-in-hand with its collection to ensure the validity of interpretation. The second phase employed the "constant comparative method" (Glaser & Strauss, 1967, p. 101), in which the coding framework and theorisations are developed in tandem. Transcribed interview data was compared with text- and video-based evidence available within the e-portfolio system to identify common categories, while correlations between these were identified through methodological triangulation of the various data sources.

CONCLUSION

Having adopted an epistemological position founded in the Deweyan tradition, this chapter has developed a theory-practice framework which is underpinned by conceptualisations from both cognitive and social constructivist theories of learning, and which is interwoven with practice-focussed literature emerging from empirical studies into assessment and feedback, uses of video in higher education, and around the needs of students affected by social communication issues and specific learning difficulties. In so doing, this chapter has highlighted gaps in the literature around the learner experience of video-enhanced assessment and feedback, and around the impact of video-enhancement strategies on the experience of learners affected by Asperger's Syndrome and dyslexia.

This chapter also considered a methodological approach to help address these gaps, identifying an opportunity to employ a series of design-based research cycles to design, develop, implement and evaluate a range of interventions designed to facilitate video-enhancement of learning, assessment and feedback. Taking forward this methodological approach, the study next moves on to follow a participatory design-based research path, collaborating with the students whom the interventions were intended to benefit, and working within a collapsed hierarchical structure to facilitate a culture of co-design, ultimately leading to the development of illustrative cases which highlight the impact of the interventions.

In the following chapter, the focus therefore turns to the realisation of the research design outlined here, discussing the iterative refinement of the video-enhancement interventions which finally combine to form an integrated and holistic model of video-enhanced learning, assessment and feedback. A discussion of the learner experience of these interventions, presenting the six illustrative cases, which emerged from the investigation are presented in Chapter 2 of this text in keeping with the earlier focus on enhanced formative assessment for all learners.

REFERENCES

Akyol, Z., & Garrison, D. R. (2011). Understanding cognitive presence in an online and blended community of inquiry: Assessing outcomes and processes for deep approaches to learning. *British Journal of Educational Technology*, *42*(2), 233–250. doi:10.1111/j.1467-8535.2009.01029.x

Akyol, Z., Garrison, D. R., & Ozden, M. Y. (2009). Online and blended communities of inquiry: Exploring the developmental and perceptional differences. *International Review of Research in Open and Distance Learning*, *10*(6), 65–83. doi:10.19173/irrodl.v10i6.765

Anderson, R., Dickey, M., & Perkins, H. (2001). Experiences with tutored video instruction for introductory programming courses. *ACM SIGCSE Bulletin*, *33*(1), 347–351. doi:10.1145/366413.364619

Anderson, T., Rourke, L., Garrison, D. R., & Archer, W. (2001). Assessing teacher presence in a computer conferencing context. *Journal of Asynchronous Learning Networks*, *5*(2), 1–17.

Asbjørnsen, A. E., Obrzut, J. E., Eikeland, O. J., & Manger, T. (2010). Can solving of wordchains be explained by phonological skills alone? *Dyslexia (Chichester, England)*, *16*(1), 24–35. doi:10.1002/dys.394 PMID:19562661

Attwood, T. (2000). *Asperger's Syndrome: A guide for parents and professionals*. London: Jessica Kingsley.

Baddeley, A. (1992). Working memory. *Science*, *255*(5044), 556–559. doi:10.1126cience.1736359 PMID:1736359

Baddeley, A. D., & Hitch, G. (1974). Working memory. In G. H. Bower (Ed.), *The psychology of learning and motivation: Advances in research and theory* (pp. 47–89). New York: Academic Press; doi:10.1016/S0079-7421(08)60452-1

Baron-Cohen, S. (2008). *Autism and Asperger's Syndrome*. Oxford, UK: Oxford University Press.

Belfer, K., & Morgan, T. (2005). Enhancing teaching presence and reducing distance using voice technologies. In *World Conference on E-Learning in Corporate, Government, Healthcare, and Higher Education* (*Vol. 2005*, No. 1, pp. 537-542). Academic Press.

Benford, P. (2008). *The use of Internet-based communication by people with autism* (PhD thesis). University of Nottingham. Retrieved from http://etheses.nottingham.ac.uk/661/

Bloxham, S., & Boyd, P. (2007). *Developing assessment in higher education*. Maidenhead, UK: Open University Press.

British Dyslexia Association. (2007). *Dyslexia research information - About Dyslexia*. Retrieved from http://www.bdadyslexia.org.uk/about-dyslexia/further-information/dyslexia-research-information-.html

Brown, A. L. (1992). Design experiments: Theoretical and methodological challenges creating complex interventions in classroom settings. *Journal of the Learning Sciences*, *2*(2), 141–178. doi:10.120715327809jls0202_2

Carey, K. (1997). Comparative disadvantage and special opportunities and the information technology revolution: General considerations and the examination of visual impairment as a case study. *Aslib Proceedings*, *4*(49), 77–81. doi:10.1108/eb051452

Chandler, P., & Sweller, J. (1991). Cognitive load theory and the format of instruction. *Cognition and Instruction*, *8*(4), 293–332. doi:10.12071532690xci0804_2

Charlop, M. H., & Milstein, J. P. (1989). Teaching autistic children conversational speech using video modeling. *Journal of Applied Behavior Analysis*, *22*(3), 275–285. doi:10.1901/jaba.1989.22-275 PMID:2793634

Cheng, G., & Chau, J. (2009). Digital video for fostering self-reflection in an ePortfolio environment. *Learning, Media and Technology*, *34*(4), 337–350. doi:10.1080/17439880903338614

Clark, J. M., & Paivio, A. (1991). Dual coding theory and education. *Educational Psychology Review*, *3*(3), 149–210. doi:10.1007/BF01320076

Cobb, P., diSessa, A., Lehrer, R., & Schauble, L. (2003). Design experiments in educational research. *Educational Researcher*, *32*(1), 9–13. doi:10.3102/0013189X032001009

Collins, A. (1992). Towards a design science of education. In E. Scanlon & T. O'Shea (Eds.), *New directions in educational technology* (pp. 15–22). Berlin: Springer. doi:10.1007/978-3-642-77750-9_2

Collins, J. L., Cook-Cottone, C. P., Robinson, J. S., & Sullivan, R. R. (2005). Technology and new directions in professional development: Applications of digital video, peer review, and self-reflection. *Journal of Educational Technology Systems*, *33*(2), 131–146. doi:10.2190/G9E9-LCQ1-H50X-4N23

Croft, N., Dalton, A., & Grant, M. (2010). Overcoming isolation in distance learning: Building a learning community through time and space. *UK Journal for Education in the Built Environment*, *5*(1), 27–64. doi:10.11120/jebe.2010.05010027

Crook, A., Mauchline, A., Maw, S., Lawson, C., Drinkwater, R., Lundqvist, K., ... Park, J. (2012). The use of video technology for providing feedback to students: Can it enhance the feedback experience for staff and students? *Computers & Education, 58*(1), 386–396. doi:10.1016/j.compedu.2011.08.025

Denzin, N. K. (2001). The reflexive interview and a performative social science. *Qualitative Research, 1*(1), 23–46. doi:10.1177/146879410100100102

Design Based Research Collective. (2003). Design based research: An emerging paradigm in educational inquiry. *Educational Researcher, 32*(1), 5–8. doi:10.3102/0013189X032001005

DeVaney, T. A. (2009). Impact of Video Tutorials in an Online Educational Statistics Course. *MERLOT Journal of Online Learning and Teaching, 5*(4), 600–608.

Dewey, J. (1929). My pedagogic creed. *Journal of the National Educational Association, 18*(9), 291–295.

Dewey, J. (1944). *Democracy and Education*. New York: Free Press.

Doolan, M. A., & Simpson, M. (2010). Engaging tutors and learners through audio supported pedagogy. In *Proceedings of the 5th Annual Blended Learning Conference*. Hatfield, UK: University of Hertfordshire.

Fleming, N. D., & Mills, C. (1992). Not another inventory, rather a catalyst for reflection. *To Improve the Academy, 11*, 137-55.

Gage, N. L. (1989). The paradigm wars and their aftermath: A "historical" sketch of research on teaching since 1989. *Educational Researcher, 18*(7), 4–10.

Garrison, D. R., Anderson, T., & Archer, W. (1999). Critical inquiry in a text-based environment: Computer conferencing in higher education. *The Internet and Higher Education, 2*(2), 87–105. doi:10.1016/S1096-7516(00)00016-6

Garrison, D. R., Anderson, T., & Archer, W. (2001). Critical thinking, cognitive presence, and computer conferencing in distance education. *American Journal of Distance Education, 15*(1), 7–23. doi:10.1080/08923640109527071

Garrison, D. R., & Kanuka, H. (2004). Blended learning: Uncovering its transformative potential in higher education. *The Internet and Higher Education, 7*(2), 95–105. doi:10.1016/j.iheduc.2004.02.001

Gibbs, G., & Simpson, C. (2004). Conditions under which assessment supports students' learning. *Learning and Teaching in Higher Education, 1*(1), 3-31.

Glaser, B. G., & Strauss, A. L. (1967). *The discovery of grounded theory: Strategies for qualitative research*. New York: Aldine Publishing Company.

Griffiths, M. E., & Graham, C. R. (2009). Using asynchronous video in online classes: Results from a pilot study. *International Journal of Instructional Technology & Distance Learning, 6*(3), 65–76.

Hardin, J. R., & Ellington, A. J. (2005). Using multimedia to facilitate software instruction in an introductory modeling course. *INFORMS Transactions on Education, 5*(2).

Hari, R., Renvall, H., & Tanskanen, T. (2001). Left minineglect in dyslexic adults. *Brain, 124*(7), 1373–1380. doi:10.1093/brain/124.7.1373 PMID:11408332

Hemmi, A., Bayne, S., & Land, R. (2009). The appropriation and repurposing of social technologies in higher education. *Journal of Computer Assisted Learning*, *25*(1), 19–30. doi:10.1111/j.1365-2729.2008.00306.x

Her Majesty's Stationery Office (HMSO). (*2001*). *The Special Educational Needs and Disabilities Act*. Retrieved from http://www.legislation.gov.uk/ukpga/2001/10/pdfs/ukpga_20010010_en.pdf

Her Majesty's Stationery Office (HMSO). (2005). *The Disabilities Discrimination Act*. Retrieved from http://www.legislation.gov.uk/ukpga/2005/13/pdfs/ukpga_20050013_en.pdf

Her Majesty's Stationery Office (HMSO). (2010). *The Equality Act*. Retrieved from http://www.legislation.gov.uk/ukpga/2010/15/pdfs/ukpga_20100015_en.pdf

Higher Education Funding Council for England (HEFCE). (2011). *National student survey: Findings and trends 2006 to 2010*. Retrieved from http://www.hefce.ac.uk/media/hefce/content/pubs/2011/201111/11_11.pdf

Higher Education Statistics Agency (HESA). (2013). *Higher education statistics for the United Kingdom, 2012-13*. Retrieved from: http://www.hesa.ac.uk/pubs/heuk

Hope, S. (2010). *Giving audio-visual feedback using Jing and GradeMark*. Retrieved from http://www.bioscience.heacademy.ac.uk/ftp/casestudies/hope.pdf

Hounsell, D. (2007). Towards a more sustainable feedback to students. In D. Boud & N. Falchikov (Eds.), *Rethinking assessment in higher education* (pp. 101–113). London: Routledge.

Hounsell, D., McCune, V., Hounsell, J., & Litjens, J. (2008). The quality of guidance and feedback to students. *Higher Education Research & Development*, *27*(1), 55–67. doi:10.1080/07294360701658765

Humfrey, P. (2010). Digital 'deviants' and the spirit of '68. *Times Higher Education*. Retrieved from http://www.timeshighereducation.co.uk/story.asp?sectioncode=26&storycode=414268&c=1

Ice, P., Curtis, R., Phillips, P., & Wells, J. (2007). Using asynchronous audio feedback to enhance teaching presence and students' sense of community. *Journal of Asynchronous Learning Networks*, *11*(2), 3–25.

Jordan, L. (2012). Video for peer feedback and reflection: Embedding mainstream engagement into learning and teaching practice. *Research in Learning Technology*, *20*(sup1), 16–25. doi:10.3402/rlt.v20i0.19192

Keller, J. M. (1983). Motivational design of instruction. In C. M. Reigeluth (Ed.), *Instructional design theories and models: An overview of their current status* (pp. 383–434). Hillsdale, NJ: Erlbaum.

Kerr, W., & McLaughlin, P. (2008). The benefit of screen recorded summaries in feedback for work submitted electronically. In F. Khandia (Ed.), *Proceedings of the CAA International Computer Assisted Assessment Conference* (pp. 153-168). Loughborough University.

King, A. (1993). From sage on the stage to guide on the side. *College Teaching*, *41*(1), 30–36. doi:10.1080/87567555.1993.9926781

Klin, A., & Volkmar, F. R. (2000). Treatment and intervention guidelines for individuals with Asperger's Syndrome. In A. Klin, F. R. Volkmar, & S. Sparrow (Eds.), *Asperger Syndrome* (pp. 340–366). New York: The Guildford Press.

Kneale, P. (1997). The rise of the 'strategic student': How can we adapt to cope. In S. Armstrong, G. Thompson, & S. Brown (Eds.), *Facing up to radical changes in universities and colleges* (pp. 119–130). London: Kogan Page.

Knight, P., & Saunders, M. (1999). Understanding teachers' professional cultures through interview: A constructivist approach. *Evaluation and Research in Education*, *13*(3), 144–156. doi:10.1080/09500799908666954

Lang, R., Shogren, K. A., Machalicek, W., Rispoli, M., O'Reilly, M., Baker, S., & Regester, A. (2009). Video self-modeling to teach classroom rules to two students with Asperger's. *Research in Autism Spectrum Disorders*, *3*(2), 483–488. doi:10.1016/j.rasd.2008.10.001

Laurillard, D. (1993). *Rethinking university teaching: A framework for the effective use of educational technology*. London: Routledge.

Laurillard, D. (2002). Rethinking university teaching: A conversational framework for the effective use of learning technologies (2nd ed.). London: Routledge Falmer.

Lee, M. J., Pradhan, S., & Dalgarno, B. (2008). The effectiveness of screencasts and cognitive tools as scaffolding for novice object-oriented programmers. *Journal of Information Technology Education*, *7*, 61–80. doi:10.28945/179

Mathieson, K. (2012). Exploring student perceptions of audiovisual feedback via screencasting in online courses. *American Journal of Distance Education*, *26*(3), 143–156. doi:10.1080/08923647.2012.689166

Mathisen, P. (2012). Video feedback in higher education. A contribution to improving the quality of written feedback. *Nordic Journal of Digital Literacy*, *7*(2), 97–116.

Mayer, R. E. (2001). *Multimedia learning*. Cambridge, UK: Cambridge University Press. doi:10.1017/CBO9781139164603

Mayer, R. E. (2002). Cognitive theory and the design of multimedia instruction: An example of the two-way street between cognition and instruction. *New Directions for Teaching and Learning*, *89*(89), 55–71. doi:10.1002/tl.47

Mayer, R. E., & Moreno, R. (2003). Nine ways to reduce cognitive load in multimedia learning. *Educational Psychologist*, *38*(1), 43–52. doi:10.1207/S15326985EP3801_6

McConnell, D. (2006). *E-learning groups and communities*. Maidenhead, UK: Open University Press.

McDowell, J. (2010). *Anarchy in the universities: Beyond the student-teacher hierarchy*. Paper presented at the Learning Futures Festival 2010. University of Leicester. Retrieved from http://eprints.hud.ac.uk/id/eprint/10802/1/McDowell_J_2010_AnarchyInTheUniversities-BeyondTheStudentTeacherHierarchy.pdf

McDowell, J. (2011). Using asynchronous video to promote learner engagement through the enhancement of assessment and feedback. In *Proceedings of the 6th International Blended Learning Conference* (pp. 69-78). University of Hertfordshire.

Meyer, J. H., & Land, R. (2005). Threshold concepts and troublesome knowledge (2): Epistemological considerations and a conceptual framework for teaching and learning. *Higher Education*, *49*(3), 373–388. doi:10.100710734-004-6779-5

Moreno, R., & Mayer, R. E. (2000). A learner-centered approach to multimedia explanations: Deriving instructional design principles from cognitive theory. *Interactive Multimedia Electronic Journal of Computer-Enhanced Learning*, *2*(2), 12–20.

National Student Forum. (2009). *National Student Forum Annual Report*. Author. Retrieved from https://webarchive.nationalarchives.gov.uk/20121106154314/http://www.bis.gov.uk/assets/biscore/higher-education/docs/n/09-p83-national-student-forum-annual-report-09

Nortcliffe, A., & Middleton, A. (2007). *Audio Feedback for the iPod Generation*. Paper presented at the International Conference on Engineering Education – ICEE 2007, Coimbra, Portugal. Retrieved from http://ineerweb.osanet.cz/Events/ICEE2007/papers/489.pdf

O'Donoghue, M., & Cochrane, T. (2010). The role of live video capture production in the development of student communication skills. *Learning, Media and Technology*, *35*(3), 323–335. doi:10.1080/17439884.2010.509350

Paivio, A. (1990). *Mental representations: A dual coding approach*. Oxford University Press. doi:10.1093/acprof:oso/9780195066661.001.0001

Palloff, R. M., & Pratt, K. (2005). *Collaborating online: Learning together in community*. San Francisco, CA: Jossey-Bass.

Rayner, C., Denholm, C., & Sigafoos, J. (2009). Video-based intervention for individuals with autism: Key questions that remain unanswered. *Research in Autism Spectrum Disorders*, *3*(2), 291–303. doi:10.1016/j.rasd.2008.09.001

Reeves, T. C. (2000). *Enhancing the worth of instructional technology research through "design experiments" and other development research strategies*. Paper presented at the Annual Meeting of the American Educational Research Association, New Orleans, LA. Retrieved from http://treeves.coe.uga.edu/AERA2000Reeves.pdf

Richardson, J. T., & Wydell, T. N. (2003). The representation and attainment of students with dyslexia in UK higher education. *Reading and Writing*, *16*(5), 475–503. doi:10.1023/A:1024261927214

Rotherham, R. (2009). *Sounds good: Quicker, better assessment using audio feedback*. Retrieved from http://sites.google.com/site/soundsgooduk/downloads/

Rumble, G. (2001). Re-inventing distance education, 1971-2001. *International Journal of Lifelong Education*, *20*(1-2), 31–43.

Rust, C. (2002). The impact of assessment on student learning: How can the research literature practically help to inform the development of departmental assessment strategies and learner-centred assessment practices? *Active Learning in Higher Education*, *3*(2), 145–158. doi:10.1177/1469787402003002004

Salmon, G. (2000). *E-Moderating: The key to teaching and learning online.* Kogan Page Limited.

Schaffer, D., Doube, W., & Tuovinen, J. (2003). Applying cognitive load theory to computer science education. In M. Petre & D. Budgen (Eds.), *Proceedings of the Joint Conference for Evaluation and Assessment in Software Engineering and the Psychology of Programming Interest Group 2003* (pp. 333-346). Academic Press.

Sekovanić, V., Vukovac, D. P., & Podbojec, Z. (2012). Usability case study of an adapted e-learning course for dyslexic students. *Proceedings of the 5th International Conference of Education, Research and Innovation* 1998-2007.

Séror, J. (2013). Show me! Enhanced feedback through screencasting technology. *TESL Canada Journal, 30*(1), 104–116. doi:10.18806/tesl.v30i1.1128

Shephard, K. L. (2002). Integrating national initiatives and local developments: streaming video to support nurse education in the UK. In *Winds of change in the sea of learning. Proceedings of the 19th Annual* Auckland, New Zealand.

Shipley-Benamou, R., Lutzker, J. R., & Taubman, M. (2002). Teaching daily living skills to children with autism through instructional video modeling. *Journal of Positive Behavior Interventions, 4*(3), 166–177. doi:10.1177/10983007020040030501

Smith, T. (2008). Integrating undergraduate peer mentors into liberal arts courses: A pilot study. *Innovative Higher Education, 33*(1), 49–63. doi:10.100710755-007-9064-6

Stake, R. (1995). *The art of case research.* Newbury Park, CA: Sage Publications.

Stannard, R. (2007). Using screen capture software in student feedback. *The Higher Education Academy English Centre Commissioned Case Studies.* Retrieved from http://www.english.heacademy.ac.uk/explore/publications/casestudies/technology/camtasia.php

Stewart, W. (2009). *Audio supported enhanced learning. JISC Final Report.* Retrieved from http://reports.jiscemerge.org.uk/Download-document/16-ASEL-Audio-Supported-Enhanced-Learning.html

Thornhill, S., Asensio, M., & Young, C. (2002). *Video streaming: A guide for educational development.* Manchester, UK: JISC Click and Go Project, UMIST.

Veronikas, S. W., & Maushak, N. (2005). Effectiveness of Audio on Screen Captures in Software Application Instruction. *Journal of Educational Multimedia and Hypermedia, 14*(2), 199–205.

Wing, L., & Gould, J. (1979). Severe impairments of social interaction and associated abnormalities in children: Epidemiology and classification. *Journal of Autism and Developmental Disorders, 9*(1), 11–29. doi:10.1007/BF01531288 PMID:155684

Woodfine, B. P., Nunes, M. B., & Wright, D. J. (2005). *Constructivist eLearning and dyslexia: Problems of social negotiation in text-based synchronous environments.* Paper presented at the 3rd International Conference on Multimedia and Information & Communication Technologies in Education (m-ICTE2005), Cáceres, Spain. Retrieved from http://www.formatex.org/micte2005/97.pdf

Woodfine, B. P., Nunes, M. B., & Wright, D. J. (2008). Text-based synchronous e-learning and dyslexia: Not necessarily the perfect match! *Computers & Education*, *50*(3), 703–717. doi:10.1016/j.compedu.2006.08.010

Yin, R. K. (2003). *Case study research: Design and methods* (3rd ed.). Thousand Oaks, CA: Sage Publishing.

Chapter 7
Towards a Dialogic Model of Video-Enhanced Learning, Assessment, and Feedback

James McDowell
Lancaster University, UK

ABSTRACT

This chapter discusses the implementation of research designed to investigate how video-enhanced learning, assessment, and feedback involves a dialogic approach and contributes to the concept of an integrated model of practice. It builds on the theoretical framework underpinned by both cognitive and social forms of constructivism. Developing design exemplars for video-enhanced learning, assessment, and feedback, three research cycles explored learner-artefact and learner-tutor interactions. The first cycle introduces instructional tutorial videos, freeing up time for formative feedback; the second cycle sees formative video-feedback situated within a conversational framework and video-enhanced assessment activities replace documentation tasks; while the third cycle introduces refinements to the frequency of feedback and regular video diaries, forming an integrated model of video-enhanced learning, assessment, and feedback.

INTRODUCTION

Background and Context

Having previously identified problems with the engagement, progression and retention of students studying for a computer games degree at a north-of-England university, the delivery of learning materials through traditional lectures was replaced by the introduction of a series of instructional tutorial videos (ITVs). These designed to free up time for formative feedback as part of a strategy of assessment for learning (Torrance, 2007), and to help establish cognitive presence (Garrison, Anderson, & Archer, 2001) within a blended learning community (e.g. Volpantesta & Frega, 2007). That intervention (DBRC1) formed

DOI: 10.4018/978-1-7998-0426-0.ch007

the first of three design-based research cycles which sought to investigate the influence of asynchronous video on the learner experience of video-enhanced learning, assessment and feedback, and this chapter discusses the work of the second and third cycles, which saw the development, refinement and integration of design exemplars for video-enhanced assessment and feedback within an integrated model underpinned by the video feedback loop outlined in the second design-based research cycle (DBRC2) and the third design-based research cycle (DBRC3), below.

THE EARLY RESEARCH WORK

Pre-Cursor Study and First Design-Based Research Cycle

The DBRC1 participant group was formed of twenty-one students, of whom five were known at the outset of the study to be affected by dyslexia. Legislative requirements on UK Higher Education Institutions (UKHEIs) are designed to ensure that students with specific learning difficulties such as dyslexia and/or Asperger's Syndrome, now known as Autism are not placed at a disadvantage in relation to their peers (e.g. Her Majesty's Stationary Office, HMSO, 2001; 2010), and an historical prevalence of students diagnosed with an autistic spectrum condition (ASC) enrolling on the course made it prudent to consider the needs of those affected by an ASC in the design of this first intervention in order to afford inclusivity for participants in subsequent research cycles. The development of the instructional tutorial videos (ITVs) was informed by an appreciation that placing excessive cognitive load (Mayer & Moreno, 2003) on the visual channel could overload the auditory channel (Attwood, 2000; Baron-Cohen, 2008), and that this carried the potential for elective social isolation due to deep immersion in predominantly visual learning materials. This took into account the needs of Alex, a student from an earlier cohort, whose Asperger's Syndrome (AS) had limited his engagement in group work prior to the introduction of computer-mediated communications (McDowell, 2015). The evaluation of DBRC1 revealed that the ITVs had made a positive impact on student engagement, retention and attainment, while also helping participants overcome threshold concepts encountered in the teaching and learning process (Mayer & Land, 2005). Further evidence which emerged through thematic analysis of students' blog entries and their responses to an anonymous questionnaire highlighted calls for greater use of video, confirming that using the medium of video had been well-received, and suggesting that asynchronous video might play a significant and valuable role in assessment and feedback.

SECOND DESIGN-BASED RESEARCH CYCLE: DBRC2

Overview of the Second Design-Based Research Cycle

The challenge for the second design-based research cycle (DBRC2) therefore was to explore and evaluate innovative approaches through which video-based activities could be introduced into assessment and feedback. DBRC1 had identified that learner-tutor dialogue and visual presentations of work-in-progress appeared to contribute to both engagement in formative tasks and, subsequently, levels of summative attainment. Consequently, the work undertaken in DBRC2 aimed broadly to enable the same group of twenty-one students who participated in DBRC1 to fulfil components of assessment through visual demonstrations of work, and to embrace a dialogic approach to formative feedback.

The development work on the two strands of video-based activities undertaken in DBRC2 was exploratory in nature, primarily aiming to establish 'proof-of-concept', and to develop *design exemplars* (cf. Hrastinski, Keller & Carlsson, 2010) on which to base a more substantive rollout over the course of a whole academic year for fuller evaluation in the third and final design-based research cycle (DBRC3).

Analysis Phase: Pedagogic Aims and Requirements Capture

Responding to both local and national calls for greater use of technology in the assessment and feedback process (National Student Forum, 2009), and cognisant of the potential for elevated rates of attrition directly attributable to a failure to provide formative feedback early in a course (Entwistle, Hounsell, Macaulay, Situnayake, & Tait, 1989; Krause, 2001), the work of DBRC2 sought to extend the use of asynchronous video by introducing technological interventions which would address key pedagogical recommendations surrounding assessment and feedback, specifically:

1. The use of video should be a participatory process (Thornhill, Asensio, & Young, 2002).
2. The format of assessment should take account of specific learning difficulties (HMSO, 2001).
3. Assessment should promote reflexive styles of learning (Crossouard, Pryor, & Torrance, 2004).
4. Students should be enabled to make improvements based on feedback received (Boud, 2000).
5. Feedback should be provided in a timely manner (Galusha, 1998; Gibbs & Simpson, 2004).
6. Feedback should be detailed and personalised to individual students (Gibbs & Simpson, 2004).
7. There should be dialogue around formative feedback (Laurillard, 2002).

Design

The learner evaluation of DBRC1 had yielded calls for greater use of video across the university experience. DBRC2 therefore aimed to respond to (i) DBRC1 participants' calls for greater use of video, (ii) institutional calls for assessment and feedback to be innovative, to inspire and motivate, and to encourage dialogue, and (iii) national calls to "enhance traditional teaching methods with new technologies" (National Student Forum, 2009).

As highlighted earlier, a tension exists between the learner-centred focus on formative feedback afforded by a model of assessment *for* learning, and an institution-centred emphasis on assessment *of* learning for the purposes of certification afforded by a model of summative assessment (e.g. Bloxham & Boyd, 2007; Torrance, 2007). Cognisant of the divergent needs of the institution and its students, one of the challenges in DBRC2 was to design appropriate interventions which would allow the effective combination of formative feedback and summative assessment. Two strands of video-based activity were therefore conceptualised, implemented and evaluated, with the use of formative video-enhanced feedback (VEF) explored before the use of formative and summative video-enhanced assessment (VEA).

The Conceptualisation and Design of a System to Facilitate Video-Enhanced Feedback

Exploring opportunities for video-enhancement of feedback, two discrete opportunities were identified as offering the potential for the provision of video-feedback that would enable learners to improve on their development work:

- **VEF Activity 1:** Short, targeted, problem-focused, formative video-feedback.
- **VEF Activity 2:** Formative video-feedback on drafts of work approaching completion.

By situating video-feedback within a *Conversational Framework* (Laurillard, 2002) and encouraging learners to participate in dialogue with tutors on specific aspects of their work, VEF Activity 1 aimed to fulfil both Pedagogic Aims 1, that use of video should be a participatory process (Thornhill, et al., 2002), and 7, that there should be dialogue around formative feedback (Laurillard, 2002), while also fulfilling Pedagogic Aim 3 through promoting learner reflexivity (Crossouard et al., 2004).

Building on the findings emerging from DBRC1 in connection with dyslexia, it was intended that replacing text-based feedback with an asynchronous, video conversation would afford learners with dyslexia the opportunity to engage more closely with feedback, thereby fulfilling Pedagogic Aim 2 by taking into account their learning difficulty (HMSO, 2001). Further, it was hoped to build on previous research into the use of video to support learners with an autistic spectrum condition, which found learners responding positively to video interventions (Rayner, Denholm, & Sigafoos, 2009), and more likely to engage in negotiation and dialogue where this was facilitated by video (Lang, Shogren, Machalicek, Rispoli, O'Reilly, Baker, & Regester, 2009).

By situating feedback within a model of assessment for learning (Torrance, 2007), all forms of formative video-feedback fulfilled Pedagogic Aim 4, that students should be enabled to make improvements based on feedback received (Boud, 2000), and Pedagogic Aim 5, that feedback should be provided in a timely manner (Galusha, 1998; Gibbs & Simpson, 2004). Further, the one-to-one nature of asynchronous learner-tutor communication ensured that feedback could be both detailed and personalised to individual students (Gibbs & Simpson, 2004), fulfilling Pedagogic Aim 6.

The Conceptualisation and Design of a System to Facilitate Video-Enhanced Assessment

Exploring opportunities for VEA, three approaches were designed which aimed to allow learners to supplement the weekly blogs documenting their development work. Against the background of an assessment model requiring both the development of a product and the production of documentation which detailed the development of that product, VEA opportunities were explored for both components of assessment. As was the case with the conceptualisation of the learner-tutor video-feedback system above, these approaches were designed to map to key pedagogic aims derived from the literature. The three video-based assessment activities are presented below in chronological order:

- **VEA Activity 1:** Formative presentations of work in progress.
- **VEA Activity 2:** Summative presentations of work completed over the year.
- **VEA Activity 3:** Summative, reflective self-assessment linking process to product.

Opportunities to make changes to the assessment model of the module were restricted by academic regulations, and, in trialling multiple approaches, it was anticipated that all participants would take part in the first and second activities, designed to replace mid-year and end-of-year presentations and demonstrations, as this activity could be substituted for traditional in-studio presentations to the class already required by the existing assessment model. The third activity could not be mandatorily imposed during DBRC2, and was therefore trialled on a voluntary basis. Each of the activities are mapped against the pedagogic aim in Table 1.

The first activity fulfilled Pedagogic Aim 1, by offering opportunities for learners' engagement with video to be a participatory process (Thornhill, et al., 2002), and, by substituting the traditional stand-up presentation with a screen capture exercise, the format of assessment fulfilled Pedagogic Aim 2 by taking account of learning difficulties (HMSO, 2001).

The second and third activities also fulfilled Pedagogic Aim 1 by offering another opportunity for the use of video as a participatory process (Thornhill et al., 2002), while creating a further opportunity for learners diagnosed with dyslexia and/or Asperger's Syndrome to avoid being disadvantaged through presenting their work without the need for direct, face-to-face contact. Additionally, by encouraging students to reflect on the development of their skills over the course of an academic year and to highlight linkages between the development and documentation work, the third activity fulfilled Pedagogic Aim 3 by promoting reflexive styles of learning (Crossouard et al., 2004). The full mapping of Pedagogic Aims against the five video-enhanced feedback and assessment activities is shown in Table 1.

Work to Support the Development of a Learner-Tutor Video-Feedback Loop

Technical work in DBRC2 sought to realise the conceptualisation above through the development of a video-feedback loop (VFL) system (see Figure 1), conceptually similar to those later described by Boud and Molloy (2013) and by Carless (2013), wherein learners were enabled to participate in a continuous and iterative dialogue around formative feedback using asynchronous video. Designed to encourage the development of a video-enhanced conversation between learner and tutor, the system allowed learners to utilise tutor responses to their work effectively by rectifying errors as they occur, thereby preventing any escalation that might lead to problems in the development of their understanding. To this end, the VFL system enabled learners to upload short video clips demonstrating their work and highlighting any problems encountered, and tutors to respond to these issues and appraise students of their progress.

Table 1. Mapping of pedagogic aims to video-enhanced feedback and assessment activities

Pedagogic Aim	VEF Activity 1	VEF Activity 2	VEA Activity 1	VEA Activity 2	VEA Activity 3
Participation	X	X	X	X	X
Inclusivity	X	X	X	X	X
Reflexivity	X	X	X		X
Improvement	X	X			
Timeliness	X	X			
Personalisation	X	X			
Dialogue	X	X			

Figure 1. The learner-tutor video-feedback loop
(McDowell, 2012)

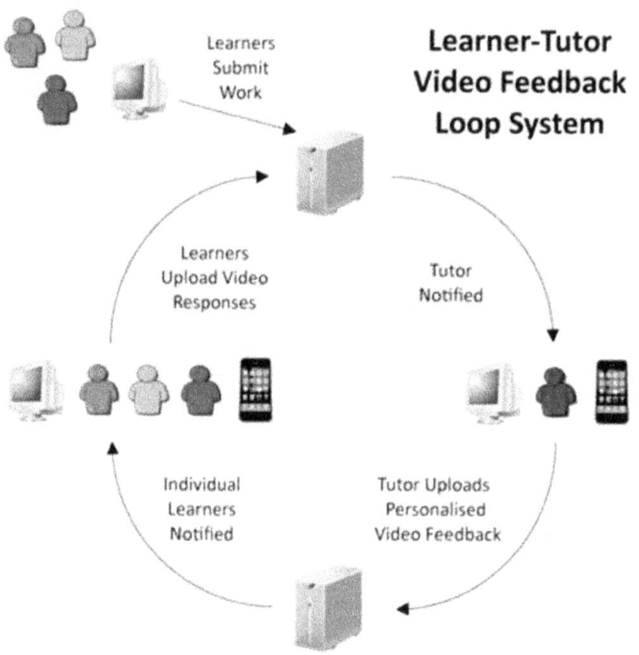

Ending an asynchronous exchange with a question encouraged learners to demonstrate that the feedback was subsequently acted upon, and to engage in dialogue around formative feedback. The design of a non-text-based approach also aimed to reduce the impact of learning difficulties such as dyslexia on engagement with feedback and, as a corollary, the assessment process.

Implementation

Findings emerging from the anonymous questionnaire and other data collection instruments used in DBRC1 were communicated to all stakeholders at the beginning of the second semester. Participants were offered the choice of continuing to receive face-to-face and online feedback as they had during DBRC1, or to raise problems encountered in their work using screencasts, and to receive personalised formative feedback through the same medium.

Implementation of Video-Enhanced Feedback Activities

Participants were given guidance on both the use of screen capture software, and on the process of engaging with the VFL, whereby they were requested to submit short videos highlighting a problem they had encountered, using the required format of a demonstration of the problem, a description of unsuccessful attempts made to overcome the problem, and an outline of a possible solution.

Implementation of VEF Activity 1: Problem-Focused, Formative Video-Feedback

The first VEF activity saw participants create short screencasts highlighting a problem encountered in their development work. Participants were also instructed to avoid submitting multiple requests for feedback in order to avoid compounded errors without first addressing the root of a problem.

In the very first example of video-feedback, a video of duration 01m:16s was uploaded to *YouTube®* by the participant, and this was played by the tutor using *Camtasia®* to capture both the screen and the tutor's head and shoulders as a picture-in-picture, as illustrated in Figures 2, 3 and 4. The problem video followed the required format, with the participant describing the problem, highlighting aspects of the code where attempts were made to correct the issue described, and demonstrating how the problem manifested itself within the compiled output. After first viewing the student's video in its entirety, the tutor then replayed the video from a key point, pausing playback to offer a short overview of the likely cause of the problem. Areas of the code on which the participant is recommended to focus their efforts were highlighted, and finally the participant was requested to attempt correction of the problem using the feedback, and to post a further video demonstrating whether and how the problem had been resolved.

The recorded video-feedback, including playback of the key aspect underpinning the student's problem and the tutor's recommendation of how to develop a solution, was of 01m:34s duration and the file size was 2.25Mb, making the file of an acceptable size to be emailed to the student and/or uploaded to *Mahara®* prior to the availability of the feedback loop application. In the follow-up video produced by the student, the problem was shown to have been resolved, and a plan for future development work outlined.

Notably, later in this VFL, the participant described seeing the tutor's image overlaid on the work as 'spooky', and as a result the inclusion of the picture-in-picture was thereafter discontinued in favour of a voiceover-only approach; this strategy was later found to be supported by the findings of work undertaken by Henderson and Phillips (2015).

Figure 2. Screenshot

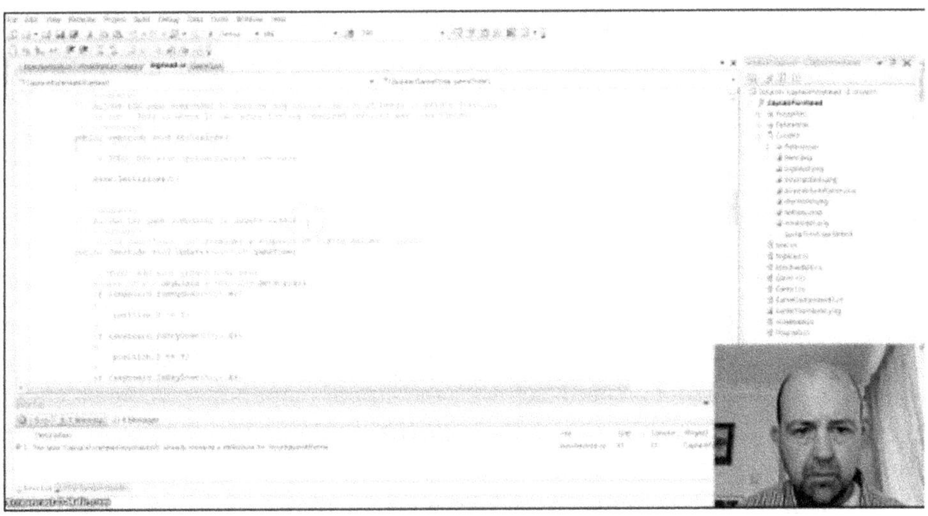

Figure 3. Student screenshot

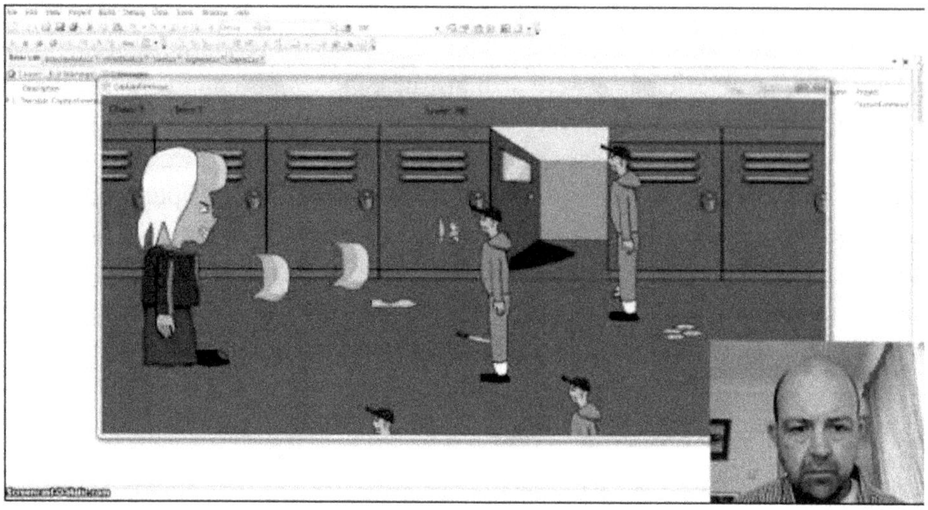

Figure 4. Screenshot of code

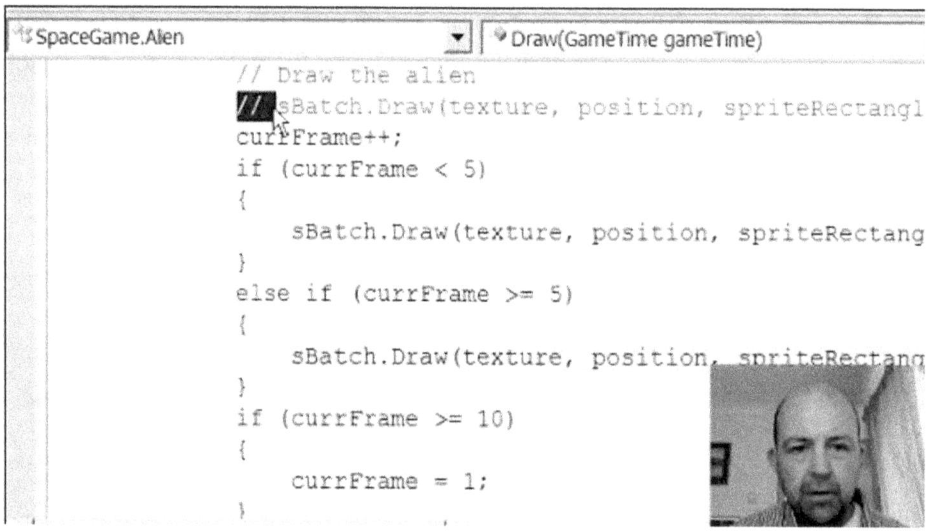

Implementation of VEF Activity 2: Formative Video-Feedback on Drafts of Work

The second VEF activity was designed to provide students with highly personalised formative feedback at a late stage in the development of an artefact, with the aim of pinpointing where amendments could be made which might lead to an improved grade in the final submission. Two weeks before the final deadline, students were requested to submit a draft of their work that was opened and run within the development environment, with audio-visual feedback captured as *Camtasia*® screencast. Following the response to the inclusion of the picture-in-picture outlined in VEF Activity 1, video-feedback was produced as a screencast-only walkthrough of the work accompanied by a voiceover, rather than including the output from the tutor's webcam, thereby lowering the risk of cognitive overload (Mayer, 2001; Mayer & Moreno, 2003).

All twenty-one participants received video-feedback on their drafts, with the duration of these videos ranging between 5-8 minutes; at a production ratio of c.1.5Mb per minute, the file size ranged between 8-12Mb. Video-feedback files were uploaded to *Mahara®* and embedded as private feedback in the e-portfolio system shared only between tutor and the individual student, thereby generating an automatic email to the student notifying them that feedback had been left.

Implementation of Video-Enhanced Assessment Activities

Participants were given guidance on the use of *Camtasia®*, and encouraged to provide a narrative account of how their skills had developed over the year. Capturing a walkthrough of their blogs and screenshots within *Mahara®* followed by a demonstration of the finished products, video summaries highlighted where problems had occurred during development, and explained how these had been overcome.

Implementation of VEA Activity 1: Formative Presentations of Work In Progress

The taught content of the module delivered over the academic year was split equally between the two semesters, with the emphasis switching from the development of design skills in Semester 1 to the development of programming expertise in Semester 2. With the assessment model for the module requiring the submission of a portfolio of work at the end of the academic year, students received feedback on design work submitted at the end of Semester 1 to which they were able to make improvements prior to a final submission at the end of Semester 2.

VEA Activity 1 required students to create a short video walkthrough of their work at a formative point in the production of the artefact, highlighting those areas in which they felt skills had been successfully developed, and acknowledging those areas in which further work remained to be completed. Lasting 2-5 minutes, these walkthrough videos were created by students when outside the studio environment, and while working in an unsupervised capacity.

In the examples shown in Figures 5 and 6, the participant used a free screencast software package, *CamStudio®*, capturing a 3-minute walkthrough that emphasised areas where unusual or interesting features had been developed, and outlining where any further work was anticipated.

Implementation of VEA Activity 2: Summative Presentations of Final Submissions

The original assessment model for the module required all students to give a presentation of their portfolio to an audience comprised of tutor and peers in the final studio session of the academic year. In previous years, a *Flip* camera had been used to record these presentations, and students later received feedback on their presentation skills in addition to the summative feedback for the module. These presentations previously provided an opportunity for students to give demonstrations of artefacts within their portfolio, with the emphasis on highlighting how and where artefacts met the requirements of the relevant product specifications, rather than on the process through which the product was developed.

Figure 5. First screenshots highlighting an example of a 'walkthrough' video of a presentation of work in progress

Figure 6. Second screenshot highlighting an example of a 'walkthrough' video of a presentation of work in progress

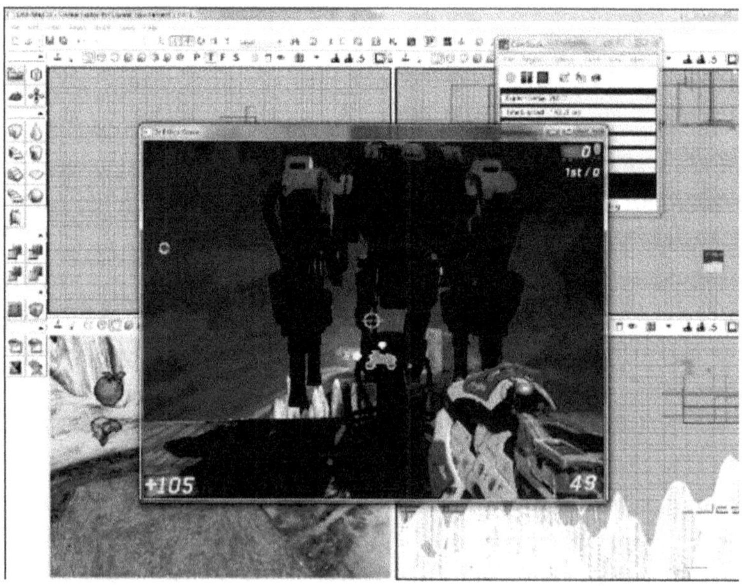

While the *Flip* camera approach captured video material that enabled tutors to provide feedback on students' presentation skills, the use of the camera could not offer the opportunity to capture the subtleties of the walkthroughs seen in on-screen demonstrations, and VEA Activity 2 sought to augment the traditional presentation with a screen capture of the final walkthrough.

Implementation of VEA Activity 3: Reflective Self-Assessment Video

The third and final intervention in this area saw the introduction of a non-mandatory VEA activity undertaken on a voluntary basis. As the exercise carried no weight towards the final grading of students' portfolios, engagement with this exercise was encouraged by framing it in a competitive format with a prize for the winner, as voted for by course tutors and peers. Participants were encouraged to work on these videos outside of scheduled studio time, and were free to choose whether to use the studio facilities after timetabled teaching activities had ended, or to work at home using their own resources.

As outlined earlier, the purpose of this activity was to encourage students to engage in a critical self-assessment of how their skills had developed by reflecting on their weekly blog posts and linking these to finished artefacts. The screenshots in Figures 7 and 8 highlight the winning entrant's use of *Camtasia®* to capture a demonstration of how development work was documented in *Mahara®* and how this connected with the development of artefacts presented in their portfolio.

Evaluation

As an exploratory cycle which aimed to identify design exemplars in the area of video-enhanced assessment and feedback, learner evaluation of the work of DBRC2 encompassed a range of data collection and analysis strategies. As highlighted earlier, a combination of the emergent nature of the research design and institutional restrictions on amendments to the assessment model meant that much of the implementation of interventions designed in this cycle was necessarily limited to prototyping activity. As a corollary, it was necessary to evaluate certain video-enhanced assessment interventions with a subset of the overall

Figure 7. First screenshot highlighting the winning entrant's linkage of weekly blogs and development work

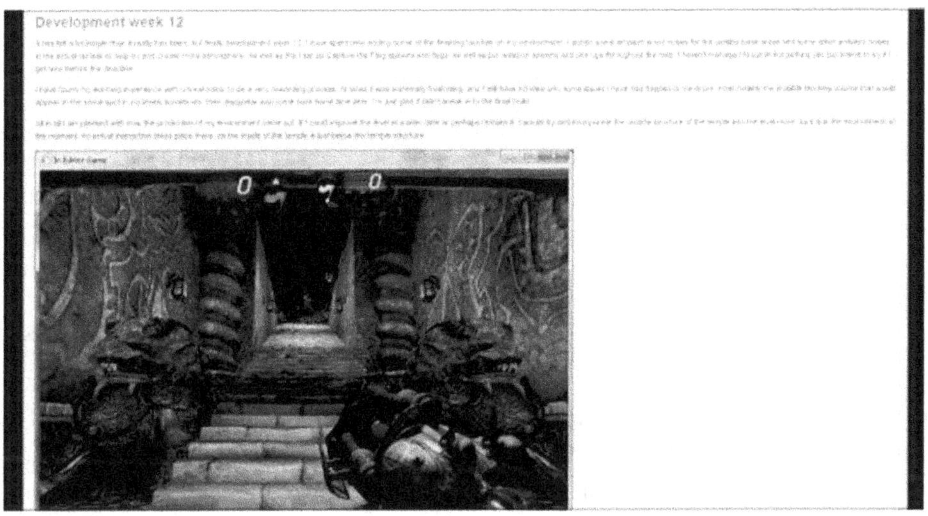

Figure 8. Second screenshot highlighting the winning entrant's linkage of weekly blogs and development work

participant group; to this end, the number of participants, the number of participants affected by dyslexia, and the number and type of sources of evidence analysed is presented on a per-sub-section basis below.

Where opportunities to introduce a video-enhanced approach to all participants within the assessment and feedback process arose, these opportunities were taken, with subsequent evaluation undertaken through the collection and analysis of data from the whole group. Where this was not possible, refinements were identified from data highlighting both positive and negative experiences of a subset of participants (cf. Wixon, Pietras, Huntwork, & Muzzey, 1996).

Evaluation of Video-Enhanced Feedback Activities

The two VEF activities were designed to fulfil several of the Pedagogic Aims outlined in the development of a framework for the evaluation of these activities, those Pedagogic Aims were used as the key criteria against which the VEF activities were judged. As outlined above, the purpose of DBRC2 was to identify design exemplars of VEF and VEA techniques, and the activities were therefore assessed against each of the relevant Pedagogic Aims.

Sources of data collected and analysed in the evaluation of VEF activities included: (i) twenty-two videos submitted by ten students initiating learner-tutor VFLs, (ii) twenty-two diagnostic feedback videos created in response to students' videos, (iii) twenty-one formative feedback videos created by the tutor, (iv) field notes created from weekly, direct observations of all twenty-one students in studio sessions, and (v) two semi-structured interviews of 30-35 minutes duration, conducted with two students whose use of VEF and VEA highlighted strong engagement and/or constituted learner-derived exemplars. As an exploratory cycle aiming to establish proof of concept and to identify design exemplars, five *a priori* categories derived from the Pedagogic Aims and outlined below directed the process of thematic analysis:

- **Improvement:** Feedback should enable improvements to be made (Boud, 2000).
- **Timeliness:** Feedback should be timely (Galusha, 1998; Gibbs & Simpson, 2004).
- **Personalisation:** Feedback should be specific to individual students (Gibbs & Simpson, 2004).
- **Dialogue:** Dialogue should take place around feedback (cf. Laurillard, 2002).
- **Inclusivity:** Assessment formats take account of specific learning difficulties (HMSO, 2001).

Evaluation of Video-Enhanced Assessment Activities

The three VEA activities were also designed to fulfil several of the Pedagogic Aims considered consistent with the approach taken with VEF, those Pedagogic Aims were used as the key criteria against which the VEA activities were judged in the development of a framework for the evaluation of these activities. As previously outlined, the purpose of DBRC2 was to identify design exemplars of VEF and VEA techniques, and the activities were therefore assessed against each of the relevant Pedagogic Aims, as was the case in the evaluation of the VEF activities. Once again, participation was voluntary, with eighteen of the twenty-one students in the class electing to create videos, including all five students affected by dyslexia.

Sources of data collected and analysed in the evaluation of VEA activities included: (i) eighteen videos of work-in-progress submitted by students, (ii) five videos offering a reflective, self-assessment of the development of skills, (iii) field notes derived from weekly direct observations of the twenty-one students' work in studio sessions, and (iv) the two semi-structured interviews of 30-35 minutes duration that focused on the uses and impact of video, conducted with the two students whose use of VEA and/or VEF formed a design exemplar. Three *a priori* categories derived from the Pedagogic Aims informed a deductive thematic analysis exercise:

- **Participation:** Use of video should be participatory (Thornhill, et al., 2002).
- **Inclusivity:** Assessment formats take account of specific learning difficulties (HMSO, 2001)
- **Reflexivity:** Assessment should promote reflexivity (Crossouard, et al., 2004).

Identifying Design Exemplars from the VEF and VEA Activities

Having identified one entry from the competition as the clear winner, there was an opportunity to investigate how this participant perceived the uses of video which had taken place in the five VEF and VEA activities, and semi-structured interviews conducted with both this participant and the student who had engaged most deeply with the VFL system were subjected to inductive thematic analysis to identify key themes. This latter student had not only engaged with the concept of the VFL, but had also identified appropriate mechanisms and technologies that enabled him to do so without the need for specialist software, enabling other members of the cohort to engage with these activities. (An illustrative case highlighting this participant's experience of VEF and VEA can be found in Chapter 2 under the pseudonym 'Keith').

FINDINGS FROM DBRC2

The evaluation of the prototype VEF and VEA activities devised as part of DBRC2 provided an indication of where and how these five interventions might make a contribution to the fulfilment of the theory-based Pedagogic Aims derived from the literature, while also highlighting design exemplars of video-enhancement of assessment and feedback for both students and academic practitioners.

Video-Enhanced Feedback

The two VEF activities were both successful as vehicles that enabled students to receive feedback which was detailed and personalised, clear and useable, and which enabled students to learn from feedback, as evidenced by refinements to students' work and through analysis of VEA artefacts.

VEF Activity 1 was found to be particularly effective as a mechanism to pinpoint the source of problems, especially in those areas highlighted as threshold concepts during DBRC1. Participants reported that problems were overcome quickly, and that misinterpretation of feedback was minimised due to the audio-visual nature of the medium. This application of animation and narration in video-feedback appeared to be consistent with the *Multimedia Principle* of the *Cognitive Theory of Multimedia Learning* (Mayer, 2001), while affording students affected by dyslexia greater opportunity to engage with feedback, and ultimately, thereby, with assessment.

VEF Activity 2 was also found to be an effective video-based intervention, and students reported that the highly personalised, pinpointed nature of the video-feedback had enabled them to make improvements to their work at a critical juncture. Further, dyslexia-affected students reported in dialogic interviews that the voiceover commentary in combination with on-screen highlighting using the cursor had enabled them to focus on and assimilate the key themes of the feedback more efficiently than a graphics-plus-text approach, and significantly faster than where feedback had been in text-only format.

For both activities, those participants who expressed a preference indicated that the inclusion of a 'talking head' was an unnecessary intrusion that served as a distraction from the feedback itself. This supported the assertions of Mayer's (2001) principles of *Redundancy* and of *Coherence*, which suggest that learning is emphasised where extraneous material is removed from the screen (Mayer, 2001), and the use of the picture-in-picture feature was therefore discontinued for the remainder of the investigation.

Video-Enhanced Assessment

The three VEA activities were all successful by default when evaluated against the themes of participation and inclusivity, however mixed results were achieved when the video-based data sources were subjected to thematic analysis for evidence of reflexivity. All participants engaged with VEA Activity 1; working away from the studio environment, participants engaged with the activity at a time of their choosing and unrestricted in their choice of tools to produce the screencasts. As argued below in connection with VEA Activity 2, the circumstances in which these screencast demonstrations of work-in-progress were produced appeared to influence the degree to which reflexivity was evident.

As the end-of-year presentations took place at a projector-enabled PC on which *Camtasia®* was installed, and participants were required simply to clip on a microphone and to click to record the screen, VEA Activity 2 was successful by default in connection with participation. Inclusivity was not a feature of this exercise, as PLSPs in place for students with a learning difficulty meant that participation was not mandatory, however the theme of reflexivity was addressed particularly poorly by the screencast exercise. In contrast to the success achieved in extra-studio settings by VEA Activity 1, the full-cohort, in-studio set and setting appeared to provide a theatre in which showcasing was significantly more prevalent, with only four of the twenty-one students demonstrating reflexivity in their presentations. It is acknowledged however that other factors may have influenced this outcome, as in addition to the combination of set and setting, (i) the timing of the exercise may have contributed to a 'last day of term' effect in which participants felt that the opportunity to make a difference was over, and (ii) closely related, an informal, celebratory atmosphere that can accompany such gatherings may have led participants to take proceedings less seriously than might have been the case had the presentations been conducted at an earlier date.

In common with VEA Activity 1, the third activity was conducted outside the studio setting, and therefore away from the whole-cohort. While the use of a competition went some way towards mitigating for students' results being unaffected by non-participation in the exercise, it was successful in ensuring that sufficient entries were submitted to warrant a prize-giving event, and VEA Activity 3 saw the production of work ranging from an exemplar piece demonstrating exceptional reflexivity, to a less successful face-to-camera format which served to highlight problems associated with that approach.

In addition to promoting reflexive styles of learning (Crossouard et al., 2004), both VEA Activity 1 and Activity 3 were found to afford students with dyslexia opportunities to participate in assessment without being disadvantaged by an insistence on working with text to document their progress with software development work. Two participants whose illustrative cases under the pseudonyms Robert and Keith can be found in the next chapter, both indicated that the opportunity to present their work visually while talking through its development led not only to enhanced clarity and coherence in the documentation process, but had also contributed to stronger levels of engagement with the assessment tasks when compared to text-based tasks in other modules.

Refinement of the Intervention and Evolution of Research Questions

The problem-solving approach in VEF Activity 1 was found to be effective in providing a boost to help students transcend threshold concepts (Meyer & Land, 2005). The success of this activity was however reliant upon students acknowledging the existence of a problem and creating a video in order to demonstrate it, and further benefits could be gained by extending this 'small picture' approach into the 'bigger picture' of the final submission. In contrast, video-feedback on drafts of work in VEF Activity 2 was found to be useful as a mechanism to enable students to improve their final grade as part of the 'bigger picture', but required students to have attained a level of competence with the development environment and to have transcended key threshold concepts at the 'small picture' level before these benefits could be realised. Despite having established two mechanisms to support both small-scale targeted and larger-scale generalised approaches to formative video-feedback, the middle ground remained under-explored, and dialogue under-exploited.

The *augmentation* of traditional presentations using video in VEA Activity 2 had achieved only limited success, and the 'showcasing' factor observed when students were engaged in synchronous, face-to-face presentations of their work highlighted how a focus on the artefacts developed served to limit the capacity for demonstrating reflexive thinking in a peer-based environment. Notwithstanding this drawback, the two other VEA activities had achieved some success in the area of formative presentations of work-in-progress and summative reflections on the relationship between product and process. In VEA Activity 1, there was evidence that students had identified and solved problems that had hindered their progress during the process of producing the video walkthrough of their work, while the exemplar piece emerging from VEA Activity 3 illustrated how using asynchronous video to document the relationship between the development of skills and competencies and a portfolio of artefacts could enable students to demonstrate a high degree of reflexivity.

For students affected by dyslexia, clear benefits were identified where participants engaged in regular dialogue around formative feedback through the medium of video, and where the weekly blog exercise was replaced with a video-based equivalent. A recurrent theme emerging from the interviews with the two participants whose illustrative cases are presented in the next chapter under the pseudonyms Robert and Keith was that increasing the frequency of video-based assessment and feedback activities within their studies promised significant benefit in terms of learner engagement. While it had not been possible to amend the assessment model of the module while in-progress during DBRC2, it was notable that the value of high-frequency formative assessment found in the literature (e.g. Gibbs & Simpson, 2004) was confirmed by these suggestions from dyslexic participants following their engagement with video-enhanced approaches to both feedback and assessment.

It was envisaged that through the introduction of a system of regular documentation using video-diaries in which students would highlight how the ITVs had been used to develop their skills, the frequency of video-feedback within the dialogic framework, and as a corollary the frequency of dialogue itself, would be increased. Against this background, the design of a refined system to support the production of regular video-diaries within the framework of the VFL was therefore taken forward for implementation in the third and final cycle.

SUMMARY OF THE SECOND DESIGN-BASED RESEARCH CYCLE

DBRC2 began by asking how the introduction of asynchronous video-enhanced assessment and feedback activities might influence the assessment and feedback process. The design, implementation and evaluation of a range of novel approaches aiming to facilitate video-enhancement of feedback and assessment in this cycle saw prototyping work that explored video-enhanced feedback and video-enhanced assessment activities lead to the introduction of five interventions that drove the research forward.

Having fulfilled the aim of designing, implementing and evaluating prototype VEF and VEA activities, one of the themes emerging from DBRC2 was that closer integration of the video-based learning, feedback and assessment components might promote greater dialogue, with the recommendation that the work of DBRC3 should focus on bringing these elements together within a holistic system. Increased frequency of both VEF and VEA activities in a symbiotic relationship was also identified as key to sustaining dialogue around feedback and assessment, and the incorporation of high frequency VEF and VEA activities within a dialogic framework therefore became a further core aim for DBRC3. Having established proof-of-concept in four of the five activities trialled, and having found that the activities appeared to contribute to the fulfilment of the seven Pedagogic Aims, the identification of areas for improvement in the effectiveness of these approaches highlighted an opportunity for a synthesis of the work of both DBRC1 and DBRC2 which would see the development of the integrated system to which we turn next.

THIRD DESIGN-BASED RESEARCH CYCLE – DBRC3

Overview of the Third Design-Based Research Cycle

As the third and final research cycle, DBRC3 focused on the refinement of the video-enhanced feedback and assessment elements introduced in DBRC2, with the aim of increasing the frequency of VEAF activities through closer integration with the ITVs developed in DBRC1 (see Figure 9). VEAF activities designed in DBRC2 were adapted and amalgamated in the process of integrating these within a holistic system of video-enhanced learning, assessment and feedback. While design and implementation remained clearly differentiated during DBRC1, where the focus was on the development of *artefacts* such as the ITVs, this distinction began to blur in DBRC2 in the context of developing *techniques* such as the video-feedback loop (VFL), where the phases of design and implementation phases became less distinct. This section therefore explores the refinement of those techniques into an integrated *system* for video-enhancement of learning, assessment and feedback.

Design Phase: Refinement of the Design

Having identified increased frequency of formative feedback as key to the work of DBRC3, it was necessary for this to be accompanied by an increased frequency of an associated formative VEA activity in order to be sustainable. The two forms of VEA activity from DBRC2 were therefore amalgamated into a single stream of VEA activity in DBRC3 for the purpose of closer integration with both the ITVs and the refined VEF activity. To this end, the design of the VEA activity was refined, replacing the weekly blog entries with a requirement that students create regular video-diary entries (VDEs) highlighting their development of an artefact specified in the tutorials, and demonstrating on a regular basis how the techniques learned were being applied to their own product as the semester progressed. Integrated

Figure 9. Illustration of the three video-enhanced learning, feedback and assessment elements

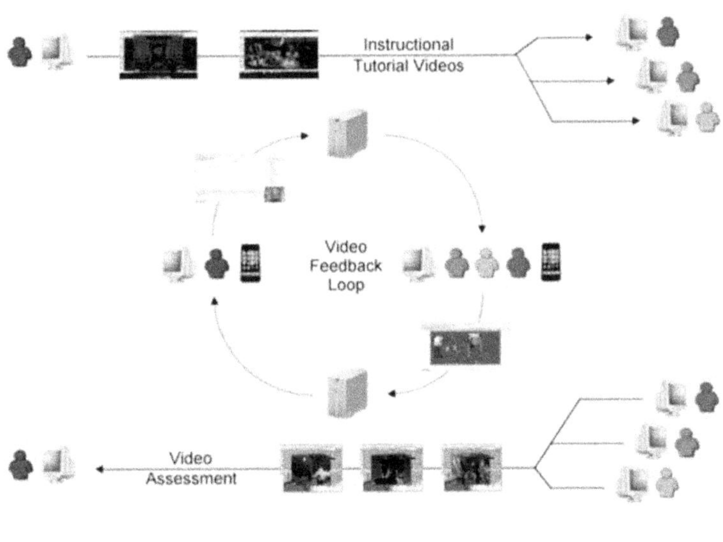

within the video-feedback loop, these VEL and VEA activities formed a holistic, tri-partite system, in which students engaged in a video-based dialogic process with the tutor around high frequency formative feedback. This system is visualised in Figure 10.

An issue identified in DBRC2 in connection with VEF was that the dialogic problem-solving approach seen in VEF Activity 1 relied on students acknowledging that a problem had been encountered and producing a screencast to demonstrate this. Further, while the production of late formative video-feedback following VEF Activity 2 appeared to enable students to make improvements to their work, it remained a possibility that without engaging with VEF Activity 1 students might subsequently find themselves unable to rectify any major or significant issues in their work within the remaining timescale.

Against this background, it was anticipated that an increased frequency of engagement in formative assessment tasks requiring the production of regular VDEs could provide the basis for increased frequency of VEF within a dialogic framework. While students would be free to work through the ITVs at their own pace, a structure for their consumption on a week-by-week basis already existed, thereby providing a platform for the production of a VDE each week, or thereabouts. Through the refinement of this activity stream into a system wherein written documentation was replaced by regular production of VDEs, it was intended that the frequency of video-feedback within the dialogic framework, and as a corollary dialogue itself, would be increased.

Implementation

In the DBRC2 case that highlighted one undergraduate student's experience of VEA, Keith expressed reservations regarding the ease with which new undergraduate students would adapt to the use of video, and in the case of DBRC3 where the creation of weekly video-diary entries was introduced, an initial 'inhibition curve' was observed in the majority of participants. Informal, in-studio conversations with participants revealed that students felt discomfort and experienced embarrassment around the process of

Figure 10. Illustration of the three holistically integrated video-enhanced learning, feedback and assessment elements

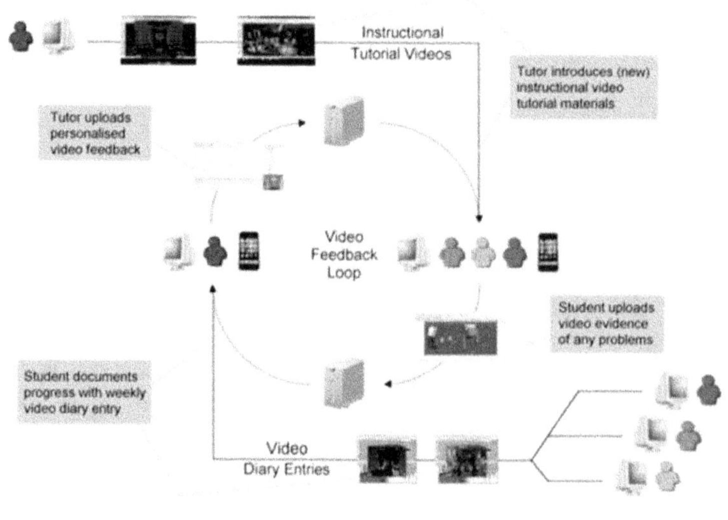

openly recording the voiceover within the studio environment while in the presence of peers. Evidence from students' VDEs indicated that most participants overcame this inhibition within 2-3 weeks, while further informal conversations suggested that students became accustomed to the activity following an initial period of adjustment, and/or began recording VDEs outside the studio.

Evaluation Phase: Data Collection and Analysis

In DBRC3, the design and implementation work focused on increasing the frequency of both video-feedback and video-assessment over the course of a full academic year, while facilitating closer integration of the three strands of video-based activities within a holistic framework. While the data collection and analysis exercises conducted in DBRC1 focused on the presentation of results which contextualised each student's use of the ITVs against the backdrop of the full cohort, DBRC3 offered an opportunity to collect and analyse data not only on the patterns of individual interaction with video activities, but also to focus in greater depth on the experience of participants with a view to the construction of those illustrative cases presented in the next chapter.

While data was collected from all participants as part of the delivery of the module, the number of participants and high volume of associated data meant that the development of illustrative cases for each student was not practical within the scope of this study. Following Stake (1995) and Yin's (2003) recommendations that cases should be identified which highlight interesting or unusual themes, while also acknowledging the need to ensure construct validity (Trochim, 2006) in the process of generating theory, a combination of representative and purposive sampling led to the selection of four cases.

The collection of a broad range of data was designed to allow the development of individual illustrative cases relevant to practitioners that could also be contextualised against the whole group, in order to strengthen both internal and construct validity of the cases. Data collection and analysis work conducted within DBRC3 was therefore designed to include:

1. A baseline audit, conducted using an online survey.
2. Questionnaire data, from an anonymous survey which collected both quantitative and qualitative data on pedagogical and technical aspects of students' experience of the holistic strategy at the mid-way point in the year.
3. Video-based data, derived from the weekly video-diary entries and video-based feedback conversations.
4. Dialogic-interview data, taken from conversational exchanges with individual students during scheduled studio-based time and supported by field notes detailing observations.
5. Interview data, collected through a series of semi-structured interviews conducted at the end of the academic year with a purposive sample of participants.

Table 2 illustrates the structure, schedule and purpose of these data collection exercises, each of which is discussed in greater detail in the respective sub-sections below.

Table 2. Structure, schedule and purpose of the six data collection exercises in DBRC3

Data Collected	Collection Schedule	Purpose
Baseline Audit Survey	Semester 1, Week 1	To assess previous experience of ITVs, VEA and VEF.
Anonymous Questionnaire	Semester 1, Week 12	To assess mid-year experience of ITVs, VEA and VEF.
Video-Diary Entries and Feedback	Semester 1, Week 1 to Semester 2, Week 12	To inform case studies through analysis of patterns of use.
In-studio Dialogic Interviews and Observations	Semester 1, Week 1 to Semester 2, Week 12	To inform the development of case studies, and to direct semi-structured interview questions.
Module Results Data	Semester 3, Week 3	To identify patterns of impact in quantitative data.
Semi-structured Interviews	Semester 3, Week 5 to Semester 3, Week 6	To inform case studies using qualitative data.

Baseline Audit

An anonymous online survey was used to perform a baseline audit of students' exposure to video-enhanced learning, feedback and assessment prior to joining the course. The survey opened by asking "Have you previously used video tutorials to support your learning?", to which responses revealed that of the twenty-two newly arrived students, four had previously used video tutorials, while the remaining eighteen had no prior experience of learning through the medium.

An external-facing 'Yes/No' question asked "Prior to starting this course, had you ever received feedback on your work in video format?" to which all twenty-two students indicated that they had not previously received feedback through the medium of video. The next question employed a multiple selection checkbox format, asking "Prior to starting the course, what was your preferred format for receiving feedback on your work?"; as illustrated in Table 3, fourteen of twenty-one respondents indicated a prior preference for receiving feedback on their work as word-processed text, two students preferred to receive feedback as an audio file, and one student preferred face-to-face feedback, while the remaining four expressed a preference for some hybrid form. Of these latter, two students highlighted a preference for a combination of text and face-to-face, while one student would have preferred to receive feedback both as an audio file and face-to-face. Although no students had indicated having previously received feedback through the medium of video in the previous question, one student suggested a mix of face-to-face and video.

Table 3. Participants' preference for format of feedback prior to starting the course (n=21)

Format	Text	Audio	Face-to-face	Video	Hybrid
Participants	14	2	1	0	4

On video-enhanced assessment, the first of two questions asked students "Prior to starting this course, had you ever used video to document the development of academic coursework?" Only one student indicated any prior experience of video-blogging, with no students having previously experienced video-enhanced assessment.

The second question asked "Prior to starting the course, what was your preferred format for documenting your work?" As illustrated in Table 4, fifteen of the twenty-one respondents indicated a prior preference for documenting their work using word-processed text, one student preferred to use a blog, while the remaining five expressed a preference for some hybrid form. Of these latter, two highlighted a preference for a combination of text and audio, two would have preferred to give spoken presentations, while one suggested a mix of text, audio and video as the ideal, despite there being no indication from the previous question that any students had previously used video to document academic work.

Anonymous Questionnaire

An anonymous online questionnaire combining Likert-scale and open-ended questions was used in DBRC3 to collect both quantitative and qualitative data on twenty-two students' experience of the VEF and VEA activities. Both are presented in separate sections below. The questionnaire was designed to maximise participation by allowing respondents the opportunity to skip questions if they wished, and while all twenty-two students completed the survey, not all respondents gave answers to all questions; the number of respondents to specific questions is indicated below in the descriptions and tables used to present these data.

Anonymous Questionnaire: VEF

This section of the questionnaire included a broad range of both internal- and external-facing questions; the internally-facing questions gathered predominantly quantitative data of a technical nature which aimed to further refine the provision of video-feedback in the second semester, whereas the collection of externally-facing data relied on a combination of multiple-choice, Likert-scale questions and open-ended questions asking about the students' experience of feedback and its effectiveness or otherwise.

In order to explore whether the integrated model of video-enhanced learning had influenced the learner experience of assessment and feedback, a series of four questions aimed first to identify students' experience of the effectiveness of video-feedback, and then to enquire whether and why students' preference for particular feedback media might have changed from that expressed at the start of the course.

Table 4. Participants' preference for format for documenting academic work prior to starting the course (n=21)

Format	Text	Blog	Text + Audio	Spoken	Hybrid
Participants	15	1	2	2	1

In the first of two selective-answer questions, students were asked "Please rate your experience of receiving video-feedback in terms of making improvements to your work?", to which available options ranged from 'Very Effective' to 'Very Ineffective' on a 5-point scale. *Figure 11 displays two bar charts illustrating students' rating of the experience of receiving video-feedback, (Chart a) and whether their feedback preference had changed since starting the course (Chart b).*

As illustrated in Figure 11, Chart (a), eighteen students indicated that receiving formative video-feedback had been either 'Very Effective' or 'Effective' in enabling them to make improvements to their work, while three students remained 'Neutral'. The second of these questions asked "Has your preferred format for feedback changed since starting the course?" offering three options: 'Yes', 'No' and 'Undecided'. As illustrated in Figure 11, Chart (b), seventeen respondents answered 'Yes', indicating that their preference had changed since starting the course, and four were 'Undecided', while no students answered 'No'.

Two follow-on open-answer questions completed this section, the first of which requested students to "Describe your experience of receiving video-feedback"; although one respondent suggested that video-feedback was *pretty much the same as any other medium of feedback*, all other comments gave positive expression to the experience. Common themes or categories emerging from inductive thematic analysis of the raw data included: SPECIFICITY; INTELLIGIBILITY; REPEATABILITY; UTILITY; MOTIVATION; and REFLEXIVITY. A frequent theme was SPECIFICITY, with respondents commenting that this form of feedback enabled them to identify the precise location of areas for improvement. Table 5 highlights the themes that emerged with examples from the raw data.

The second of these two final questions in this section asked students to "Tell us about any change in your feedback preference since starting the course". While only one of the four students who remained 'Undecided' responded to this open-answer question, commenting that "*I do not really have a preference about my form of feedback*", inductive thematic analysis of the responses received from the seventeen respondents whose preference changed saw the emergence of categories including: UTILITY, EASE OF USE, MIXED MEDIA, SPECIFICITY, SIMPLICITY, FREQUENCY, as seen in Table 6.

Figure 11. Bar charts illustrating (a) students' rating of the experience of receiving video-feedback, and (b) whether their feedback preference had changed since starting the course

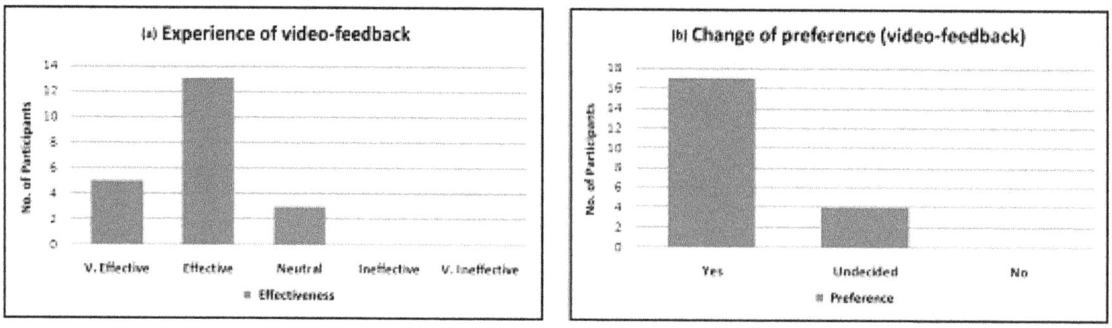

Table 5. Categories emerging and examples of raw data from open-ended question on the experience of video-feedback

Category	Example(s) from Raw Data	Frequency
SPECIFICITY	*Shows the exact location I need to edit.*	12
INTELLIGIBILITY	*It's a lot easier to understand what to improve when it's shown to you directly.* *I now know how I could improve my level.*	9
REPEATABILITY	*They have helped me with a few things I forgot about which help me remember what they were.*	7
UTILITY	*I could see what the problems were with my work and fix them straight away for example.*	9
MOTIVATION	*It helps me want to improve work.*	3
REFLEXIVITY	*Helped to see another person's point of view on my work.*	5

Table 6. Categories emerging and examples of raw data from open-ended question on changes in feedback preference

Category	Example(s) from Raw Data	Frequency
SPECIFICITY	*When making things that [users] interact with, it is a lot better to have video-feedback to show the exact points in which there needs to be an improvement.* *I like that I can see exactly where I have gone wrong and what I need to do.*	10
EASE OF USE	*Video-feedback gives a more visual approach so I can see exactly what has gone wrong.*	7
MIXED MEDIA	*Very easy to use and effective, Its better having audio and visual aspects at the same time to explain something.*	5
UTILITY	*Video-feedback is a very quick and helpful way of receiving feedback and so it is preferable now.* *I prefer being told directly how to improve things, so the video format is very helpful.*	8
SIMPLICITY	*It is just so much simpler to open a video and hear someone speak then it is to read. Put it down to laziness if you want, but that's how I feel. . ..*	6

Anonymous Questionnaire: VEA

Figure 12 displays two bar charts that illustrate the frequency distribution of production of video-diary entries (Chart a) and location of storage (Chart b). The first of two selective-answer questions asked "How frequently have you produced a video to document your work?", and as illustrated in Figure 12 Chart (a), while only ten of the twenty-one respondents had produced video-diary entries at least as frequently as once per week during the first semester, fifteen students had produced entries at least fortnightly, while a total of eighteen students reported having produced a video-diary at least once per month.

Two students produced only one video entry during the semester, correlating with a prolonged period of sickness absence in one case and particularly low levels of motivation across the course in another, while one student who did not produce any video-diary entries self-identified in the studio session as unwilling to use either the blogging tool or video, preferring to produce a report using *MS-Word®*.

The second of these selective-answer questions asked "Where did you store your video-diary entries?", and as illustrated in Figure 12(b), while twelve of the twenty respondents had uploaded the files to the eportfolio system, *Mahara®*, a further six had uploaded their entries either to *YouTube®* or to both *YouTube®* and *Mahara®*, while one student stored videos in a *SkyDrive®* account, and one elected to

Figure 12. Bar charts illustrating (a) frequency distribution of production of video-diary entries, and (b) location of storage

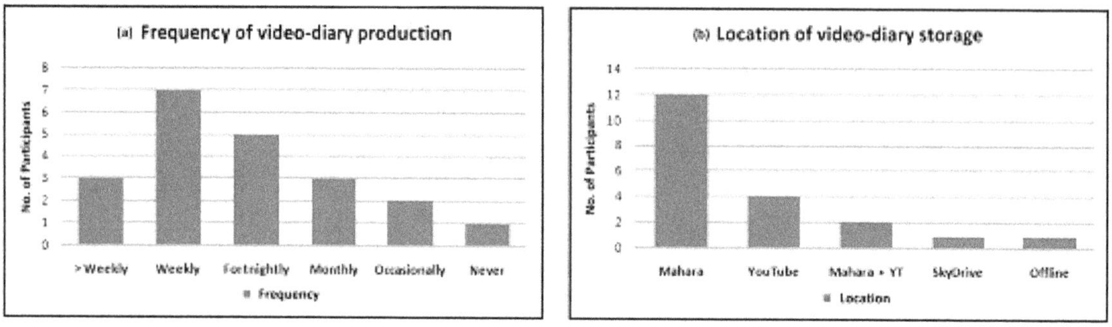

keep the diary entries offline. *Mahara®* offers opportunities to embed video content stored remotely on services such as *YouTube®*, and video-diary entries produced by an overwhelming majority of students (90%) could therefore be viewed within their e-portfolio, with the tutor's screencast video-feedback response to the issues highlighted in the VDEs uploaded to individuals' feedback area.

The next selective-answer question asked students "Please rate your experience of using video-diaries to document your work?", and again, it was evident that in this case students may be undecided and/ or have legitimate reasons for wishing to avoid committing to one side or the other of a dichotomous question, and a 5-point scale from 'Very Effective' to 'Very Ineffective' included a 'Neutral' option. As illustrated in Figure 13(a), fifteen students indicated that the video diaries had been either 'Very Effective' or 'Effective', and three students chose 'Neutral', while two students felt that video diaries had been 'Ineffective', and one selected 'Very Ineffective'. Notably, this distribution mirrored that found for the frequency of production, and while the anonymous nature of the questionnaire might be thought to undermine grounds for any deductive claims to have found an identity relationship between the authors of these responses, tracking of each participant's anonymous ID codes highlighted that these answers were consistently selected by the same three participants. In particular, evidence from tracking the ID codes

Figure 13. Bar charts illustrating (a) distribution of students' experience of producing video-diary entries, and (b) whether their documentation preference had changed since starting the course

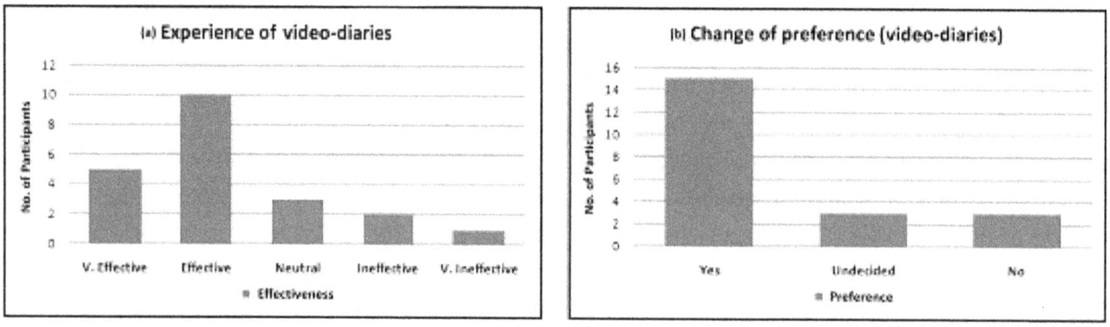

revealed that: (i) three students indicated having produced a video-diary entry on a monthly basis, and that these three students remained 'Neutral' regarding their experience; (ii) two respondents indicated having produced an entry only occasionally, and these same two respondents found the production of these entries to be 'Ineffective'; while (iii) one student did not produce any video-diary entries, and the same student felt their experience of using video-diaries was 'Very Ineffective'.

Further indications of this apparent relationship could be seen in the responses to the final question in this section that asked "Has your preferred format for documenting your work changed since starting the course?" *Figure 13 displays two bar charts that illustrate the distribution of students' experience of producingf video-diary entries (Chart a) whether their documentation preference had changed since starting the course (Chart b).*

Fifteen of twenty-one respondents signalled a change in their preferred method for documenting work since the start of the course, while three respondents remained 'Undecided', and a further three indicated 'No' change in preferred method, as illustrated in Figure 13, Chart (b). While this numeric data might appear to suggest that approximately three-quarters of this cohort of students valued the VEA intervention, clearly around one student in seven remain undecided, and a further one in seven appear to reject video-diaries as a mechanism to document development work.

A final open-ended, follow-on, question asked students to "Describe your experience of using video-diaries", and the qualitative data collected in response to this question enabled triangulation of the analysis thus far conducted. Inductive thematic analysis of these data saw the emergence of similar categories to those revealed in the analysis of the VEF data in DBC3; these categories are highlighted alongside examples from the raw data in Table 7.

Table 7. Categories emerging and examples of raw data from open-ended question regarding experience of video-diaries

Category	Example(s) from Raw Data	Frequency
UTILITY	*Its better than doing word docs it's just I have never had the option to do video before.* *I'm not very used to recording my work by video but it hasn't been too hard to adapt to this.*	10
COMPARABILITY	*I was able to easily compare one week to the other so I could tell if I had been wasting my time or not.*	3
REFLEXIVITY	*It's been OK since I did log my work. However, I wish that I was more in-depth on my problems and solutions.*	6
EASE OF USE	*My preference has changed due to the fact it's easier and you can include more information.* *I find it easier to show work rather than describing it in text.*	12
DYSLEXIA	*This makes it easier to say what I mean as I find it difficult to explain in written form as I change words that I know but can't spell.* *This is better and much easier for me as I have dislaxyer so videoing is far easier.*	3
SPEED	*Video documenting is much easier to do than in a text document, because it's faster, and you're giving a visual view, on your work, so there's not much explanation to do.* *I find it has summarized my work up in a few minutes.*	9

SUMMARY OF THE THIRD DESIGN-BASED RESEARCH CYCLE

In refining the design and frequency of the video-based activities introduced in DBRC2, the development of a holistic model of asynchronous video-enhanced learning, assessment and feedback saw the integration of these three strands of video-based activities situated within a dialogic framework. Instructional tutorial videos continued to scaffold the development of learners' skills in software development, while a synthesis of problem-focused and final-draft approaches to formative video-feedback evolved into a single iterative feedback activity existing in symbiosis with regular video-based formative assessment that amalgamated formative presentations of work in progress and reflective self-assessment videos. Integrated within the VFL, these three strands of video-based activity were implemented with and evaluated by a group of 22 students over the course of a full academic year.

An initial online survey confirmed the extent of students' previous exposure to video in learning, assessment and feedback, and provided a baseline audit from which to explore students' early responses to the intervention. An anonymous questionnaire formed a mid-year checkpoint through which to examine the development of the group's impressions of both the holistic approach and the individual video-based component activities. While the findings of this interim evaluation suggested that video was not unanimously preferred as the primary medium through which to engage with assessment and feedback activities, it was evident that the intervention was both generally well received and recognized as offering particular benefits to those learners affected by dyslexia and/or Asperger's Syndrome.

At the culmination of the academic year, a series of semi-structured interviews yielded detailed qualitative data on the impact of the intervention on the in-depth experience of individual students, and through both purposive and representative selection the findings from these data collection exercises informed the development of further illustrative cases. Supplementing those introduced in DBRC2, thick descriptions of the stories of four students illustrate how the integrated model for video-enhanced learning influenced their experience of and engagement with the assessment and feedback process.

CONCLUSION

While the positive impact on individual academic performance was one common theme noted by the subjects of the illustrative cases, the key findings stemming from these accounts were in the association of the intervention with increased reflexivity, enhanced learner autonomy and the emergence of metacognitive self-awareness. These narratives also highlighted the ways in which students affected by dyslexia and Asperger's Syndrome, found opportunities to approach assessment tasks using video-based techniques that reduced or eliminated disadvantages experienced with traditional assessment, while enabling meaningful engagement with detailed and personalised feedback through ongoing dialogic exchanges.

Chapter 2 of this text, in keeping with the earlier focus on enhanced formative assessment for all learners, reports on case study students. Thus, in that chapter, the focus shifts to the learner experience of the VEF and VEA interventions (as referred to above), drawing on and presenting raw data from semi-structured interviews to present six illustrative cases, with two exemplar activities emerging from DBRC2, and a further four from the evaluation of DBRC3. Across these narratives, three cases highlight the impact of VEA and VEF on the experience of students affected by dyslexia, one discusses the case of a student diagnosed with Asperger's Syndrome, while the remaining cases discuss the experience of two neurotypical participants.

REFERENCES

Attwood, T. (2000). *Asperger's syndrome: A guide for parents and professionals*. London: Jessica Kingsley.

Baron-Cohen, S. (2008). *Autism and Asperger's Syndrome*. Oxford University Press.

Bloxham, S., & Boyd, P. (2007). *Developing assessment in higher education*. Open University Press.

Boud, D. (2000). Sustainable assessment: Rethinking assessment for the learning society. *Studies in Continuing Education*, *22*(2), 151–167. doi:10.1080/713695728

Boud, D., & Molloy, E. (2013). Rethinking models of feedback for learning: The challenge of design. *Assessment & Evaluation in Higher Education*, *38*(6), 698–712. doi:10.1080/02602938.2012.691462

Carless, D. (2013). Trust and its role in facilitating dialogic feedback. In D. Boud & E. Molloy (Eds.), *Feedback in higher and professional education: Understanding it and doing it well* (pp. 90–103). London: Routledge.

Crossouard, B., Pryor, J., & Torrance, H. (2004). Creating an alternative assessment regime with online formative assessment: Developing a researcher identity. In *European Conference on Educational Research* (pp. 22-25). Academic Press.

Entwistle, N. J., Hounsell, C. J., Macaulay, C., Situnayake, G., & Tait, H. (1989). *The Performance of Electrical Engineers in Scottish Education. Report to the Scottish Education Department, Centre for Research on Learning and Instruction Department of Education*. University of Edinburgh.

Galusha, J. (1998). *Barriers to learning in distance education*. Hattiesburg, MS: The University of Southern Mississippi (ERIC Document Reproduction No. ED 416 377). Retrieved from http://files.eric.ed.gov/fulltext/ED416377.pdf

Garrison, D. R., Anderson, T., & Archer, W. (2001). Critical thinking, cognitive presence, and computer conferencing in distance education. *American Journal of Distance Education*, *15*(1), 7–23. doi:10.1080/08923640109527071

Gibbs, G., & Simpson, C. (2004). Conditions under which assessment supports students' learning. *Learning and Teaching in Higher Education*, *1*(1), 3-31.

Henderson, M., & Phillips, M. (2015). Video-based feedback on student assessment: Scarily personal. *Australasian Journal of Educational Technology*, *31*(1), 51–56. doi:10.14742/ajet.1878

Her Majesty's Stationary Office (HMSO). *(2001)*. *The Special Educational Needs and Disabilities Act*. Retrieved from http://www.legislation.gov.uk/ukpga/2001/10/pdfs/ukpga_20010010_en.pdf

Her Majesty's Stationary Office (HMSO). (2010). *The Equality Act*. Retrieved from http://www.legislation.gov.uk/ukpga/2010/15/pdfs/ukpga_20100015_en.pdf

Hrastinski, S., Keller, C., & Carlsson, S. A. (2010). Design exemplars for synchronous e-learning: A design theory approach. *Computers & Education*, *55*(2), 652–662. doi:10.1016/j.compedu.2010.02.025

Lang, R., Shogren, K. A., Machalicek, W., Rispoli, M., O'Reilly, M., Baker, S., & Regester, A. (2009). Video self-modeling to teach classroom rules to two students with Asperger's. *Research in Autism Spectrum Disorders*, *3*(2), 483–488. doi:10.1016/j.rasd.2008.10.001

Laurillard, D. (2002). Rethinking university teaching: A conversational framework for the effective use of learning technologies (2nd ed.). London: Routledge Falmer.

Mayer, R. E. (2001). *Multimedia Learning*. Cambridge University Press. doi:10.1017/CBO9781139164603

Mayer, R. E., & Moreno, R. (2003). Nine ways to reduce cognitive load in multimedia learning. *Educational Psychologist*, *38*(1), 43–52. doi:10.1207/S15326985EP3801_6

McDowell, J. (2012). Cultivating a viral community of practice to drive institutional enhancement through the promotion of video-enhanced learning, feedback and assessment. In *Proceedings of the 7th International Blended Learning Conference*. University of Hertfordshire.

Meyer, J. H., & Land, R. (2005). Threshold concepts and troublesome knowledge (2): Epistemological considerations and a conceptual framework for teaching and learning. *Higher Education*, *49*(3), 373–388. doi:10.100710734-004-6779-5

National Student Forum. (2009). *National Student Forum Annual Report*. Retrieved from http://www.bis.gov.uk/assets/biscore/higher-education/docs/n/09-p83-national-student-forum-annual-report-09

Rayner, C., Denholm, C., & Sigafoos, J. (2009). Video-based intervention for individuals with autism: Key questions that remain unanswered. *Research in Autism Spectrum Disorders*, *3*(2), 291–303. doi:10.1016/j.rasd.2008.09.001

Stake, R. (1995). *The art of case research*. Newbury Park, CA: Sage Publications.

Thornhill, S., Asensio, M., & Young, C. (2002). *Video streaming: A guide for educational development*. Manchester, UK: JISC Click and Go Project, UMIST.

Torrance, H. (2007). Assessment as learning? How the use of explicit learning objectives, assessment criteria and feedback in post-secondary education and training can come to dominate learning. *Assessment in Education: Principles, Policy & Practice*, *14*(3), 281–294. doi:10.1080/09695940701591867

Trochim, W. M. K. (2006). *Construct validity*. Web Centre for Social Research Methods. Retrieved from http://www.socialresearchmethods.net/kb/constval.php

Volpentesta, A. P., & Frega, N. (2007). Developing a blended-learning community in a university setting. *International Journal of Web Based Communities*, *3*(2), 134–150. doi:10.1504/IJWBC.2007.014076

Wixon, D. R., Pietras, C. M., Huntwork, P. K., & Muzzey, D. W. (1996). Changing the rules: A pragmatic approach to product development. In D. Wixen & J. Ramey (Eds.), *Field methods casebook for software design* (pp. 57–89). New York: John Wiley & Sons.

Yin, R. K. (2003). *Case study research: Design and methods* (3rd ed.). Thousand Oaks, CA: Sage Publishing.

Chapter 8
Course Development:
Feedback and Video to Improve Teaching Capacity

Beverly Dann
University of the Sunshine Coast, Australia

ABSTRACT

This chapter discusses the impact of feedback as dialogue and video use in a new science methods course that was created to meet new governmental requirements. National demands increased the evidence required by preservice teachers to demonstrate quality teaching and learning in initial teacher education programs impacting on program and course design. This led to increasing reflective opportunities and demonstrating knowledge of subject content for preservice teachers in one small metropolitan university. The ability of 32 preservice teachers to reflect on teaching strategies and their subject knowledge as part of a video teaching assignment with peers in a university classroom is described and discussed. Results indicated high participation in dialogue and dialogic feedback. Personal reflections revealed preservice teachers' understanding of teaching strategies, pedagogies, and subject knowledge improved with support. This has implications for final year preservice teacher progress where they need to demonstrate graduate requirements to transition into the profession.

INTRODUCTION

Initial Teacher Education (ITE) programs in Australia respond to regulatory bodies such as the Australian Institute for Teaching and School Leadership (AITSL) that shape and guide the education programs and the professional learning of education Preservice Teachers (PSTs). AITSL policies are designed to prepare preservice teachers to become quality teachers who know the subject matter, can apply appropriate teaching strategies to engage all learners and can reflect on their practice and on student learning effectively. The intentions are worthy and understandable; however, these policies create pressure at the university level to ensure all PSTs meet these expectations and high standards. Therefore, the course

DOI: 10.4018/978-1-7998-0426-0.ch008

design, course activities and assessment need careful consideration. The use of effective feedback during course activities and assessment including the use of video to reflect on teaching performance can have a positive impact on developing quality preservice teachers (Star & Strickland, 2008).

BACKGROUND

In a Bachelor of Primary Education program in Queensland, Australia, new courses needed to be designed to become part of a specialization or subject-focus. This new design consideration was a national government level response to continuing poor standardized test results in Australia, such as those from the National Assessment Program-Literacy and Numeracy (NAPLAN) (Adams, 2012). Each university was able to decide on their specialization focus based upon the needs of the local schools. This particular university provided two subject choices. Preservice teachers (PSTs) had the option of choosing English or Science as their subject-focus. A subject-focus has four courses (semester long classes) that gradually develop the PSTs' teaching capacity in that subject area. The four subject-focus courses were taken in the second and third years with the final course planned for second semester of the third year. These four courses were sequenced in a developmental manner so that more familiar concepts such as biology were taken first to develop confidence and basic science teaching skills. The last two courses covered content such as physical and chemical science that most PSTs had little to no background in and space science where many misconceptions were held. The developmental nature of the science courses and the feedback provided throughout the courses were expected to support preservice teacher learning for greater success in the final subject-focus course. Feedback in these science courses was embedded in collaborative activities whereby PSTs switched between involvement as a learner and involvement as a teacher. Discussions were at the heart of the feedback process.

Feedback during the four science courses took three forms; feedback through inquiry activities, feedback through dialogue and feedback on assessment. Feedback through inquiry activities involved testing out PSTs' own ideas and identifying if their conceptions were accurate or needed adjusting. PSTs could see outcomes for themselves and would have opportunity to explain/discuss what happened and therefore led to a change in misconceptions. Feedback through dialogue refers to the conversations between tutor and PSTs during tutorials. These conversations identified PST understanding and attempted to build upon that understanding. PSTs could freely add to the conversation in many ways, e.g.: describing observations and experiences, attempting to explain observations, challenging others' ideas, and so on. Feedback on assessment was in the form of written comments that identified concepts they did understand and those they were still trying to understand. Clarification of the concepts in context were made as needed. For example, within a tutorial, a model would be revisited to clarify remaining misconceptions. This information allowed PSTs to continue to act on their learning needs.

The final subject-focus course had specifications from the Australian Institute for Teaching and School Leadership (AITSL) that expected the course to provide opportunity for PSTs to demonstrate 'expert' level knowledge of subject content and subject specific teaching strategies. For example, the year 5 Earth and Space curriculum contains: modelling the relative size of and distance between Earth, and other planets in the Solar System and the Sun. These are very abstract concepts and have commonly been taught through the creation of paper mâché placed on balloons to represent the planets and the Sun. These are often painted to represent the planets then hung from the ceiling in their order from the

Sun. Unfortunately, this well remembered activity only contributes to misconceptions rather than correct scientific conceptions since the relative sizes are far from accurate and the distances cannot easily be demonstrated to a scale within the classroom.

The difficulty in planning the course was knowing what "expert" looks like for a novice teacher and how it can be demonstrated in a course. This caused several questions about course design to emerge.

Some of the concerns included:

- How would content knowledge and science teaching strategies be assessed?
- How would the course and assessment connect to classroom experiences?
- How would feedback and feed forward be maximized in the course?
- Would it include teaching segments or videos?
- Should PSTs reflect on their experiences?

In addition, the CC wanted to provide support and practice for some of the mandated requirements within the final science course. Therefore, the course design, activities and assessment were carefully considered to fulfill that purpose. The chosen course design emphasized tutor modeling of science teaching strategies while teaching content and providing feedback as dialogue during tutorial activities. Further, videos of preservice teachers' teaching became a focus of the summative assessment for the course, therefore shaping the formative processes employed. This chapter describes and evaluates the fourth science course for the subject-focus; the innovations, challenges and the learning outcomes as they related to the involvement of feedback as dialogue and video use in the assessment process.

LITERATURE REVIEW

Research since the 1960's has identified primary teachers as lacking confidence in teaching science (Appleton, 2006; Fitzgerald, Dawson, & Hackling, 2013; Murphy, Neil, & Beggs, 2007) and lacking knowledge of the science concepts (McConnell, Parker, & Eberhardt, 2013; Mulholland & Wallace, 2005; Vail Lowery, 2002) to teach the subject. This impacts on preservice teachers' science coursework while at university and science teaching skills after being employed to teach (Appleton, 2003; McConnell et al., 2013; Skamp & Mueller, 2001). These reasons contributed to the decision at this university to make science a subject-focus option. Studies (Nicol, 2010; Sadler, 2010) have reported that engagement in learning and receiving regular and appropriate feedback improves knowledge and understanding of coursework, however, many university classrooms can still be described as didactic and lacking in feedback that is understandable by the student (Price, Handley, Millar, & O'Donovan, 2010). A didactic approach to teaching and learning can lead PSTs to become disengaged, thus negatively influence learning. Boud and Molloy (2013) support this in their call to increase the co-construction of knowledge through feedback to engage PSTs in their own learning. Doing so would increase understanding of task components and encourage discussion, a two-way dialogue, which could lead to improved feedback and clarity.

The importance of PSTs having the opportunity to engage in dialogue about their performance is well established in the literature as fundamental to teaching and learning. Pask (1976) introduced 'conversational framework' and described it as dialogue between the teacher and the student that is focused on a topic or goal. Wells (1999) described dialogue as creating meaning out of the interaction between different voices and perspectives that could assist in the feedback process to engage learners in personal learning. Laurillard (2002) built upon these to describe four characteristics of a feedback process that encourages dialogue as:

- **Adaptive:** Contingent on PSTs' needs.
- **Discursive:** Rich in two-way communication.
- **Interactive:** Linked to actions related to a task or goal.
- **Reflective:** Encourage reflection on the goal-action-feedback cycle.

Laurillard (2002) believed the purpose of dialogue was to help PSTs understand concepts and ideas and to apply their understanding in learning tasks. Her framework prescribed a form of interaction between a teacher and a student that she believed would result in effective learning and could be viewed as a description of the feedback process when both the teacher and the student make a coordinated effort to communicate. Therefore, feedback was a dialogical process and not a delivered message.

The purpose of dialogue is to help PSTs understand ideas and concepts. This requires ongoing and dynamic interactions between PST and supervising teacher/university tutor to draw attention to the learners' degree of understanding and adjust the conversation to lead to successful learning (Wood, Wood, & Middleton, 1978). Further, Price et al. (2010) noted a need for focus on the 'relational dimension of feedback' that can improve student engagement and student judgment-making on the feedback. They suggest that the feedback provided can help to shape the expected relational dimension between the tutor and the student. The degree to which the feedback "must be accepted or can be debated" implies the relational aspect of communication between the tutor and the student (Price, et al., 2010, p 280). Therefore, feedback is viewed as interactive and student-centred whereby both parties contribute to the topic of the conversation and each contribution is valued.

Several aspects about learning science can engage preservice teachers to improve knowledge and understanding about teaching science. These include: interaction and engagement with content, collaborative activities that challenge ideas (Gokhale, 1995), deep conversations and effective feedback (Carless, 2016). Research clearly identifies the importance of feedback as impacting on learning (Boud, 2015; Carless, 1998; Rushton, 2005). Feedback is considered fundamental to curriculum design where diverse forms of continuous feedback should be employed to ensure participation and discussion that clarifies concepts and expectations (Boud & Molloy, 2013). The creation and design of the final science course therefore incorporated (1) content knowledge, (2) Meta-cognitive skills, (3) teaching pedagogy, (4) assessment, (5) feedback, (6) reflection and (7) video use to ensure a rigorous course. The aspects of each of these are discussed below.

Content Knowledge

Poor content knowledge and lack of confidence impact on the teaching strategies chosen by preservice teachers when teaching science (Appleton, 2002, 2003; McConnell et al., 2013). Although they learn about a myriad of strategies for teaching in multiple subject areas and contexts, practicing them in the classroom is where they begin to see the impact of their strategies on student learning. When PSTs take mathematics courses, they learn about mathematics strategies and pedagogies and likewise for other subject areas. However, a major concern with novices is that they use the same strategies for all subject areas and all contexts (Appleton, 2002; McConnell et al., 2013). Thus, there is a need to teach and discuss strategies that are specific to science teaching and learning.

Studies have shown that PSTs who take science methods courses in which they practice the pedagogy of teaching along with the science content, do improve in confidence, science knowledge and the ability to teach science with appropriate and effective teaching strategies for science learning (Kawalkar & Vijapurkar, 2013; Skamp & Peers, 2012). Important for curriculum design is that the more exposure PSTs have to science learning and teaching, the more confident they become. This relates directly to the student ability to see themselves and expose their practices in relation to content and allows them to identify content knowledge within their teaching.

Metacognitive Skills

The use of subject specific language and pedagogical skills are a vital component of the PST course. It is essential for them to build the meta-language of the subject area and in turn impacts on their metacognitive skills. Metacognitive skills are essential for lifelong learning or sustained learning as posited by (Veenman, Van Hout-Wolters, & Afflerbach, 2006). It is the process of thinking purposefully, to be self-aware and to self-assess and reflect on personal learning and development (Joseph, 2009). If quality teachers are the goal, then appropriate teaching, feedback and assessment that teach and promote metacognitive skills (Joseph, 2009) are essential in university courses. These may include providing practice to recognize what is not understood, time to reflect on coursework and feedback and to use structured sessions to practice self-assessing and setting goals. Research suggests that as preservice teachers' metacognitive abilities improve, they will be able to achieve higher levels in course work and teaching skills (Martinez, 2006). In addition, metacognitive skills have a positive effect on self-reflecting and learning. Studies (Joseph, 2010; Lambert, 2000) have shown that PSTs who are able to assess their knowledge and examine their cognitive abilities are more successful than PSTs who lack introspective skills. Moreover, it is well established that when the dialogue of a lesson is high in the use of the metalanguage of the subject, students' learning is more effective (Department of Education and Training, DET, 2003). Geoghegan, O'Neill, & Petersen, 2013; Rowe, 2005). Therefore, adding metacognitive learning as part of tutorial instruction in higher education is essential and necessary for the teaching profession. Further, the inclusion of video viewing of oneself teaching may support PSTs ability to improve metacognition skills.

Pedagogy: Engagement

Engaging learning situations can aid in the development of science knowledge and science pedagogies (Haefner & Zembal-Saul, 2004). The idea of engaging learning experiences is not just for primary and high school teaching but should also be employed in university classrooms. Learner engagement in higher education has been the foci of many studies, most recently Compton, (2016) and Ali, Ahmed and Rose (2018) have suggested that the introduction of a digital form of feedback can impact greatly on such engagement.

Experiential and explorational learning is about engaging with natural and social phenomena through science processing skills: questioning, investigating, collecting data and applying problem solving skills (Haefner & Zembal-Saul, 2004; Smith et al., 2007). These science-processing skills include science content that can be learned while engaging in real-life learning situations.

This type of engagement allows for dialogue and collaboration that help to develop understanding of the science content and also the understanding of the science process skills (Akerson, Cullen, & Hanson, 2009). Effective learning experiences provide opportunity for learners to discuss observations and make comparisons and connections to earlier learning. When PSTs work together to plan, test out their hypotheses, collect data, discuss and try to explain their observations they develop collaboration skills (Haefner & Zembal-Saul, 2004). These collaboration skills or co-constructed knowledge opportunities (Boud & Molloy, 2013) enable them to problem solve and involve others who may have different opinions. Working with others to solve problems can be challenging but is part of workplace expectations and so developing these skills is essential (Martinez, 2006) for PSTs to be able to move forward in their education to be life-long learners.

Pedagogy: Learning Constructively

Constructivism is a learning theory that is based on the PSTs having an active role in constructing new knowledge both socially and individually (Hussain, 2012; Vail Lowery, 2002). Real-life learning situations are the beginning of learning constructively. When the classroom is designed for learning constructively, the teacher begins with discovering what the PSTs already know and understand about a topic. Then each lesson is sequentially built from the starting point and catered to lead the student to a greater and deeper understanding of the topic. Social constructivism allows learners to challenge their personal ideas thus helping to address misconceptions (Masui & De Corte, 2005). It will also cement ideas further when evidence supports their thinking and understanding.

When ideas or concepts are new and there are no prior experiences to link to the learning experience then understanding of the new experience can be more fully developed through social interactions and dialogue with an experienced person in the field (Dann, 2016). Learning as a solo event and as a monologic practice, one directional, is less effective (Masui & De Corte, 2005; Paris & Winograd, 1990). This aligns to the conceptual ideas expressed in chapter 14 where Dann and O'Neill propose the need for increased dialogic practices and a more effective learning pathway. Working alone or individually does not allow for the challenges of varying viewpoints. Nor does it allow for development of deep understanding and co-constructed knowledge. Predominately, PSTs who work in a social constructivist learning environment develop skills to demonstrate knowledge of new concepts, are able to form an argument for or against concepts, have a range of knowledge about the concepts and have multiple experiences to link to the concepts (Leach & Scott, 2002). Social constructivism provides more breadth and depth of learning than monologic practices.

This social constructivist design to develop learning is not age dependent. Adults will learn and thrive in a constructivist environment too (Hussain, 2012). Since learning constructively is highly social and depends on communication and effective dialogue, it would be a valuable addition to university classrooms.

Assessing Learning

Assessing PSTs' learning and skills takes evidence and time to make accurate judgements about what has been learnt and what has not been learnt. PSTs need to demonstrate both content knowledge of the subject and application of strategies and skills. Formative assessment of knowledge and application of

Figure 1. Feedback process
(From Dann & Dann, 2018, p. 78)

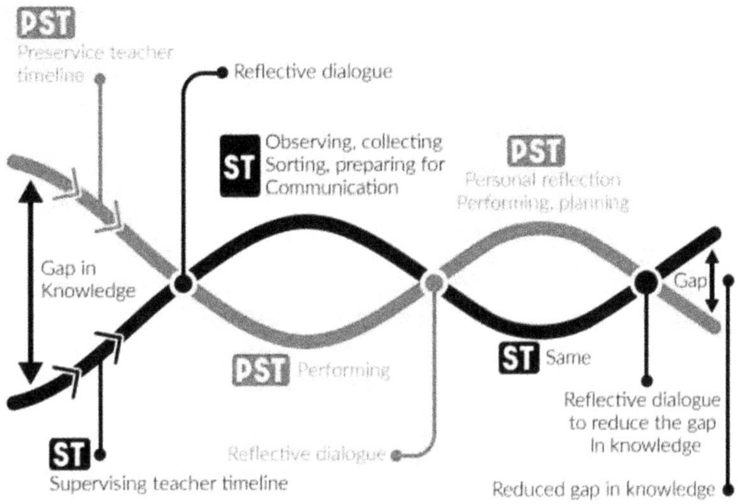

skills means that evidence is being collected along the learning journey. The evidence determines the next steps in the learning journey and provides feedback to both the learner and the teacher (Richardson, Dann, Dann, & O'Neill, 2018).

Research by Dann and Dann (2018) identified a feedback process that supervising teachers and preservice teachers went through when working together during a school placement (Figure 1). Steps one and two are mostly independent actions. To complete the process, Step three (3), a collaborative action, involves the preservice teacher and supervising teacher in reflective dialogue that focuses on how the evidence from (1) compares with the relevant standards as recorded in (2). This is the feedback step and the dialogue should be designed to help preservice teachers to reflect on their practice and learning and would be expected to be rich in the metalanguage of the pedagogy and teaching standards in focus. Thus, the term 'feedback' or 'feedback process' referred to this three-step process: (1) collect evidence, (2) sort against the standards and (3) share and engage in reflective dialogue (p. 78).

Within a classroom context, this feedback process would prompt teachers to identify where changes needed to be made with the instruction and would also ensure that gaps in knowledge (Rushton, 2005; Dann and Richardson, 2015) and PSTs' misconceptions (Allen, 2014) are being addressed in a timely manner. Further to this, the PSTs are provided feedback to help them understand how they are progressing and where they need to continue working. The university classroom must demonstrate similar knowledge and application outcomes but usually in less time. Formative assessment and formative feedback in the university setting is often limited with less opportunity for feedback due to a shortened class schedule. Nevertheless, university PSTs, like young students, demonstrate better outcomes when formative feedback and dialogue are part of their learning process (Nicol, 2010).

Feedback and Feed Forward

To be effective 'feedback' in any learning situation needs to lead to feed forward opportunities, particularly with regard to formative assessment. Feed forward is the ability to use the feedback in a similar situation to improve skills and understanding in assessed learning situations (Nicol, 2010). The purpose of feedback is to help PSTs to monitor, evaluate and regulate their learning. For example, when PSTs have been learning to develop an investigation, the teacher provides examples, PSTs practice developing an investigation with the teacher, and eventually they complete one without assistance. In this example, formative feedback most likely occurred along the way as dialogue that created a shared understanding of the task and content. Even more feedback can occur with the final piece as written feedback. This feedback can be used to develop another investigation. Therefore, when the combined written and dialogic feedback are considered and internalized by the PST, then feed forward is in process as s/he makes improvements to create a new investigation that heeds the feedback advice provided (Nicol, 2010). This would likely result in a second investigation completed without assistance that was of better quality than the first. In this instance, the student is improving their investigation skills and has a greater chance of remembering those skills into the future. Sometimes feedback just sits on a piece of work, like the first investigation and nothing further is done with it, thus allowing the feedback to remain in the past (Carless, 2016). Effective learning occurs when teachers provide PSTs with multiple feedback opportunities so they may learn and improve their knowledge and skills. This would include conversations to ensure there is understanding of the feedback that will result in change in the future. Therefore, feedback as dialogue in science is a vital aspect of practice to improve preservice teacher learning and teaching.

Reflecting

Reflecting on work and performance is part of the skills all people use for their workplace and daily encounters. It is not meant to be a fleeting thought but rather a way to check oneself and engage in continuous improvement (Joseph, 2009). A definition created by Black and Plowright (2010) suggests:

reflection is the process of engaging with learning and/or professional practice that provides an opportunity to critically analyze and evaluate that learning or practice. The purpose is to develop professional knowledge, understanding and practice that incorporates a deeper form of learning which is transformational in nature and is empowering, enlightening and ultimately emancipatory (p. 246).

Further, Thompson and Pascal (2012) extend this by breaking reflective practice down into four components; blending theory and practice, active learning, participative learning and challenging dogma. Loughran's (2002) study supports this and points out the value in continually connecting reflection with the meaning and significance of the teaching and learning practice. The problems preservice teachers are confronting need attention. As preservice teachers practice reflecting on experiences, they gradually enhance their meaning making to inform their practice (Loughran, 2002).

Teacher education regards reflective capacity highly (Australian Institute for Teaching and School Leadership, 2019). It is integrated into education courses; assessment and part of the government requirements to successfully complete an education program and become a practicing teacher (Australian Institute for Teaching and School Leadership, 2019). This is understandable when teaching practice involves regular and thoughtful reflection on teaching performance and on student development of learning outcomes (Thompson & Pascal, 2012). However, preservice teachers need support to learn to reflect deeply and effectively (Thompson & Pascal, 2012). As novices, they often have little experience with purposeful reflection and tend to provide descriptions of events instead (Baecher, Kung, Jewkes, & Rosalia, 2013). This is a skill that requires practice with experienced educators who can help PSTs identify essential aspects of teaching and learning on which to reflect and then lead them to improve their teaching and learning skills.

Video and Video Reflection

The use of video for self-reflection may positively influence preservice teachers' development (Baecher, et al., 2013) and allow them to gradually get to know themselves (Gaudin & Chaliès, 2015). Self-reflection helps PSTs to explore and consider different perspectives and develop teacher identity (Rogers & Scott, 2008). However, self-reflection is not without biases, particularly as a novice. Preservice teachers may view themselves in a favorable light and over-rate themselves or they may be more modest and under rate themselves (Sitzmann, Ely, Brown, & Bauer, 2010). This results in an inaccurate representation of themselves and may negatively impact on personal teaching skills and ultimately their teacher identity. Osipova, Prichard, Boardman, Kiely, and Carroll (2011) found that preservice teachers' personal conceptions of themselves were more accurate when they were initially presented with specific descriptions and criteria to reflect upon. Further, Baecher et al. (2013) added that viewing teaching models and assessing that practice against criteria prior to self-assessing enabled preservice teachers to experience dissonance that challenged their beliefs and standards about teaching. Dissonance can be uncomfortable to experience as it challenges preconceptions. However, this experience positively influenced preservice teachers' self-reflections about their own teaching (Baecher et al., 2013; Rosaen, Lundeberg, Cooper, Fritzen, & Terpstra, 2008). Equipping preservice teachers with tools for meaningful self-reflection may support preparation for the performance expectations required to demonstrate achievement of standards in the teaching profession (Baecher et al., 2013).

Additionally, Gaudin and Chaliès (2015) reported that the role of the tutor as facilitator and the importance of video selection enhanced teacher learning through "video-mediated" activities (p 44). The facilitator was responsible for drawing out the learning outcomes being viewed in order to reflect on the practice, compare with other teaching videos and lead a meaningful discussion. These components assist preservice teachers to 'notice' and analyze teaching practice thus leading them towards improved critical reflection skills.

Quality teachers are reflective teachers and know the content and skills required of the profession. Therefore, use of video in course assessment may help to develop metacognitive skills required for deep reflection.

TRADITIONAL TEACHING APPROACH VS PROGRESSIVE APPROACH

A traditional approach to teaching implies that a teacher directs PSTs to learn through memorization and recitation techniques that do not develop problem-solving skills (Cimer, 2007; Mayer, 2002). A traditional approach also implies that teachers are the giver of knowledge and PSTs are the receivers of information and through this style of teaching and learning, PSTs are expected to master the content. Further, a traditional approach assumes all PSTs are at the same level of understanding (Cambridge Centre for Sixth-form Studies, 2017). Hence, it does not address diverse learner needs.

Alternatively, more progressive and modern teaching is viewed differently. Modern teaching focuses on the individual learner needs, is more activity based with questions, explanations, demonstrations and collaboration (Wenger, 2009). The goal is to help PSTs make connections to improve their memory and knowledge. In addition, a progressive approach develops metacognitive skills more effectively and aligns well with teaching and learning expectations within the Australian Curriculum.

Preservice teachers felt less capable to teach science than other subjects (Osborne, Simon, & Collins, 2003). The use of engaging and active learning, video, reflection and feedback as dialogue opportunities may begin to change preservice teachers' efficacy and capacity to teach science and to become effective teachers. These were the challenges and the aims for the new science subject-focus course in an initial teacher education program.

This chapter encourages research that supports the use of feedback as dialogue that feeds forward in the vital role of formative assessment in learning, combined with video to support critical reflection to improve personal reflections and teaching practice. Personal reflection is a lifelong skill that empowers a person to notice and to improve. The ability to reflect effectively can impact on many areas of one's life and can be transformational (Boud, Keogh, & Walker, 2013). In teacher education, the teachers' ability to reflect on their own practice and on their PSTs' learning is pivotal and can ultimately impact on the learning environment and the ability for PSTs to attain an adequate year's growth. Further, constructivist learning empowers the learner and provides a learning structure for the teacher. Not only does it begin with a student's prior knowledge and allow for development of understanding through dialogue and experiences, it creates a learning environment in which PSTs can acquire deep learning that leads to transfer of knowledge into other contexts. Constructivist learning maintains that all learners can learn, participate and develop skills at their own level. The teachers' ability to reflect and the use of constructivist learning approach is important for the range of learners found in today's classrooms. Thus, university PSTs can enhance their learning of difficult topics through engaging experiences and exploration, dialogue and collaboration.

The CC took note of the PSTs' responses and understanding of the weekly topics and teaching strategies. These were used to plan for the next session. Re-teaching and conversations were based on this weekly data. Group work and inquiry activities allowed opportunities for the CC to sit with the groups and have further conversations and address the learning needs of the groups. Anecdotal notes from weekly classes and any individual meetings and CC reflections became part of the data for this chapter. Concerns and assessment results contributed to the data discussed in the following section.

There are limitations around the data since the CC is also the writer of the chapter and uses personal reflections and perspective to describe and interpret the data from the course. However, while this is a personal reflection of the course, discussions with colleagues who also teach in the science courses and share student assessment and anecdotal notes occurred to ensure the CC was interpreting the course data as bias free as possible. As a practitioner in education, the CC endorses and regularly participates in reflection of her own teaching practice.

COURSE INFORMATION AND DISCUSSION

Developing and writing a university education course takes careful planning as there are many internal university requirements and external accreditation requirements that need to be met. The key concerns for the primary subject-focus courses were the expectation that it provides opportunity for PSTs to demonstrate 'expert' knowledge of content and teaching strategies suitable to the subject area. This was a concern for the CC because many of the PSTs who enrolled into primary teaching at this university have reported during class discussions that they have already developed a fear of and a dislike towards science and would prefer not to teach it if they had a choice. Statements such as these have impacted on course design for the science courses.

Participants

The PSTs who chose science as their subject-focus indicated they enjoyed science. Comparatively, the numbers in science were 32 and those in English were 75 aligning with the research (Osborne, Simon & Collins, 2003) that reported most PSTs prefer subjects other than science. However, these PSTs elected to be in the science course thus leading to some assumptions about them. For instance; they enjoy science, they have greater confidence to teach science than the other 75 PSTs, and they have some science content knowledge. These assumptions were confirmed by many of the PSTs during class engagement and observed through weekly participation. Further, there were never more than two PSTs absent from a tutorial with most weeks having 100% attendance.

The purpose of the course was to have PSTs demonstrate 'expert' content knowledge and teaching strategies for teaching science. The CC decided to incorporate modeling, feedback as dialogue and video through a collaborative and constructive learning approach as the research supports these for effective teaching and learning. The final science subject course included science content for space science (F-8) and earth science, specifically magnetism and electricity.

Modeling Strategies

Modeling is used to teach concepts and performance skills within a context (Haston, 2007). Science content was taught while demonstrating science teaching strategies. The CC modeled appropriate teaching strategies for teaching science each week during tutorials. The strategy chosen was based upon the content to be taught. For example, an inquiry approach was used when PSTs needed to compare materials and decide upon the most appropriate material for a particular use. E.g.: How many C batteries will light bulb X? What happens to the brightness of bulbs when more bulbs are added to a parallel circuit? The demonstration strategy clarified misconceptions about the cause of seasons by using equipment to demonstrate the Earth's rotation around the Sun. Models were used as a strategy to teach concepts for abstract ideas such as the Solar System, planet size and planet rotations. A discovery approach was used to encourage trial and error, testing out and to justify results. EG: Which materials can contain an electrical charge? Using these materials, how many ways can you make a complete circuit? Direct teaching was used to advance learning from concrete to symbolic abstract as with electrical diagrams. Cooperative learning groups were used to enhance group activities; whereby each person was allocated a role for purposeful contribution (Johnson & Johnson, 2008).

Table 1. Course overview of topics and strategies

Week	Topics - A Constructivist Approach	Strategy Modelled or Taught
1	Prior knowledge activity introduction to teaching strategies/theories	Discrepant event 24-hour daylight Earth and torch
2	Space: seasons and day and night	Modelling
3	Space: orbits and rotation of planets, size of planets	Use of models and cooperative group activity Fruit and seeds for relative size of planets
4	Space: Moon phases. eclipses	Daily drawings
5	Magnetism and static electricity and a simple circuit	Discovery approach
6	Circuits in a series and in parallel	Analogy Cooperative group activity
7	Planning and practice teaching	feedback
8	Planning and practice teaching	feedback
9	Video teaching / peer feedback	
10	Video teaching / peer feedback	

Early in the course PSTs had identified that they were familiar with most of the strategies on the list of strategies provided. Nevertheless, the CC modeled a strategy each week and led open discussions about its effectiveness for science learning. However, the CC soon realized that the PSTs did not understand how to enact the strategies, nor did they know how to match strategies to the content to be taught. Further, after PSTs had definitions of the strategies, they continued to have difficulty aligning with content and applying correctly. They were unable to link prior knowledge to the definitions. This feedback to the CC influenced the direction of the course and the type of feedback as dialogue enacted with the PSTs. As a result, the CC incorporated more modeling of strategies and added explicit discussion about the strategies and why they were chosen for the content (Louca, Zacharia, & Constantinou, 2011). This type of teaching needed to be repeated each time a strategy was used. This approach helped PSTs identify the strategies observed and those that they were demonstrating themselves and discuss why it was the preferred choice over other strategies. Further, it became powerful feedback to improve their understanding of teaching strategies. This purposeful and deliberate thinking helped build the metacognitive skills required for effective teaching.

Feedback

Feedback as dialogue occurred during tutorials and written feedback was provided after assessments. During tutorials modeling of strategies took place and the teaching of science content took place. It was important for PSTs to have conversations on both aspects. Feedback was often in the form of clarifying and re-teaching. For example, re-teaching through use of a model to correct misconceptions about the angle of the Earth and its rotation around the Sun, thus breaking information down into smaller portions that built understanding and connected to their prior knowledge and learning experiences (Alesandrini & Larson, 2002; Saunders, 1992). The tutorials were designed to be sequential and build on knowledge from the previous activities.

Therefore, the use of explicit conversations along with materials to demonstrate, model and test out built preservice teacher knowledge and understanding constructively (Masui & De Corte, 2005; Skamp & Peers, 2012). The CC had 2-3 boxes of materials and equipment for teaching within the tutorial room each week. Therefore, when it was necessary to break down information or use a different method/material to reteach concepts, the opportunity and materials were there.

While the course was very hands-on and interactive, it had a strong dialogue component that allowed strong and explicit links between curriculum content, language and teaching strategies to be made (Craig, Sullins, Witherspoon, & Gholson, 2006). Examination of this dialogue 'in the moment' was a powerful form of feedback that led to improved personal understanding of science content and understanding of strategies to teach science. For example, after a few weeks of modeling and discussing teaching strategies, the CC provided a learning experience and at the completion of the learning experience asked PSTs to identify the teaching strategy used and why they thought so. The learning experience was to gather a range of materials from the boxes and identify which were able to contain static electricity and explain their thoughts. There was no other instruction or guidance provided. After the experience and discussion around their findings, the CC asked PSTs to identify the teaching strategy and a reason for their choice. Some of the choices were (1) inquiry because they were answering a question, (2) cooperative learning because they were working in groups and (3) discovery because they were finding things out for themselves without teacher guidance. Each of these choices was discussed to identify what constitutes each strategy or theory that they chose. PSTs then re-chose the strategy that was most applicable. After the discussion, the whole group was able to identify the correct strategy. In this situation, discovery approach.

The feedback dialogue improved the way PSTs viewed the learning process for their future teaching careers. Research has clearly identified that effective teachers know their content, how to teach it and understand how their PSTs learn (Zientek, 2007). It is essential then that PSTs understand the impact of dialogue and collaboration to build metacognitive skills, especially since good metacognitive skills lead to lifelong learning (Joseph, 2009; Paris & Winograd, 1990).

Peer to Peer Feedback

Studies have shown that peer feedback conjures a range of opinions from highly negative to positive. (For example, peers tend to be skeptical that some peers will be able to provide adequate feedback particularly with written tasks (Patchan & Schunn, 2016), however, research indicated that even low level PSTs can provide some kind of beneficial feedback for their peers especially when several peers are providing feedback to the one student (Patchun & Schunn, 2016). In video situations, after PSTs moved past feeling confronted and vulnerable, they were able to improve through cycles of practice, reflection, feedback and revision (Bower, Cavanagh, Moloney, & MingMing, 2011).

Preservice teachers had the opportunity to provide feedback to peers during tutorial learning activities. One example is when PSTs were asked to get a small light bulb to light with the materials provided; a bulb, a battery a wire. This allowed them thinking and reflection time including practice with providing feedback to others. Especially when they believed it was not possible to light the bulb with only one wire. Further, it allowed the CC to intervene as needed to provide further feedback or to make immediate clarification and correction if misconceptions were identified. Preservice teachers were organized into groups of 4-6 each week. They completed activities as groups and demonstrated to the class their results and conclusions. Such as identifying weather a series in a circuit or in parallel would produce the best light for a bedroom. Other groups were asked to comment on their results and conclusions. Most of them were somewhat comfortable speaking out in class and giving opinions (I think . . .), challenging each other (Are you sure? I found on the web . . .) and even correcting each other (I disagree. When I did . . . then . . . happened).

This type of feedback and dialogue between PSTs and the CC can impact on the development of knowledge and skills in teaching. Placing the PSTs in a position to be interactive learners, just as PSTs would be in their future classrooms, allows the PSTs to consider how learning takes place and how understanding is developed. Research has identified that PSTs will understand what teaching looks like and how learning feels when they are actively participating in discussions about the teaching and learning environment (Joseph, 2010). These are important and valuable aspects of the course design.

Video and Reflection

The final four weeks was designed to be a collaborative time for the groups to plan a mini unit and lessons that would be taught by the group to the class. By this time PSTs had been exposed to new science content, modeling of appropriate teaching strategies and conversations about linking content with strategies that enhanced learning. This type of engagement and dialogue enabled PSTs to address misconceptions and to improve understanding for personal improvement (Baecher et al., 2013). Therefore, demonstrating subject knowledge and effective use of strategies became a summative assessment through video.

Each person in the group took one of the lessons to teach to another group of their peers which was videoed on their own devices, saved as a YouTube® then added to Playposit®. Playposit® is a platform that allows for the creation of interactive video to enhance teaching and learning. Within Playposit® PSTs had the capacity to view their performance and reflect on their performance regarding the science content and the teaching strategies employed. The Playposit® program allowed for annotation directly with the video. Therefore, PSTs could add annotations at specific moments during their teaching that then allowed the CC to view exactly what the PST was referring too. This could be a pathway for further dialogue that could lead to feed forward. However, in this course it was the summative assessment.

Additionally, immediately after teaching their lesson, peers could provide verbal or written feedback to the student teaching the lesson. Many of the PSTs took advantage of this within their planning groups at the end of the class. Those who felt comfortable sharing their video did so, however, many also shared conversations about aspects they felt should be a discussion point. For example, four of the six working groups discussed similar areas outside of the course focus. These included: voice, gesturing, handwriting, ability to see what was written on the board and use of praise. Clearly, these groups had been collaborating to make decisions. This preservice teacher response is similar to the findings by Baecher et al. (2013). Conversation with peers provided dialogue around different points of view and provided time to interpret and analyze the teaching.

Studies have shown that PSTs require practice to become critical reflectors of their teaching practice and for student learning (Gaudin & Chaliès, 2015). The PSTs in previous studies have shown that PSTs are less critical of themselves when viewing teaching videos but will be more critical of others they do not know (Gaudin & Chaliès, 2015). The PSTs displayed similar responses with their reflections. At the start of the video reflection work PSTs asked the CC to check their reflections. Many were retelling the teaching rather than discussing its effectiveness and its accuracy. At this point the CC provided feedback to help PSTs reflect more critically. While the feedback improved the reflections, it was clear that practice noticing and reflecting on personal knowledge and skills will be needed in future classes.

Tripp and Rich (2012) found that providing practice with viewing other teachers' teaching and discussing the teaching against specific criteria helped to develop preservice teacher skills in noticing what good practice looks like and sounds like. Additionally, Baecher et al. (2013) identified that providing multiple practice with videoed teaching opportunities including feedback helped PSTs to improve their reflective ability prior to the final video teaching assessment. These practice opportunities helped to increase the metacognitive skills of the PSTs (Joseph, 2009; Loughran, 2002).

The Learning Process

The feedback process described in Figure 1 is similar to the learning process experienced in the science course. PSTs started the course with a pretest that identified content knowledge. This revealed the gaps in knowledge (step 1, individual). Each week, new learning experiences and dialogue between the tutor and the PSTs took place that would essentially begin to narrow the gaps (step 3, collaborative). This led to regular practice with and improvement in cognitive skills such as reading, thinking, remembering, learning and reasoning. These critical skills enable PSTs to process information and learn to evaluate, analyze, make comparisons and understand cause and effect. Most cognitive skills are learned. Therefore, thinking and cognitive skills can be improved with practice (Kusschau, 2003).

DEVELOPING QUALITY TEACHERS

The course design was unique and attempted to improve PSTs' metacognitive abilities so they could demonstrate 'expert' level content knowledge and teaching strategies for teaching primary science. This new design does not follow traditional approaches to teaching at university. Instead, it sought a progressive approach that would develop 21st Century skills. Therefore, the final science course encouraged learning through modern strategies that included collaboration, demonstrations, active learning, dialogue, questions and explanations of the content that helped to develop the PSTs' metacognitive skills.

Additionally, the final science course provided opportunity for PSTs to engage with the activities in order to support their personal understanding of how to teach with progressive strategies. The use of video to view themselves and each other and reflect on their personal teaching skills as well as those of their peers is an important part of the course design to develop reflective practitioners. These are essential components as the PSTs are the next generation of teachers and will have an impact on the future of education.

Nevertheless, it could be argued that incorporating both traditional and modern teaching approaches could be an effective way to develop metacognitive skills. There is evidence that there is a place for rote learning (Mayer, 2002), such as knowing and recalling number facts, times tables and vocabulary. Regardless, learning situations should increase participation, include activities that promote collaboration, questioning, testing out, discussions and problem solving. These are the types of activities that lead to lifelong skills that are needed for any profession that young people may pursue. The following image attempts to illustrate the course design.

Figure 2. Course pedagogical design

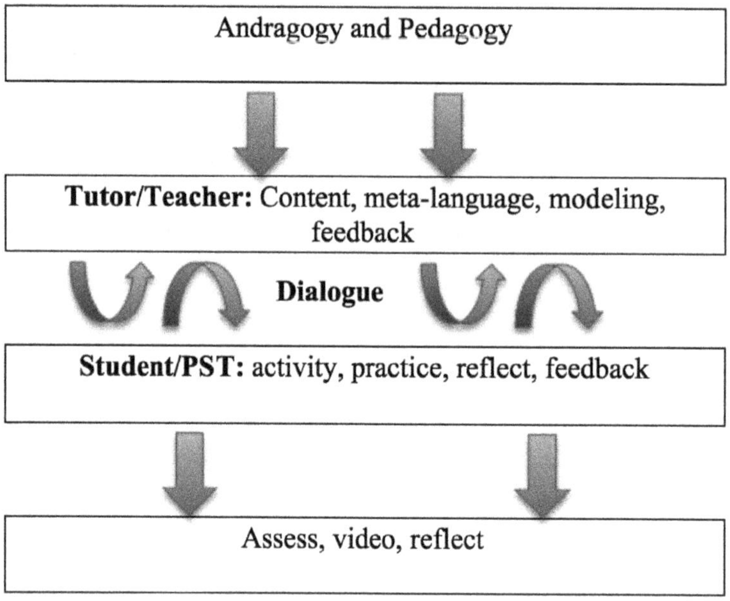

This model summarizes the approach to the science course design. It attempts to identify the overall use of modeling pedagogy while applying andragogy for the adult learners. This, in itself, is complex but it meant that the CC had to deliver and model one or the other as appropriate for the purpose. In addition, the CC would ask PSTs to think or act with different hats such as how would a teacher respond or what might a child conclude after this activity? This style of teaching considers and respects both the adult learner and the fact that these PSTs must learn appropriate pedagogy for their future teaching. Therefore, the course design emulated a school classroom whereby the interactions cannot be preplanned but responded to dialogue, or content needs as they arose. This allowed for feedback and dialogue to address immediate concerns and link to previous experiences thus establishing grounds for feed forward and application of concepts in other contexts. Teaching is typically fluid with dialogue interplay moving from various points as dictated by the group while at the same time moving PSTs forward in their learning to be successful in the course content and assessment.

This final science subject course included key strategies that improved learning and preparation for teaching and will be maintained within the course. They include: modeling, feedback as dialogue and the use of video for reflection on teaching skills. Modeling was very important as it revealed to PSTs how these strategies look in action and they began to understand their purpose. PSTs began to consider the better strategy for given situations. The dialogue that supported the strategies acted as feedback for PSTs therefore became another essential component of the learning process.

On the other hand, there were aspects to the course that needed modifying to improve the depth of engagement and the outcome to demonstrate 'expert' level. The video was a powerful indicator of teaching abilities, but PSTs needed further practice to develop skills in noticing key behaviors and language around effective teaching prior to using it as assessment. A modification could include videos of teachers teaching that are discussed in classes and key strategies, language and behaviors of the teachers identified (Baecher et al., 2013). Helping PSTs to notice effective teaching skills and then practice them may lead PSTs to improve their personal metacognitive skills.

Learning can be defined as the acquisition of knowledge or skills through study, experience or being taught. Teaching is the process of attending to people's needs, experiences and feelings, and making specific interventions to help them learn particular things. This final science course must contain opportunity for PSTs to both learn and teach. Both are very different, and both are essential. As teachers, they will need to understand how learning occurs and they will need to know how to teach effectively. These are two important characteristics of quality teachers.

CONCLUSION

This unique course design was pivotal in creating a learning environment that responded to learner needs in the moment by incorporating dialogue to provide feedback and feed forward opportunities that dealt with misconceptions in the learning process and would propel student learning forward. The video reflections were evidence of the positive impact the dialogue made on student learning. Further, the use of video helped PSTs to conceptualize their learning for themselves in a self-reflective manner. This course design allowed PSTs to develop their personal metacognitive skills and their science content knowledge by incorporating feedback into every class, dialogue to clarify concepts and interactive group activities to practice the concepts.

The literature and the science course experience led the writer to believe that initial teacher training and higher education courses need to embed specific pedagogy that improves feedback and metacognitive skills. Research supports embedding feedback into course design through activities such as modeling, dialogue, video and problem solving through collaboration. These activities can improve PSTs' metacognitive skills that will ultimately influence their future teaching practice. It is important to foster these integral life skills, so PSTs can lead the way in creating enjoyable, interesting, and useful learning environments in their own classrooms.

REFERENCES

Adams, R. (2012). *NAPLAN reporting – measures and models for reporting gain over time*. Retrieved from http://www.coagreformcouncil.gov.au/excellence/docs/improvement/CI_NAPLAN_reporting_2012.pdf

Akerson, V. L., Cullen, T. A., & Hanson, D. L. (2009). Fostering a community of practice through a professional development program to improve elementary teachers' views of nature of science and teaching practice. *Journal of Research in Science Teaching*, *46*(10), 1090–1113. doi:10.1002/tea.20303

Alesandrini, K., & Larson, L. (2002). Teachers bridge to constructivism. *The Clearing House: A Journal of Educational Strategies, Issues and Ideas*, *75*(3), 118–121. doi:10.1080/00098650209599249

Ali, N., Ahmed, L., & Rose, S. (2018). Identifying predictors of students' perception of and engagement with assessment feedback. *Active Learning in Higher Education*, *19*(3), 239–251. doi:10.1177/1469787417735609

Allen, M. (2014). *Misconceptions in primary science*. McGraw-Hill Education.

Appleton, K. (2002). Science activities that work: Perceptions of primary school teachers. *Research in Science Education, 32*(3), 393–410. doi:10.1023/A:1020878121184

Appleton, K. (2003). How do beginning primary school teachers cope with science? Toward an understanding of science teaching practice. *Research in Science Education, 33*(1), 1–25. doi:10.1023/A:1023666618800

Appleton, K. (2006). Science pedagogical content knowledge and elementary school teachers. *Elementary science teacher education: International perspectives on contemporary issues and practice*, 31-54.

Australian Institute for Teaching and School Leadership. (2019). *Australian professional standards for teachers*. Retrieved from https://www.aitsl.edu.au/teach/standards

Baecher, L., Kung, S.-C., Jewkes, A. M., & Rosalia, C. (2013). The role of video for self-evaluation in early field experiences. *Teaching and Teacher Education, 36*, 189–197. doi:10.1016/j.tate.2013.08.001

Black, P. E., & Plowright, D. (2010). A multi-dimensional model of reflective learning for professional development. *Reflective Practice, 11*(2), 245–258. doi:10.1080/14623941003665810

Boud, D. (2015). Feedback: Ensuring that it leads to enhanced learning. *The Clinical Teacher, 12*(1), 3–7. doi:10.1111/tct.12345 PMID:25603699

Boud, D., Keogh, R., & Walker, D. (2013). Promoting reflection in learning: A model. In D. Boud, R. Keogh, & D. Walker (Eds.), *Reflection: Turning experience into learning* (pp. 18–40). London: Routledge Falmer. doi:10.4324/9781315059051

Boud, D., & Molloy, E. (2013). Rethinking models of feedback for learning: The challenge of design. *Assessment & Evaluation in Higher Education, 38*(6), 698–712. doi:10.1080/02602938.2012.691462

Bower, M., Cavanagh, M., Moloney, R., & Dao, M. M. (2011). Developing communication competence using an online Video Reflection system: Preservice teachers' experiences. *Asia-Pacific Journal of Teacher Education, 39*(4), 311–326. doi:10.1080/1359866X.2011.614685

Cambridge Centre for Sixth-form Studies. (2017). *Teaching methods: Traditional vs modern*. Retrieved from https://www.ccss.co.uk/news/traditional-vs-modern-teaching/

Carless, D. R. (1998). A case study of curriculum implementation in Hong Kong. *System, 26*(3), 353–368. doi:10.1016/S0346-251X(98)00023-2

Carless, D. R. (2016). Feedback as Dialogue. Encyclopedia of Educational Philosophy and Theory. doi:10.1007/978-981-287-532-7_389

Cimer, A. (2007). Effective teaching in science. A review of literature. *Journal of Turkish Science Education, 4*(1), 20–44.

Compton, J. (2016). Flipped learning with Turkish students: Issues of dissonance to possible harmonization. *The Global eLearning Journal, 5*(22), 1-7.

Craig, S. D., Sullins, J., Witherspoon, A., & Gholson, B. (2006). The deep-level-reasoning-question effect: The role of dialogue and deep-level-reasoning questions during vicarious learning. *Cognition and Instruction, 24*(4), 565–591. doi:10.12071532690xci2404_4

Dann, B. (2016). *Implementation of the new Australian Curriculum: Science: A study of the experiences of a range of Queensland schools and teachers* (Doctoral thesis). University of the Sunshine Coast, Queensland, Australia.

Dann, C., & Dann, B. (2018). Exemplifying formative assessment practices in the field through casc study: Concepts, issues and challenges. In T. Richardson, B. Dann, C. Dann, & S. O'Neill (Eds.), *Formative assessment practices for pre-srevice teacher practicum feedback: Emerging research and opportunities* (pp. 76–112). IGI Global. doi:10.4018/978-1-5225-2630-8.ch003

Department of Education and Training (DET). (2003). *Quality teaching in NSW public schools: A classroom practice guide.* Department of Education and Training, Professional Support and Curriculum Directorate.

Fitzgerald, A., Dawson, V., & Hackling, M. (2013). Examining the beliefs and practices of four effective Australian primary Science teachers. *Research in Science Education, 43*(3), 981–1003. doi:10.100711165-012-9297-y

Gaudin, C., & Chaliès, S. (2015). Video viewing in teacher education and professional development: A literature review. *Educational Research Review, 16*, 41–67. doi:10.1016/j.edurev.2015.06.001

Geoghegan, D., O'Neill, S., & Petersen, S. (2013). Metalanguage: The 'teacher talk' of explicit literacy teaching in practice. *Improving Schools, 16*(2), 119–129. doi:10.1177/1365480213493707

Gokhale, A. (1995). Collaborative learning enhances critical thinking. *Journal of Technology Education, 7*(1), 22–30. doi:10.21061/jte.v7i1.a.2

Haefner, L., & Zembal-Saul, C. (2004). Learning by doing? Prospective elementary teachers' developing understandings of scientific inquiry and science teaching and learning. *International Journal of Science Education, 26*(13), 1653–1674. doi:10.1080/0950069042000230709

Haston, W. (2007). Teacher modeling as an effective teaching strategy. *Music Educators Journal, 93*(4), 26–30. doi:10.1177/002743210709300414

Hussain, I. (2012). Use of constructivist approach in higher education: An instructor's observation. *Creative Education, 3*(02), 179–184. doi:10.4236/ce.2012.32028

Johnson, R. T., & Johnson, D. W. (2008). Active learning: Cooperation in the classroom. **教育心理学年報**, *47*, 29-30. doi:10.5926/arepj1962.47.0_29

Joseph, N. (2009). Metacognition needed: Teaching middle and high school students to develop strategic learning skills. *Preventing School Failure, 54*(2), 99–103. doi:10.1080/10459880903217770

Joseph, N. (2010). Practical intelligence: Do students have it? Helping them develop it through metacognitive learning. *Language Arts Journal of Michigan, 26*(1), 38–42. doi:10.9707/2168-149X.1016

Kawalkar, A., & Vijapurkar, J. (2013). Scaffolding science talk: The role of teachers' questions in the inquiry classroom. *International Journal of Science Education, 35*(12), 2004–2027. doi:10.1080/09500693.2011.604684

Kusschau, R. (2003). *Understanding psychology.* Columbus, OH: Glencoe and McGraw Hill Co. Inc.

Lambert, M. (2000). Using cognitive and metacognitive learning strategies in the classroom. *Preventing School Failure, 44*, 81–82. doi:10.1080/10459880009599788

Laurillard, D. (2002). Rethinking university teaching: A conversational framework for the effective use of learning technologies (2nd ed.). London: Routledge Falmer.

Leach, J., & Scott, P. (2002). Designing and evaluating science teaching sequences: An approach drawing upon the concept of learning demand and social constructivist perspective on learning. *Studies in Science Education, 38*(1), 115–142. doi:10.1080/03057260208560189

Louca, L. T., Zacharia, Z. C., & Constantinou, C. P. (2011). In quest of productive modeling-based learning discourse in elementary school science. *Journal of Research in Science Teaching, 48*(8), 919–951. doi:10.1002/tea.20435

Loughran, J. J. (2002). Effective reflective practice: In search of meaning in learning about teaching. *Journal of Teacher Education, 53*(1), 33–43. doi:10.1177/0022487102053001004

Martinez, M. (2006). What is Metacognition? *Phi Delta Kappan, 87*(9), 696–699. doi:10.1177/003172170608700916

Masui, C., & De Corte, E. (2005). Learning to reflect and to attribute constructively as basic components of self-regulated learning. *The British Journal of Educational Psychology, 75*(3), 351–372. doi:10.1348/000709905X25030 PMID:16238871

Mayer, R. E. (2002). Rote versus meaningful learning. *Theory into Practice, 41*(4), 226–232. doi:10.120715430421tip4104_4

McConnell, T. J., Parker, J. M., & Eberhardt, J. (2013). Assessing teachers' science content knowledge: A strategy for assessing depth of understanding. *Journal of Science Teacher Education, 24*(4), 717–743. doi:10.100710972-013-9342-3

Mulholland, J., & Wallace, J. (2005). Growing the tree of teacher knowledge: Ten years of learning to teach elementary science. *Journal of Research in Science Teaching, 42*(7), 767–790. doi:10.1002/tea.20073

Murphy, C., Neil, P., & Beggs, J. (2007). Primary science teacher confidence revisited: Ten years on. *Educational Research, 49*(4), 415–430. doi:10.1080/00131880701717289

Nicol, D. (2010). From monologue to dialogue: Improving written feedback processes in mass higher education. *Assessment & Evaluation in Higher Education, 35*(5), 501–517. doi:10.1080/02602931003786559

Osborne, J., Simon, S., & Collins, S. (2003). Attitudes towards science: A review of the literature and its implications. *International Journal of Science Education, 25*(9), 1049–1079. doi:10.1080/0950069032000032199

Osipova, A., Prichard, B., Boardman, A. G., Kiely, M. T., & Carroll, P. E. (2011). Refocusing the lens: Enhancing elementary special education reading instruction through video self-reflection. *Learning Disabilities Research & Practice*, *26*(3), 158–171. doi:10.1111/j.1540-5826.2011.00335.x

Paris, S., & Winograd, P. (1990). How metacognition can promote academic learning and instruction. In B. Jones & L. Idol (Eds.), *Dimensions of thinking and cognitive instruction* (pp. 15–52). Lawrence Erlbaum Associates.

Pask, G. (1976). *Conversation theory: Applications in education and epistemology*. Amsterdam: Elsevier.

Patchan, M., & Schunn, C. (2016). Understanding the effects of receiving peer feedback for text revision: Relations between author and reviewer ability. *Journal of Writing Research*, *8*(2), 227–265. doi:10.17239/jowr-2016.08.02.03

Price, M., Handley, K., Millar, J., & O'Donovan, B. (2010). Feedback: All that effort, but what is the effect? *Assessment & Evaluation in Higher Education*, *35*(3), 277–289. doi:10.1080/02602930903541007

Richardson, T., Dann, B., Dann, C., & O'Neill, S. (2018). *Formative assessment practices for pre-service teacher practicum feedback: Emerging research and opportunities*. IGI Global. doi:10.4018/978-1-5225-2630-8

Rogers, C., & Scott, K. (2008). The development of the personal self and professional identity in learning to teach. In M. Cochran-Smith, S. Feiman-Nemser, D. McIntyre, & K. Demers (Eds.), *Handbook of research on teacher education* (pp. 732–755). New York: Routledge.

Rosaen, C. L., Lundeberg, M., Cooper, M., Fritzen, A., & Terpstra, M. (2008). Noticing noticing: How does investigation of video records change how teachers reflect on their experiences? *Journal of Teacher Education*, *59*(4), 347–360. doi:10.1177/0022487108322128

Rowe, K. (2005). *Teaching reading: Report of the national inquiry into the teaching of literacy*. Canberra: Department of Education, Science and Training. Retrieved from http://www.dest.gov.au/nitl/documents/report_recommendations.pdf

Rushton, A. (2005). Formative assessment: A key to deep learning? *Medical Teacher*, *27*(6), 509–513. doi:10.1080/01421590500129159 PMID:16199357

Sadler, D. (2010). Beyond feedback: Developing student capability in complex appraisal. *Assessment & Evaluation in Higher Education*, *35*(5), 535–550. doi:10.1080/02602930903541015

Saunders, W. L. (1992). The constructivist perspective: Implications and teaching strategies for science. *School Science and Mathematics*, *92*(3), 136–141. doi:10.1111/j.1949-8594.1992.tb12159.x

Sitzmann, T., Ely, K., Brown, K. G., & Bauer, K. N. (2010). Self-assessment of knowledge: A cognitive learning or affective measure? *Academy of Management Learning & Education*, *9*(2), 169–191. doi:10.5465/amle.9.2.zqr169

Skamp, K., & Mueller, A. (2001). Student teachers' conceptions about effective primary science teaching: A longitudinal study. *International Journal of Science Education*, *23*(4), 331–351. doi:10.1080/09500690119248

Skamp, K., & Peers, S. (2012). *Implementation of science based on the 5E learning model: Insights from teacher feedback on trial Primary Connections units.* Paper presented at the The 43th Annual Conference of the Australasian Science Education Research Association (ASERA), University of the Sunshine Coast, Australia.

Smith, T. M., Desimone, L. M., Zeidner, T. L., Dunn, A. C., Bhatt, M., & Rumyantseva, N. L. (2007). Inquiry-oriented instruction in science: Who teaches that way? *Educational Evaluation and Policy Analysis*, *29*(3), 169–199. doi:10.3102/0162373707306025

Star, J. R., & Strickland, S. K. (2008). Learning to observe: Using video to improve preservice mathematics teachers' ability to notice. *Journal of Mathematics Teacher Education*, *11*(2), 107–125. doi:10.100710857-007-9063-7

Thompson, N., & Pascal, J. (2012). Developing critically reflective practice. *Reflective Practice*, *13*(2), 311–325. doi:10.1080/14623943.2012.657795

Tripp, T., & Rich, P. (2012). Using video to analyze one's own teaching. *British Journal of Educational Technology*, *43*(4), 678–704. doi:10.1111/j.1467-8535.2011.01234.x

Vail Lowery, N. (2002). Construction of teacher knowledge in context: Preparing elementary teachers to teach mathematics and science. *School Science and Mathematics*, *102*(2), 68–83. doi:10.1111/j.1949-8594.2002.tb17896.x

Veenman, M., Van Hout-Wolters, B., & Afflerbach, P. (2006). Metacognition and learning: Conceptial and methodological considerations. *Metacognition and Learning*, *1*(1), 3–14. doi:10.100711409-006-6893-0

Wells, G. (1999). *Dialogic inquiry: Toward a sociocultural practice and theory of education*. Cambridge, UK: Cambridge University Press. doi:10.1017/CBO9780511605895

Wenger, E. (2009). Communities of practice. *Communities*, *22*, 57.

Wood, D., Wood, H., & Middleton, D. (1978). An experimental evaluation of four face-to-face teaching strategies. *International Journal of Behavioral Development*, *1*(2), 131–147. doi:10.1177/016502547800100203

Zientek, L. R. (2007). Preparing high-quality teachers: Views from the classroom. *American Educational Research Journal*, *44*(4), 959–1001. doi:10.3102/0002831207308223

Chapter 9
Student Perceptions of Screencast Video Feedback for Summative Assessment Tasks in the Creative Arts

Josh McCarthy
University of South Australia, Australia

ABSTRACT

This chapter evaluates the use of screencast video feedback for summative assessment tasks in the creative arts and analyzes the advantages and disadvantages of such a format when compared to traditional feedback techniques. In 2017, in the second-year course Narrative Animation at the University of South Australia, video feedback was trialed for summative assessment tasks, in an attempt to improve students' understanding of their academic performance. Thirty-seven students participated in the course and received a five-minute feedback video for each of their three submissions. The video feedback provided to students during the course was evaluated at the end of the semester in the form of two online surveys, allowing participating students with the opportunity to critically reflect on the learning experience. The findings of the study disseminate the learning benefits afforded by the video feedback model and provide insight into the varying attitudes of both students and staff.

INTRODUCTION

This chapter evaluates the use of screencast video feedback for summative assessment tasks in the creative arts, and analyses the advantages and disadvantages of such a format when compared to traditional feedback techniques. More specifically, video feedback was trialed in the course *Narrative Animation* at the University of South Australia, in Australia, in order to improve students' understanding of their academic performance. Student evaluations of the two previous iterations of the course identified feedback as an area that could be improved in the future, specifically the provision of feedback for summative assessment tasks. As such, in 2017 alternate forms of feedback provision were researched and explored.

DOI: 10.4018/978-1-7998-0426-0.ch009

The second-year course, which sits in the Bachelor of Media Arts program, included three animated video summative assessment tasks, and it was hypothesized that video-based feedback would provide students with a clearer understanding of their academic performance within the course by providing more detailed and relevant commentary on their work. Within this hypothesis, the aim of the case study was to establish the advantages and disadvantages of a screencast video feedback model. To achieve this aim, a series of research questions were established:

- Does screencast video feedback enable better understanding of academic performance than other feedback methods, such as written, verbal or audio?
- Does screencast video feedback enable better understanding of academic performance within different student demographics?
- Are screencast video feedback files accessible: can students access and view video feedback easily?
- Are screencast video feedback files easy for staff to produce and distribute in a timely manner?
- What are the workload implications on course staff, for providing screencast video feedback?

BACKGROUND

Summative Assessment and Feedback in Higher Education

Summative assessment involves evaluating or examining a student's academic performance, and reviews their progress at a specific time within a course or program, through feedback and a grade (Harlen & James, 1997; Scriven, 1967; Taras, 2005). The focus of summative assessment is on the outcome of a task, such as an exam or assignment, and seeks to monitor educational outcomes (Knight, 2002; Shepard, 2005). Summative assessment is considered as *assessment of learning*, differing from formative assessment which can be defined as *assessment for learning* (Black, Harrison, Lee, Marshall, & Wiliam, 2004).

It is accepted in the research literature that assessment is vital to student learning in higher education, and that feedback is a core component of the assessment process in terms of evaluating and also elevating student academic performance (Biggs, 2003; Bloxham & Boyd, 2007; Gibbs & Simpson, 2004; Harris, Brown, & Harnett, 2014; Henderson & Phillips, 2015; Lunt & Curren, 2009; Orsmond & Merry, 2011). Moreover, high quality and timely feedback, both in formative and summative form, are crucial features for enhancing effective student learning and in developing strong relationships between staff and students (Crook, Mauchline, Maw, Lawson, Drinkwater, Lundqvist, Orsmond, Gomez & Park, 2012; Irons, 2008). Henderson and Philips (2015) point out however, that feedback is a broad term and that both its design and perceived benefits depend on its context, mode and delivery. High quality feedback can improve the learning experience, and has the capacity to motivate students, and facilitate stronger academic performance, whereas poor quality feedback can lead to a weaker learning experience and dishearten students.

Gould and Day (2013) note, the use of effective feedback by teachers provides the foundations for learner autonomy and a framework for high achievement. Effective feedback is essential as the scaffolding that enhances learning. Merry and Orsmond (2008) suggest that meaningful feedback enables students to achieve goals to a greater extent than they would without it, while Nicol and MacFarlane-Dick (2006) propose that feedback should relate to performance in terms of goals, criteria and expected standards, and should also be timely, detailed and specific.

Despite the research literature clearly emphasizing the importance of high quality and timely feedback, it has been recognized globally that there are significant pitfalls in terms of delivering such feedback, as well as engaging students in the assessment-feedback process (Bevan, Badge, Cann, Willmott, & Scott, 2008; Crook et al., 2012; Merry & Orsmond, 2008). For many students, feedback can often be provided in a manner which they feel is too late to be useful, too vague, unclear or inconsistent (Glover & Brown, 2006; Weaver, 2006). Feedback can often be misunderstood, particularly by international students - via illegible handwriting or language differences (Price, 2007; Walker, 2009) - or simply ignored (Mutch, 2003). While for staff, the provision of feedback can be a highly repetitive process (Henderson & Phillips, 2015) and is frequently time consuming, particularly with large class sizes, making the delivery of 'timely' feedback a hard task to achieve. To combat these potential pitfalls, researchers have identified many different design characteristics that should be utilized in order to improve the value of feedback for both formative and summative assessment (Boud & Molloy, 2013; Evans, 2013; Gibbs & Simpson, 2004; Mory, 2003). More specifically, Mory (2003) outlined four areas where feedback for summative assessment supports learning:

- Feedback can be considered as an incentive for increasing response rate and or accuracy.
- Feedback can be regarded as a reinforcer that automatically connects responses to prior stimuli (focused on correct responses).
- Feedback can be considered as information that learners can use to validate or change a previous response.
- Feedback can be regarded as the provision of scaffolds to help students construct internal schemata and analyze their learning processes.

While, Gibbs and Simpson (2004) outline six key drivers where feedback for summative assessment positively influences student performance:

- Feedback is sufficient in detail.
- Feedback is focussed on students' performance, on their learning, and on the actions under students' control, rather than on the students themselves.
- Feedback is timely.
- Feedback is appropriate to the aim of the assessment and its criteria.
- Feedback is appropriate in relation to students' conception of learning, of knowledge and of the discourse of the discipline.
- Feedback is attended to and acted upon in future assessment tasks.

Utilizing Different Feedback Models for Summative Assessment

Feedback is central to the learning experience, and providing written remarks on students' assignments is commonly viewed as a key feature of feedback processes for assessment in higher education (Nicol, 2010). Despite this, researchers have highlighters several difficulties students face in relation to understanding such feedback (Blayney & Freeman, 2004; Duncan 2007; Higgins, Hartley & Skelton, 2002; Merry &

Orsmond, 2008). Duncan (2007) notes that written feedback can often focus on mechanical aspects of the submission, such as spelling and grammar, rather than concentrating on the core of the work. Duncan also found that written comments could be vague; illegible if handwritten; inconsistent in quality and quantity across a range of tutors; and could focus on positive and encouraging comments at the expense of clear advice on how to improve the quality in subsequent work. Duncan's findings are consistent with numerous prior investigations that have analyzed students' perceptions of teacher comments (Crook et al., 2012; Lizzio & Wilson 2008; Orsmond, Merry, & Reiling 2005; Poulos & Mahony, 2008).

In response to this, several alternatives to traditional feedback mechanisms, for both summative and formative assessment, have been explored. The use of audio for feedback delivery has become popular in higher education over the past decade, and in several cases, has been found to be of significant value to students for a range of reasons (Emery & Atkinson, 2009; Francis, Shannon & Murison, 2008; Ice, Curtis, Phillips, & Wells, 2007; King, McGugan, & Bunyan, 2008; Merry & Orsmond, 2008; Nortcliffe & Middleton, 2008; Ribchester, France, & Wheeler, 2007; Rotherham, 2008). Merry and Orsmond (2008) found that students appreciated audio feedback because it was perceived as being of good quality; was easier to understand; had more depth; and was more personal than written feedback. Within the same study, associated staff found audio feedback particularly valuable to explain complex ideas, and by adjusting the volume or tone of their voice they could highlight specific points, and consequently more understanding could be gained from the spoken word than written text. Ribchester et al. (2007) noted that audio feedback facilitated increased discussions between staff and students as it was perceived to be more personal and involved. This idea is supported by Ice et al. (2007) who found students felt audio feedback was more supportive and caring than written feedback and reported a greater understanding of comments. As well, Middleton, Nortcliffe and Owens (2009) found that audio feedback was seen as conveying more than just words – tone, pronunciation and emphasis all added to the depth of the communication. Furthermore, Emery and Atkinson (2009) noted that one minute of audio feedback was equivalent to approximately 100 words and provided a more in-depth and detailed review of the students' work than written feedback. In their study, they found that audio feedback was also more likely to include suggestions for how to improve work, rather than just identifying the problems, because it was faster to narrate such issues rather than write them down.

Video feedback techniques, while not as common as written and audio formats, have also been successfully incorporated into teaching and learning in higher education in recent times (Abdous & Yoshimura, 2010; Abrahamson, 2010; Bracher, Collier, Ottewill, & Shephard, 2005; Ferguson, 2011; Guadagnoli, Henderson & Phillips, 2015; Holcomb & Davis, 2010; McCarthy, 2015; Richardson, Dann, Dann & O'Neill, 2018). Video feedback can be designed and utilized in different ways to suit the assessment task. For visual-based assessment tasks, a screencast video can enable the assessor to review the student's submission on screen, allowing for more directed feedback. Screencast video feedback can enable the assessor to demonstrate where a student has either achieved or fallen short of an assessment goal, by visually referring to specific elements within the student's submission. For text-based assessment tasks, video can be used to record the assessor, allowing for the provision of more personalized feedback. Like audio feedback, video has the capacity to convey the assessor's tone and emphasis, providing further clarity to their comments. Video feedback however moves beyond this, with the visual medium allowing the assessor to also include expression and gesture within their communication. Cann (2007) notes that videos have a broad acceptance amongst students, and can offer a richer format for feedback provision than audio or written techniques. As video is a visual medium. It has the potential to enhance learning in different ways to other technologies, including the potential for demonstrations (i.e. seeing as opposed to being told how to improve subsequent coursework (Abrahamson, 2010; McCarthy, 2015). Video feedback files also provide a permanent record, which can be stored online and replayed at the students' convenience, as opposed to handwritten feedback forms which can be lost or damaged.

MAIN FOCUS OF THE CHAPTER

Establishing a Feedback Model for Summative Assessment

The case study presented in this chapter was conducted in 2017 within the course *Narrative Animation* at the University of South Australia in Australia. The course trialed the use of screencast video feedback for summative assessment tasks and included 37 students. *Narrative Animation* is a second-year course which explores storytelling, design and performance within 3D computer animation. The course sits within a broad Media Arts degree, which explores film, animation, visual effects and digital media. The course profile reads:

Narrative Animation introduces students to an intermediate level of 3D modelling, lighting, texturing, rigging, animating and rendering concepts, techniques and processes. During the lecture series, students will enhance their understanding of the fundamental principles of animation specific to character design and performance. During the practical series, students will develop their design and performance skills in Maya, including character modelling, pelting, rigging, walk and run cycles, pose-to-pose and straight-ahead animation, and facial animation. The assessment tasks will provide students with the opportunity to combine these skills and techniques to develop a short story; and design, build and animate a character within a narrative.

The video feedback model utilized within the course for the three summative assessment tasks, was derived from the models generated by Gibbs and Simpson (2004) and Mory (2003), as well as key considerations from Evans (2013), and Boud and Molloy (2013). Whilst designing the screencast video feedback model for summative assessment in the course, it was established that, to successfully support student learning and outcomes, summative feedback was required to be timely, detailed, and most importantly, relevant to the assessment criteria. The key considerations for these principles are outlined in more detail in Table 1.

Within the course, there were three summative assessment tasks, in the form of assignments, situated at different stages during the semester. All of which required a video submission from the students. The first piece of summative assessment was submitted by students in the fourth week of the course, and the assessment process started with the assessor first looking at all of the submissions, and then assessing the students one by one. The method of assessing each student's work took approximately 25 to 30 minutes, and included viewing the student's submission several times (a 15-second video of a 3D character) and handwriting or typing notes and comments related to the submission in regards to the assessment criteria. When this had been completed, and a clear idea of the student's performance had been established, the assessor opened *Camtasia Studio®* and recorded narrated, visual feedback, which included critiquing the student's video in *VLC Player®*, for a period of between three and five minutes.

Table 1. Key principals established for providing screencast video feedback for summative assessment tasks within the course Narrative Animation.

Feedback Principle	Key Considerations
Timely	Feedback should be provided to students within a short timeframe, ideally within ten days of the submission date;
Detailed	Feedback should take advantage of the video format by highlighting specific elements within the students' submissions for discussion and evaluation; Feedback should take advantage of the video format by providing examples of how the student could improve their work in future tasks, including external examples from websites such as *YouTube* and *Vimeo*; Feedback should effectively utilize narration by emphasizing tone, expression and pronunciation to provide further clarity to the students regarding their performance;
Relevant	Feedback should be focused on the student's academic performance within the task; Feedback should be direct and address the specific assessment criteria of the task at hand; Feedback should be constructive, and provided as an incentive to the student to improve in future tasks, and to reinforce their design concepts and ideas; Feedback should be appropriate in terms of the student's conceptions of learning and knowledge;

Once the feedback had been recorded, it was then reviewed; edited or re-recorded if necessary; exported as an .mp4 file; and uploaded to the course assessment website. Once all submissions had been graded and uploaded, the videos were then returned to students. The second and third summative assessment tasks, due in weeks eight and thirteen respectively, followed a similar process.

The Student Experience

The student experience within the course was evaluated online, in the form of two anonymous surveys at the conclusion of the semester. The first survey established a demographic breakdown of the cohort, and allowed students to consider and compare the video feedback they received during the semester, with other forms of feedback they had received in previous courses. This included evaluating the perceived advantages and disadvantages of the video feedback; determining whether it was more beneficial for their learning; considering issues such as accessibility, comprehension, engagement, and knowledge transfer; and indicating which type of feedback they'd like to receive in future courses. 31 students from the cohort participated in the survey, resulting in a response rate of 84%. Participants were given the opportunity to assess the video feedback in the form of Likert-scale statements and open-ended questions, enabling the provision of both quantitative and qualitative data. Within the cohort, 60% of students were male, 40% were female; 78% were local students, and 22% were international students. Being a second-year course, there were no school leavers in the cohort - 78% were aged between 19 and 24, and 22% were aged 25 and over. The cohort was made up of 81% Media Arts students and 19% from Computing and Multimedia. Student demographic breakdowns from the participating cohort are shown in Table 2.

The survey queried students about their experiences with feedback for summative assessment tasks in previous courses at university, and asked them to evaluate the effectiveness of each type of feedback they had received. Written feedback, either in comment (84%) or assessment criteria matrix/rubric (74%) form was the most common type of feedback received. Other types of feedback received in previous courses included video feedback (26%); individual verbal feedback (26%); audio feedback (19%); and group verbal feedback (19%). When asked, which type of prior feedback provided the most insight into their academic performance, the responses were varied due to the range of different experiences within

Table 2. Student demographics within the participating cohort. The survey yielded a response rate of 84%

Demographic	Number of Students in the Course	Percentage of Total Cohort	Number of Respondents	Percentage of Respondents Within Each Demographic
Number of respondents	37	100	31	84
Gender				
Male	22	60	17	77
Female	15	40	14	93
Student Type				
Local Student	29	78	25	86
International Student	8	22	6	75
Age				
19-24	29	78	26	90
25-34	4	12	3	75
35+	3	10	2	67
Program Type				
Media Arts	30	81	25	83
Computing and Multimedia	7	19	6	86

the cohort, however seven of the eight students who had received video feedback in the past identified it as the most effective. Students' experiences with previous types of feedback received for summative assessment tasks, and their perceived effectiveness, are outlined in Table 3.

Having established the students' previous experiences with feedback for summative assessment at University, the survey then focused on their attitudes towards the provided feedback within *Narrative Animation*. Students were asked to assess the accessibility of the video feedback they received during the

Table 3. Students' experience with feedback for summative assessment tasks in previous courses at University. Students listed as many types of feedback as they had received.

Feedback Type	Percentage of Students to Experience Feedback	Perceived as Most Effective (All Respondents)	Perceived as Most Effective (Those Who Experienced the Feedback)
Written (comments)	84 (25 respondents)	32 (10)	40 (10)
Written (assessment criteria matrix/rubric)	74 (23)	39 (12)	52 (12)
Video	26 (8)	23 (7)	88 (7)
Individual verbal	26 (8)	3 (1)	13 (1)
Audio	19 (6)	3 (1)	17 (1)
	19 (6)	0 (0)	0 (0)

course, as well as assess the perceived quality of the feedback. Ninety percent of students indicated they found the video feedback files easy to access and view, with 7% undecided and 3% finding it difficult. One student indicated that they couldn't access the feedback via their mobile device – "couldn't view video through mobile devices so needed computers" – however several others noted that they could – "I easily could access them from my phone, which was handy."

All students (100%) indicated that the video feedback was easy to comprehend and provided insight into their academic performance within the assessment tasks – "it's almost like personal interaction"; "video feedback is great, the feedback is easy to understand and link to the project"; "the video feedback was very in-depth, and was clear on what sections needed improvement." Student responses regarding the accessibility and perceived effectiveness of the video feedback are outlined in Table 4.

When asked to discuss the benefits of the video feedback format, students identified more in-depth analysis of their performance, and a more personalized approach as the key advantages:

- *You can see exactly what the teacher is commenting on at that moment. There is no deciphering what they mean in a comment.*
- *Visual is far more direct and effective.*
- *The benefits were that it was much more descriptive than other methods and also provided visual cues in regards to what section of the assignment was being spoken about. This level of feedback helped me know where I needed to improve for future assessments.*
- *Actually, feeling like the lecturer read my assignment instead of just issuing a mark without comment.*
- *More engaging and personal compared to written feedback. Also helps with discussing certain visual elements, such as a certain point in a video or animation.*
- *It was very helpful. Even though the videos weren't long, they allowed me to follow along with the marking process and actively see what's wrong or good with my work.*

Students identified some limitations with the video feedback format, which included having to watch the entire video before receiving their grade for the assessment task, as well as some accessibility issues:

- *Extremely nerve racking when finding out if you passed or not. Also slightly disheartening when you know what you submitted wasn't up to par.*
- *Not able to quickly view your grade and have to download and watch the whole video.*
- *There is no written record.*
- *Can't watch the video if no internet is available when outside.*

Table 4. Mean response and broad agreement data related to Likert-scale statements in the survey. The survey used a 5-point Likert scale from 1 (strongly disagree), to 3 (undecided), to 5 (strongly agree).

Topic	Mean Response	Percentage Broad Agreement
The video feedback files were easy to understand and provided insight into my academic performance within the assessment tasks.	4.77	100
The video feedback files were easy to access and view.	4.52	90

When asked to discuss potential improvements to the video feedback, students were predominantly happy with the existing format, although some reiterated that they would like to receive a grade *prior* to the provision of feedback:

- *I know it may stop some students watching their feedback, but could we please get a mark or some sort of indication of what mark we will be getting before the video?*
- *A quick view of the grade outside of the video would be great.*

While another student noted a written component, such as an assessment criteria matrix or rubric, used in conjunction with the video would be an improvement:

- *Maybe certain marking criteria could be placed side by side with video feedback to make the marking process more understandable.*

At the conclusion of the survey, students were asked how the video feedback received in *Narrative Animation* compared to their previous experiences in other courses. 100% of respondents indicated the video feedback was an improvement, with 44% responding it was better, and 56% responding it was much better (using a five-point Likert scale) – "the video feedback is much better for my personal development, and a clear understanding of why I received the grades that I did." When asked whether they would like to receive video feedback for summative assessment tasks in future courses, 94% of students responded positively, while 6% remained undecided identifying the assessment itself should inform the feedback format:

- *It depends on the course. It works for this one because of the submitted video content, but if an essay was marked the same way then it wouldn't be useful at all.*

Further insight into the learning experience was provided by the standard, university-distributed course evaluation survey, which was open to students during the final three weeks of the course. This survey yielded a response rate of 61%, and included topics relating to the quality of the delivered course. 100% of respondents indicated they had a clear idea of what was expected of them during the course, and 94% indicated they received feedback that was constructive and helpful, a considerable improvement from the previous two iterations of the course in 2015 and 2016, when written feedback was employed for all summative assessment tasks. 95% of respondents indicated they were satisfied with the quality of the course, often referring to the video feedback as a key indicator – "the video feedback for the assignments was also extremely helpful." Student responses to the post-semester course evaluations from 2015 to 2017 are outlined in Table 5.

Table 5. Mean response and broad agreement data from the end of semester, university-distributed course evaluation. The survey used a 5-point Likert scale from 1 (strongly disagree), to 3 (undecided), to 5 (strongly agree). The 2017 survey yielded a response rate of 61%.

Topic	2015 (Response Rate = 35%)		2016 (Response Rate = 34%)		2017 (Response Rate = 61%)	
	% Broad Agreement	Mean Response	% Broad Agreement	Mean Response	% Broad Agreement	Mean Response
Overall, I was satisfied with the quality of the course.	85	4.15	77	4.00	95	4.53
I have a clear idea of what is expected of me in this course.	85	4.31	100	4.46	100	4.68
The assessment items assisted my learning in this course.	85	4.23	85	4.23	100	4.58
I have received feedback that is constructive and helpful.	77	4.00	85	4.00	94	4.56

SOLUTIONS AND RECOMMENDATIONS

The two surveys conducted at the conclusion of the semester provided much insight into the student learning experience, and the perceived effectiveness of the screencast video feedback format for summative assessment tasks. The video feedback was well received by all students, and there were no significant variances of attitudes within the different demographics in the student cohort. The principle focus of this case study was to improve the quality of feedback provided to students for summative assessment tasks within the course *Narrative Animation*. The surveys both indicated that this was achieved, and that the method of feedback delivery not only improved students' understanding of their academic performance, but also had a positive impact on their overall learning experience. Students consistently referred to the screencast video feedback as more engaging, personal, dynamic and effective than written feedback, and every student identified it as being the best form of feedback they had received at university.

There were some key issues to arise with the video feedback that also need to be addressed for future iterations of the model, from both student and staff perspectives. The majority of students found the video feedback files easy to access and view, however at least one student was unable to view them on their phone. Providing feedback in a timely manner was identified as one of the key attributes of high quality feedback at the start of this study, and the student's inability to quickly view the feedback via their phone was a concern. That said, several students identified that they *were* able to use their phone to both view and store their feedback files, thus making the problem harder to qualify. To tackle such accessibility issues, it is important to ensure that video feedback files are relatively small in size so they can be downloaded and stored easily, but also of a common file type. The videos in this case study were exported as mp4 files, using the H264 codec, recognized as being one the most common video formats available.

Another issue raised by a small group of students was the lack of a visible grade prior to viewing the video feedback file, and that watching the video itself could be 'nerve racking' without first knowing their mark. Providing a grade at the start of the video, and then following that up with more detailed commentary related to the assessment criteria may be a good approach to take in future iterations. Allowing students to see their grade without even opening the feedback file may results in some students not bothering to view the feedback file at all. Lastly, some students identified that, while the video format allowed more in-depth analysis of their work, it didn't provide them with a *'written record'* of their performance. Providing a written assessment criteria matrix or rubric, in conjunction with the video feedback file, would alleviate this concern, and provide further clarity regarding the student's performance against each key criterion.

From a teacher's perspective, there were significant advantages and disadvantages to arise from the case study, regarding both pedagogy and workload. Course staff found the provision of feedback in the 2017 iteration of the course, was more time consuming than previous years, but also more rewarding. The process of creating the video files was more involved than the previously used text-based feedback forms, including loading the student's work on screen, recording the feedback, editing or re-recording the video if necessary, and then exporting the video in .mp4 format. On average, it took between 25 and 30 minutes to provide a three to five-minute video feedback to each student. The video files also took significantly longer to upload to the course's learning management system and distribute to students, due to their file size – the feedback files ranged in size from five to fifteen megabytes. Varying time frames associated with delivering different feedback models clearly has an impact on a teacher's capacity to incorporate them into a curriculum. With a small class size, it is much easier to experiment and innovate with different feedback techniques, and spend more time with formats such as video.

If the class size is large however, there may not be the opportunity to utilize such techniques, given the desire, as well as the expectation, to provide *timely* feedback. While there was an increased timeframe to delivering the video feedback in this course, it also promoted within the teachers, a stronger sense of engagement with, and enthusiasm towards marking. A summary of the advantages and disadvantages of the screencast video feedback model utilized within *Narrative Animation* is provided in Table 6.

Table 6. A summary of the advantages and disadvantages of the screencast video feedback model trialed in Narrative Animation.

Advantages	Disadvantages
• Students find the video feedback more engaging and dynamic. • -Students may perceive the video feedback as more personal. • Vocal tone, pronunciation and emphasis can be utilized within the narrated feedback to improve students' understanding of feedback. • Students find the video feedback more detailed and that it allows for greater comprehension of their academic performance. • Teachers can be very specific about certain aspects of the student's work by combining narration and visual cues. • Teachers can easily incorporate external elements, such as videos from *YouTube*® or *Vimeo*®, into the feedback. • Teachers may become more engaged in the marking process. • Students have a digital record of their academic performance within an assessment task, which they can download and store. • Provided the video feedback file is correctly encoded when rendered, it is viewable on many devices, including desktop computers, laptops, tablets and phones.	• Video feedback files can be comparatively large, when compared to word of pdf documents. • The process of recording and editing the feedback file can increase staff workload. • Due to the larger file size, the video feedback file can be slower to upload and distribute to students. • Like any other digitally supplied feedback format, video feedback files require computer access and (initial) internet access to download and view. • Students do not have a written record of their academic performance within an assessment task. • If the video feedback file is not optimally exported (file type; video codec), it may not be accessible to students on mobile devices. • While some suitable screen recording software packages are free, such as *CamStudio*®, more advanced programs can cost up to $179 for a license, such as *Camtasia Studio*®.

The process of narrating feedback while viewing the students' work on screen, enabled the provision of far more detailed commentary focused on core concepts and issues, which can often be hard to explain easily, or at least quickly, in text-based form. Instead of being limited to a handful of short comments and an assessment criteria matrix, the teachers were able to engage with the student's work and provide more expansive and directed feedback. Furthermore, it was easy to incorporate external elements into the feedback provision – a good character design or an animation precedent from *YouTube®* or *Vimeo®* could be quickly integrated into the feedback to provide further clarity. While ultimately more time consuming, it proved to be much easier to simultaneously narrate and visualize a problem and resolution related to a submission in video format, rather than writing extensive comments and providing an assessment criteria rubric. The process of discussing the students' work while looking at it on screen, often pausing at certain points and rewinding and replaying their submission, led to a more engaging experience for the staff, who felt they were providing valued advice to the students, rather than simply trying to justify a grade.

FUTURE RESEARCH DIRECTIONS AND CONCLUSION

Feedback is an essential component to both formative and summative assessment in higher education, and academics will continue to explore new models of feedback delivery, in order to improve the student learning experience. The screencast video feedback trialed in this small-scale case study, proved to be successful in enhancing students' understanding of their performance within the assessment tasks. While the feedback model was successful in providing further clarity on academic performance, and popular with students, the case study also illustrated that it is not the answer to all summative assessment tasks. Students identified a desire for teachers to match the feedback model to the purpose and context of the assessment tasks at hand; in essence the assessment tasks themselves should guide the feedback format that is utilized. When adopting a feedback model, it is important for educators to take into consideration several factors, including the field of study; assessment type (formative or summative); assessment format (visual, aural, written); the class size; the student type (age; local or international; visual / hearing impaired); and available staff and student resources (software; hardware; internet access). For the students participating in this study, the screencast video feedback was considered more beneficial than other formats because it provided more in-depth analysis of their academic performance in assignments, which themselves were video-based. The feedback model matched the format of the assessment. For assignments or other assessment tasks which are text-based, such as essays, text or audio feedback models may be much more appropriate. However, video feedback that captures the assessor discussing a student's text-based submission, and utilizes tone, emphasis, expression and gesture would still remain highly suitable.

The screencast video feedback in *Narrative Animation* was also appropriate in this instance because students had access to the internet, through a range of devices, and could collect their feedback files quickly and easily. This is not the case with many student cohorts in developing countries around the world, where internet access, or even access to a computer, may be limited. That said, when time and resources allow for it, video feedback has proven to be a highly valued format by students working in visual-based fields of study. The positive responses from both staff and students towards the screencast video feedback utilized in *Narrative Animation*, encourages more exploration into the field. Future research will delve into refining the video feedback format to further strengthen students' understanding of their academic performance; and streamlining the assessment process to improve teachers' workloads.

REFERENCES

Abdous, M., & Yoshimura, M. (2010). Learner outcomes and satisfaction: A comparison of live video-streamed instruction, satellite broadcast instruction, and face-to-face instruction. *Computers & Education, 55*(2), 733–741. doi:10.1016/j.compedu.2010.03.006

Abrahamson, E. (2010). *Assessment through video-feedback on an undergraduate sports rehabilitation programme. Higher Education Academy [HEA] Case Study.* Retrieved from http://www.heacademy.ac.uk/assets/hlst/documents/case_studies/147_abrahamson_video-feedback.pdf

Bevan, R., Badge, J., Cann, A., Willmott, C., & Scott, J. (2008). Seeing eye-to-eye? Staff and student views on feedback. *Bioscience Education, 12*(1), 1–15. doi:10.3108/beej.12.1

Biggs, J. B. (2003). *Teaching for quality learning at university: What the student does.* Maidenhead, UK: Society for Research into Higher Education & Open University Press.

Black, P., Harrison, C., Lee, C., Marshall, B., & Wiliam, D. (2004). Working Inside the Black Box: Assessment for Learning in the Classroom. *Phi Delta Kappan, 86*(1), 9–21. doi:10.1177/003172170408600105

Blayney, P., & Freeman, M. (2004). Automated formative feedback and summative assessment using individualised spreadsheet assignments. *Australasian Journal of Educational Technology, 20*(2), 209–231. doi:10.14742/ajet.1360

Bloxham, S., & Boyd, P. (2007). *Developing effective assessment in higher education: A practical guide.* Maidenhead, UK: Open University Press.

Boud, D., & Molloy, E. (2013). *Feedback in higher and professional education.* New York: Routledge.

Bracher, M., Collier, R., Ottewill, R., & Shephard, K. (2005). Accessing and engaging with video streams for educational purposes: Experiences, issues and concerns. *ALT-J. Research in Learning Technology, 13*(2), 139–150. doi:10.3402/rlt.v13i2.10990

Cann, A. (2007). Podcasting is dead. Long live video! *Bioscience Education, 10*(1), 1–4. doi:10.3108/beej.10.c1

Crook, A., Mauchline, A., Maw, S., Lawson, C., Drinkwater, R., Lundqvist, K., ... Park, J. (2012). The use of video technology for providing feedback to students: Can it enhance the feedback experience for staff and students? *Computers & Education, 58*(1), 386–396. doi:10.1016/j.compedu.2011.08.025

Duncan, N. (2007). 'Feedforward': Improving students' use of tutors' comments. *Assessment & Evaluation in Higher Education, 32*(3), 271–283. doi:10.1080/02602930600896498

Emery, R., & Atkinson, A. (2009). *Group assessment feedback: 'The good, the bad and the ugly'.* Paper presented at the 1st Annual Meeting of a Word in Your Ear, Sheffield, UK.

Evans, C. (2013). Making sense of assessment feedback in higher education. *Review of Educational Research, 83*(1), 70–120. doi:10.3102/0034654312474350

Francis, R., Shannon, S., & Murison, S. (2008). *Delivery of online digital feedback and assessment for design and engineering (architectural) students.* Paper presented at the 46th Annual Conference of the Architectural Science Association, Griffith University. Retrieved from http://anzasca.net/wp-content/uploads/2014/08/p201.pdf

Gibbs, G., & Simpson, C. (2004). Conditions under which assessment supports learning. *Learning and Teaching in Higher Education, 1*(1), 3–31.

Glover, C., & Brown, E. (2006). Written feedback for students: Too much, too detailed or too incomprehensible to be effective? *Bioscience Education, 7*(3), 1–15. doi:10.3108/beej.2006.07000004

Gould, J., & Day, P. (2013). Hearing you loud and clear: Student perspectives of audio feedback in higher education. *Assessment & Evaluation in Higher Education, 38*(5), 554–566. doi:10.1080/02602938.2012.660131

Harlen, W., & James, M. (1997). Assessment and learning: Differences and relationships between formative and summative assessment. *Assessment in Education: Principles, Policy & Practice, 4*(3), 365–379. doi:10.1080/0969594970040304

Harris, L., Brown, G., & Harnett, J. A. (2014). Understanding classroom feedback practices: A study of New Zealand student experiences, perceptions, and emotional responses. *Educational Assessment, Evaluation and Accountability, 26*(2), 107–133. doi:10.100711092-013-9187-5

Henderson, M., & Phillips, M. (2015). Video-based feedback on student assessment: Scarily personal. *Australasian Journal of Educational Technology, 31*(1), 51–66. doi:10.14742/ajet.1878

Higgins, R., Hartley, P., & Skelton, A. (2002). The conscientious consumer: Reconsidering the role of assessment feedback in student learning. *Studies in Higher Education, 27*(1), 53–64. doi:10.1080/03075070120099368

Ice, P., Curtis, R., Phillips, P., & Wells, J. (2007). Using asynchronous audio feedback to enhance teaching presence and students' sense of community. *Journal of Asynchronous Learning Networks, 11*(2), 3–25.

Irons, A. (2008). *Enhancing learning through formative assessment and feedback. Key guides for effective teaching in higher education.* Abingdon, UK: Routledge.

King, D., McGugan, S., & Bunyan, N. (2008). Does it make a difference? Replacing text with audio feedback. *Practice and Evidence of Scholarship of Teaching and Learning in Higher Education, 3*(2), 145–163.

Knight, P. (2002). Summative assessment in higher education: Practices in disarray. *Studies in Higher Education, 27*(3), 275–286. doi:10.1080/03075070220000662

Lizzio, A., & Wilson, K. (2008). Feedback as assessment: Students' perceptions of quality and effectiveness. *Assessment & Evaluation in Higher Education*, *33*(3), 263–275. doi:10.1080/02602930701292548

Lunt, T., & Curran, J. (2010). 'Are you listening please?' The advantages of electronic audio feedback compared to written feedback. *Assessment & Evaluation in Higher Education*, *35*(7), 759–769. doi:10.1080/02602930902977772

McCarthy, J. (2015). Evaluating written, audio and video feedback in higher education summative assessment tasks. *Issues in Educational Research*, *25*(2), 153–169.

Middleton, A., Nortcliffe, A., & Owens, R. (2009). *iGather: Learners as responsible audio collectors of tutor, peer and self-reflection.* Paper presented at the A Word in User Ear – Audio Feedback Conference. Retrieved from http://research.shu.ac.uk/lti/awordinyourear2009/docs/Middleton-Nortcliffe-Owens-iGather_final.pdf

Mory, E. H. (2003). Feedback research revisited. In D. H. Jonassen (Ed.), *Handbook of research for educational communications and technology* (pp. 745–783). New York: Macmillan.

Mutch, A. (2003). Exploring the practice of feedback to students. *Active Learning in Higher Education*, *4*(1), 24–38. doi:10.1177/1469787403004001003

Narciss, S. (2008). Feedback strategies for interactive learning tasks. In J. M. Spector, M. D. Merrill, J. Van Merrienboer, & M. P. Driscoll (Eds.), *Handbook of research on educational communications and technology* (3rd ed.; pp. 125–143). New York: Erlbaum.

Nicol, D. (2010). From monologue to dialogue: Improving written feedback processes in mass higher education. *Assessment & Evaluation in Higher Education*, *35*(5), 501–517. doi:10.1080/02602931003786559

Nicol, D., & Macfarlane-Dick, D. (2006). Formative assessment and self-regulated learning: A model and seven principles of good feedback practice. *Studies in Higher Education*, *31*(2), 199–218. doi:10.1080/03075070600572090

Nortcliffe, A., & Middleton, A. (2008). A three year case study of using audio to blend the engineer's learning environment. *Engineering Education*, *3*(2), 45–57. doi:10.11120/ened.2008.03020045

Orsmond, P., & Merry, S. (2008). Students' attitudes to and usage of academic feedback provided by audio files. *Bioscience Education*, *11*(3). doi:10.3108/beej.11.3

Orsmond, P., & Merry, S. (2011). Feedback alignment: Effective and ineffective links between tutors' and students' understanding of coursework feedback. *Assessment & Evaluation in Higher Education*, *36*(2), 125–136. doi:10.1080/02602930903201651

Orsmond, P., Merry, S., & Reiling, K. (2005). Biology students' utilisation of tutors' formative feedback: A qualitative interview study. *Assessment & Evaluation in Higher Education*, *30*(4), 369–386. doi:10.1080/02602930500099177

Poulos, A., & Mahony, M. (2008). Effectiveness of feedback: The students' perspective. *Assessment & Evaluation in Higher Education*, *33*(2), 143–154. doi:10.1080/02602930601127869

Price, M. (2007). Should we be giving less written feedback? Centre for Bioscience Bulletin, 22(9).

Ribchester, C., France, D., & Wheeler, A. (2008). Podcasting: A tool for enhancing assessment feedback? In E. O'Doherty (Eds.), *The Fourth Education in a Changing Environment Conference Book 2007* (pp. 119-136). Santa Rosa, CA: Informing Science Press.

Richardson, T., Dann, B., Dann, C., & O'Neill, S. (2018). *Formative assessment practices for pre-service teacher practicum feedback: Emerging research and opportunities*. Hershey, PA: IGI Global. doi:10.4018/978-1-5225-2630-8

Rotheram, B. (2009). *Sounds good.* JISC Final Report. Retrieved from. http://www.jisc.ac.uk/publications/reports/2009/soundsgoodfinalreport.aspx

Scriven, M. (1967). The methodology of evaluation. In R. Tyler, R. Gagne, & M. Scriven (Eds.), Perspectives on curriculum evaluation (pp. 39-83). Chicago: Rand McNally.

Taras, M. (2005). Assessment – summative and formative – some theoretical reflections. *British Journal of Educational Studies*, *55*(4), 466–478. doi:10.1111/j.1467-8527.2005.00307.x

Walker, M. (2009). An investigation into written comments on assignments: Do students find them usable? *Assessment & Evaluation in Higher Education*, *34*(1), 67–78. doi:10.1080/02602930801895752

Weaver, M. (2006). Do students value feedback? Student perceptions of tutors' written responses. *Assessment & Evaluation in Higher Education*, *31*(3), 379–394. doi:10.1080/02602930500353061

Chapter 10
Mobile Devices Contribute to Feedback Processes

Beverly Dann
University of the Sunshine Coast, Australia

ABSTRACT

The use of mobile devices such as mobile phones and tablets in education is a problematic field of research that fits within the scope of assessment, mobile technologies, dialogic practices, and more broadly, feedback. This small pilot study investigated how supervising teachers incorporated a mobile device in the form of a video-enabled app into practicums to promote feedback in the form of dialogue and record achievements in alignment with requisite criteria. It further investigated the role of the app in the dialogic feedback process and the interactions between supervising teachers and preservice teachers when they undertake practical performance reviews. The findings showed that using mobile devices aids the dialogic practices of preservice teachers and leads to better outcomes. Despite the evidence, systemic organizational intent will be needed to reinforce the benefits and encourage adoption.

INTRODUCTION

This chapter sits within complex, interconnected, yet discrete fields of knowledge that embrace concepts of higher education assessment (Bennett, Dawson, Bearman, Molloy, & Boud, 2017; Boud & Falchikov, 2005, 2006; Denton & McIlroy, 2017) and include formative assessment models (Nicol & Macfarlane-Dick, 2006; Rushton, 2005) for self-regulated learning (Carver, 2017), feedback (Dowden, Pittaway, Yost, & McCarthy, 2013; Molloy & Boud, 2014), mentoring (Hudson, 2010; Mena, García, Clarke, & Barkatsas, 2015) and the duality of practice-based learning supervision (Tillema, Smith, & Leshem, 2011). The researchers sought alignment between these areas of knowledge and the emerging capabilities of mobile devices such as iPads, Tablets, and mobile smart phones as part of the solution for pedagogical assessment issues in university teaching programs. These types of mobile devices had heretofore not been used in higher education assessment, so this study materialized as an exploration of their impact on feedback processes in the context of assessments, further elaborated in the following sections.

DOI: 10.4018/978-1-7998-0426-0.ch010

BACKGROUND

Practitioners in the field of education at a small regional university conceptualized a systematic, iterative approach to assessment feedback (Bennett, Dawson, Bearman, Molloy, & Boud, 2017) with school-based staff and preservice students as partners. They concluded that mobile devices could be part of the solution to the feedback and assessment issues being faced at the time (Dann & Allen, 2015).

Universities have recently been investigating the use of varied assessment and feedback practices (Carless & Boud, 2018) to improve teaching and learning. Unfortunately, the role of technology in higher education assessments has yet to be defined (Nicol & Milligan, 2006). This coincides with a need for enhanced clarity and transparency with regard to communicating assessment feedback (Juwah, Macfarlane-Dick, Matthew, Nicol, Ross, & Smith, 2004), and for feedback in initial teacher education to more closely reflect the learning intentions of school placements. At the time, the disparate nature of the feedback (Knight, 2002) being provided by supervising teachers has signalled a lack of reliability and validity in assessments (Dann & O'Neill, 2018; Stellmack, Konheim-Kalkstein, Manor, Massey, & Schmitz, 2009). These circumstances have led to the creation of a video-enabled application, subsequently developed into a Tablet and iPad application (App) for teacher candidates at a regional Australian university (Willis, Dann, Jones, Lowe, & Toohey, 2012). In a later study (Dann & Allen, 2015) found that supervising teachers' feedback in placement was confined to areas of their personal interest, so with this in mind, the App was designed to extend that approach by integrating clear and transparent university criteria for measuring performance in placements. Thereafter, Dann and Dann's (2018) research focus shifted to the processes and strategies employed in formative assessments, based on noticing (Cowie, Harrison, & Willis, 2018; Sherin, 2017) during practicum experiences, and later, tutorial presentations. Noticing became an important aspect of this research because noticing refers to the heightened awareness and aptitude of supervising teachers to observe, in order for their actions and reactions to effectively respond to their students' thinking, interests and needs (Cowie et al., 2018). Thus, the impetus for learning can be revealed and nurtured.

Akkerman and Meijer (2011) and Carless (2016) argued that noticing in the classroom should be supported by engagement in dialogic practices between supervising and preservice teachers. The dialogic process refers to conversation between the parties to attain understanding through dialogue with another, while at the same time, advancing personal thinking and reasoning. It goes beyond superficial interactions, and allows teachers and preservice teachers to challenge existing practices and test their capabilities in a safe and supportive environment (Cochran-Smith & Lytle, 2009; Cochran-Smith & Villegas, 2014; Parker, Patton, & O'Sullivan, 2016). Deglau, Ward, O'Sullivan, and Bush (2006) argued that dialogue is essential for teachers to understand and challenge policies, including for example, the national teaching standards in Australia, against which preservice teachers are assessed and that supervising teachers are required to interpret.

LITERATURE REVIEW

Assessment in teacher education has increasingly been influenced by rapidly evolving mobile technologies, allowing for video collection and accessibility to real-time viewing on mobile devices in the form of mobile phones, tablets and ipads. Ally and Wark (2018) urged educators to consider the impact of mobile devices because they tend to exert technological influences. Empirical evidence suggests that students favor the use of mobile devices in feedback processes (O-K & Hong-Fa, 2014), yet there has been limited research on their impact, and how, if at all, they improve learning outcomes and student satisfaction.

Video feedback in teacher education is a valuable tool for assessing practical performance as a demonstration of theoretical knowledge. Efforts to bridge the gap between theory and practice have generated significant empirical research and knowledge of constructivist, student-centered approaches that empower students in higher education programs. Boud and Molloy (2013) labelled feedback by "telling" as Mark 1 feedback, describing it as one-sided and ineffective. On the other hand, they defined Mark 2 feedback as sustainable and effective processes for empowering students and placing them in the center of the learning process. Four out of the seven Mark 2 processes specify: (a) learners' participation in activities that promote self-regulation; (b) developing dispositions in learners to seek feedback; (c) providing opportunities for students to produce work; and (d) designing assessments with nested tasks that allow for feed forward (Boud & Molloy, 2013, p. 707).

For mobile devices to positively impact on student feedback and align with Mark 2 (Boud & Molloy, 2013) characteristics needed consideration of issues such as: professional learning opportunities available to students, mediation and coaching processes, the ability of students and coaches to notice when observing student performance, students' ensuing reflective practices, and the role of assessment tools and feedback.

Research on professional learning and development in education has historically focused on practicing teachers rather than preservice teachers; but in the past decade, Masats and Dooly (2011) used video case studies for training, in their efforts to guide preservice teachers' professional development and improve their digital competencies. Darling-Hammond (2010) proposed broadening the intentional use of targeted feedback within teacher education programs, and sought systemic changes to ensure preservice teachers were given opportunities to observe quality teaching in action. She understood the importance of watching pedagogical practices to deepen understanding of the complicated nuances of teaching (Darling-Hammond, 2010; Hammerness, Darling-Hammond, Bransford, Berliner, Cochran-Smith, McDonald, & Zeichner, 2005). The development of teacher-mentor skills is vital as they tend to follow an ad hoc response that would include what they perceive as important to focus on rather than allowing the practicum criteria to lead the shaping of learning opportunities and feedback to the practicing preservice teacher (Dann & Allen, 2015; Dann & Dann, 2018).

Coaching

Coaching is a continuous process embedded in the ethos and context of a school, and an effective way for teachers to help one another and induct new teachers into the profession. Studies have shown that coaching contributes to learning new strategies and using them more effectively, for demonstrating transference of skills to other contexts, and improving the ability of students to reflect on practice and build collegial relationships (Anderson, Feldman, & Minstrell, 2014). Coaching can also be a valuable practice for assisting and supporting preservice teachers within a new school, and as this study showed, has the potential to positively influence the learning of preservice teachers and the support provided by supervising teachers during practicums.

Noticing

Noticing begins with knowledge and close observation of what is being viewed. Jacobs, Lamb, and Philipp (2010) identified three interrelated skills for noticing children's thinking in mathematics as: attending to, interpreting and responding, all of which rely on prior knowledge and experience combined with professional knowledge for making judgements about what is being observed. Hatch, Shuttleworth, Jaffee, and Marri (2016) found that preservice-teacher transformation resulted from discussions with an experienced partner when attempting to bridge the gap in knowledge between "abstract bodies of professional knowledge and the 'craft knowledge' of practitioners (p. 274)". They also found scaffolding a necessary component for encouraging preservice teachers to challenge their conceptions. The authors used videos of experienced high-school teachers to deepen preservice teachers' understanding of key pedagogical strategies in social studies, and then apply what they had learned in a student teaching seminar. Hatch et al. (2016) noted that effective preservice teacher transformation was borne out of dialogue with another person when scaffolding was included. Dann & Allen (2015) also highlighted the evidence of scaffolding during the use of mobile video.

Scaffolding tools may include checklists, examples, or even criteria sheets. According to the work of Grainger, Heck, and Carey (2018), students appreciate having a tool that demonstrates assessment criteria in a specific manner. In addition, students were found to value high-quality exemplars for assessment tasks as useful tools for improving their understanding of the expectations. However, the authors did not investigate the use or suitability of video as one such tool for demonstrating assessment criteria in a more specific manner.

Reflective Practice

Reflection is a common practice in education and encouraged by external Australian accrediting bodies (Simon, 2013). Reflection after viewing a video of one's own or others' performance is a more recent phenomenon. Blomberg, Sherin, Renkl, Glogger, and Seidel (2013) investigated 67 teachers' responses after analyzing their own teaching videos, and found they noticed more about certain components of teaching and learning and were "less self-reflective regarding critical incidents". Kleinknecht and Schneider (2013) also studied the effect on teachers after viewing and analyzing videos of themselves and other teachers' classroom instruction, and found that teachers were "more deeply engaged in analysis of problematic events" when viewing others. The authors concluded that personalizing the video was a powerful tool, requiring prearrangements and scaffolding in order to bring about changed and transformed practice. Dann's (2014) use of an App to introduce scaffolding for preservice and supervising teachers was purposely aimed at improving dialogue and teaching practice in a more personalized manner.

Baecher, Kung, Jewkes, and Rosalia (2013) adopted a different approach to assisting preservice teachers develop reflective skills. They placed volunteers into two groups: Group 1 was shown video models and evaluation rubrics prior to their practicum, while Group 2 received descriptions of desired lessons and evaluation rubrics. The preservice teachers in Group 1 achieved higher scores in self-evaluation and understanding of the expectations, indicating that video played a significant role in the learning outcomes. The authors deduced this was the result of focusing on reflective practices that included peer discussion, professor-led discussions, case studies and multi-media (Christ, Arya, & Chiu, 2017). They also identified infrequent and varied use amongst tutors and attributed this to high teaching loads and whether or not individual discipline areas and colleagues supported the use of video in their classes.

THE FEEDBACK MODEL

Dann and Dann's (2018) feedback/feed forward model (Figure 1) emerged from a constructivist action research study, and illustrates the feedback process between supervising and preservice teachers on school placements. The diagram shows the dialogic pattern which gave preservice teachers greater understanding of their own pedagogical practice and subject knowledge, resulting in a reduction in their knowledge gap.

The findings of this study showed a combination of four feedback components: (a) professional learning; (b) mediation/coaching; (c) noticing; and (d) reflective practice boosted the ability of students to bridge the gaps in their knowledge and had a sustained impact on their learning. The use of video to increase students' abilities to self-assess was examined in a study by Baecher et al. (2013), and showed

Figure 1. Feedback model for supervising and preservice teachers

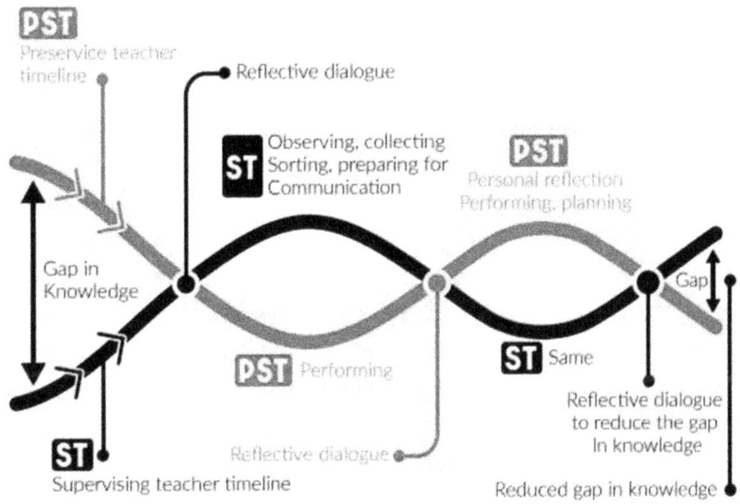

higher self-evaluation scores for students who were shown video as part of the feedback process (Baecher et al., 2013; Dann & Dann, 2018).

In Dann and Dann's (2018) study, feedback referred to dialogue, in written, verbal and recorded form, between preservice and supervising teachers during the learning process. The dialogue had all the characteristics of feed forward, i.e. positive, constructive, engaging interactions, that on an ongoing basis, make a developmental impact on learning. The model was created after observing preservice and supervising teachers' interactions during school placements, using a video-enabled App on an ipad as part of the feedback process (Dann & Dann, 2018). On the left-hand side of Figure 1, the two prominent snaking lines represent preservice teachers' knowledge of a particular subject (e.g., behavior management, teaching strategy, early reading, content knowledge and so on) at the start of their placement. As the lines move to the right they cross one another, representing points of discussion and reflection between preservice and supervising teachers. In the diagram the two lines that snake through the diagram move apart, indicating planning and teaching by the preservice teacher while the supervising teacher observes and takes notes and videos. As this cycle is repeated, preservice teachers improved their knowledge and practice and the gap began to close. The cycle of reflective dialogue and planning in Figure 1 shows feed forward in action, and demonstrates the impact of dialogue and its significance for the development of preservice teachers in practicum situations (O'Neill & Dann, 2018).

Guides or scaffolding tools, like the video-enabled App used in Dann and Dann's (2018) study, serve to steady the focus on specific aspects or assessment criteria, thereby facilitating growth in particular areas. The constructed guides of Brouwer, Besselink, and Oosterheert (2017) sit at the Reflective dialogue intersection in Figure 1, where dialogue takes place between the preservice teacher and another, more experienced individual. The preservice teacher and student dialogue sits inside the preservice teacher's (PST) performing space.

According to Brouwer et al. (2017), assessments that include peer discussion and viewing a video can be used to raise domain-specific behaviors. Constructed guides help to develop competence by eliciting contributions from learners in the form of dialogue, in turn, improving their performance. Kleinknecht and Schneider (2013), having found they created a much deeper analysis of events, viewed guides as essential and referred to them as prearranged scaffolding that assists the viewing process. In combination with video, therefore, guides can contribute significantly to the development of preservice teachers as effective beginners in the field.

This study took into account previous research on feedback in higher education, and expanded the literature with evidence of video and mobile devices such as mobile phones, tablets and ipads as having a positive impact on feedback with a particular focus on formative dialogue in professional practice, communication and the craft of teaching. The varied and inconsistent approaches to methods of feedback and formative data continue to cause disparity in summative judgements and require further attention in the teacher education space. Both preservice teachers and mentors are affected by the complexity of summative judgements (Klenowski, 2013), and video feedback has been shown to mitigate inconsistent outcomes. Without a clear understanding of the criteria and learning outcomes, neither supervising teacher nor preservice teacher can respond adequately in their roles. This examination of the impact of video via mobile devices on areas such as active teaching, communication, body language and other pedagogical nuances, provides valuable insights.

School placements are designed to meet the specific content and skills requirements of government registration bodies. These skills are sequenced over a four-year university education program that allows for practice and development over time. Students are assessed on specific components that gradually increase in difficulty and demand, culminating in a final placement. This requires supervising teachers to align their assessment of preservice teachers with university expectations – the video-enabled App developed by Dann and Allen (2015) was used in this small pilot study to enable easy alignment with university criteria and allowed preservice teachers to participate in a unique way.

This research answered a call by Bennett et al. (2017) to rethink feedback so that it becomes part of the learning process, emerging over time as a "designed sequence of development" (Boud & Molloy, 2013, p. 710). The guiding objectives of the study were to allow for deeper and sequenced development of teaching skills by exploring dialogic practice using a video-enabled App and assessment criteria during a school placement. The initial inquiry was followed by a larger action research study into how preservice and supervising teachers incorporated the App into their daily classroom activities, and their attempts to meet the assessment requirements of a one-month school placement. A structured and systematic formative feedback system formed a critical component of the App, while the video feedback promoted positive dialogue and ultimately enhanced the performance of preservice teachers. The structured aspect of the App refers to the school placement criteria that are embedded in the App along with the ability to add comments, goal setting, and timelines. It also has the ability to rate the PST on a scale of 1- 4 against each criterion. As video or photos are taken, they can be dropped into a section beside the chosen criteria as evidence of performance. Videos also form a valuable component for dialogue between the PST and the ST.

METHODS

The study used a pragmatic theoretical perspective, concerned with making connections between theory and practice, with a strong focus on understanding the "real world". This approach suggests knowledge only has meaning when applied to real-world problems, such as the practicum issues addressed in this study. A pragmatic approach also accommodates both qualitative and quantitative methods (Feilzer, 2010; Plano Clark & Creswell, 2008), however, the results of this study will report on the qualitative data.

This research took the form of a pilot study to identify how supervising teachers and preservice teachers would employ the use of a video-enabled App to assist in streamlining the practicum experience with the expected learning outcomes and criteria. The pairs of preservice and supervising teachers who participated in the study decided for themselves how best to use the system in their particular contexts after being supplied with the App, training, and an assurance of ongoing support. The investigation into the impact of the App on the practicum experience was underpinned by the research question: *How do preservice teachers and supervising teachers incorporate the use of mobile technologies for practicums and how does this impact on pedagogy and assessment of the professional experiences of pre-service teachers?*

Study Context

The study took place in Queensland, Australia, in one regional independent school with both primary and high school sections. The school had 1314 students and partnered with the university in school placements by regularly supervising preservice teachers. Highly regarded in the community and well known for turning out Year 12 graduates with outstanding results, the school had an Index of Community Socio-Educational Advantage (ICSEA) value of 1152, while the average for Australia was 1000, indicating the majority of students were from well-educated families (ACARA).

Participants

Participants included two experienced teachers and two preservice teachers in their third year of a four-year Bachelor of Education degree. Supervising teacher 1 (ST 1) had 15 years' experience and supervising teacher 2 (ST 2) had 21 years' teaching experience. Both had been teaching year 2's for over five years at the time of the study and participated as supervising teachers for the past 10 years.

Data Collection

Data collection took place over four weeks. Data collection tools used in this mixed methods action research included semi-structured interviews and post surveys. In addition, volunteer participants undertook analysis of the data captured by the system during use. Using personal devices meant that students' feedback data was self-selected and personally driven, thereby containing more value and purpose.

Central themes were derived only from the interviews and system data since the survey was ineligible due to the small number of participants. Table 1 shows a timeline of the study.

The two researchers participated as university academic and coordinator for the project and therefore had first-hand knowledge of the participants' experiences. The researchers' extensive teaching backgrounds (50 years combined) also contributed to interpretation of the data and the experiences of participants.

Table 1. Timeline of the study

Month	Activity
August	• School visit and introduction of study. • Researchers meet the mentors at the school. • Researchers provide supervising teachers and preservice teachers with a research information sheet. • Schedule next meeting.
September	• Visit school for pre-interviews with supervising teachers about their supervising practice. • Demonstrate the App system to supervising teachers and allow them to play with it for the next month to get used to it.
October	• Students were introduced to the features of the App so they could participate in its use during placements and support their supervising teacher if needed.
November	• Post interviews and surveys

Analysis

Key to the analysis was ongoing correlation and triangulation of the data. Data included interviews with supervising teachers and preservice teachers, surveys for all participants and the App system analytics. The system analytics could confirm, usage times, frequency, length of time in use, sections of comments, amount of video, types of comments and more.

NVivo was used for coding and theming, as well as initial independent manual coding (Kemmis, McTaggart, & Nixon, 2014) of the interview transcripts. Transcript themes broadly aligned with the research question and comparisons were constantly made to provide benchmarks and discussion points for subsequent theming and understanding of the data. Both researchers checked for alignment to clarify themes and interpretations. Unfortunately, due to low survey responses during end of year activities, these were not incorporated into the results.

The rich, qualitative data were captured chronologically as digital evidence of the system's use, and provided insightful storylines about the school context and the nature of the relationships between the preservice and supervising teachers. The videos afforded a window into the events, attitudes, conversations, perceptions, changes in performance, and the ways in which the pairs decided to incorporate the digital device into their feedback processes.

FINDINGS

The principal aim of this study was to investigate how supervising and preservice teachers incorporated a video-enabled App into their practicums to assist with dialogue and measuring of achievements against a clear set of 28 criteria. ST1 was eager to use the mobile device and App and did so several times a week in a variety of ways. ST2 needed more prescription around the times and frequency of using the system, and her lack of confidence with ICTs caused her to use it infrequently. In contrast to the regular use of ICT in ST1's class, it was nowhere near as predominant in ST2's class; "regular" being defined as using ICT for teaching and/or learning twice or more times per week. This may have influenced the varied responses of supervising teachers. It should also be noted that the data collected from the App comprised only the feedback of ST1, who used the system regularly. ST2's App data were not analyzed because she only used it three times. When asked why she used the App so infrequently, ST2 replied: "I think it is a great idea but I'm not very comfortable with Apps and computers. There is so much to do in the classroom anyway, and with a university student to assess, I just don't see how I can fit it all in".

Both preservice teacher (PST) participants were interested in using the mobile device in placements. Initially, however, as they were getting to know the school environment, they preferred not to rock the boat, instead, going with the flow of the classroom and their supervising teachers. For this reason, PST2 did not encourage her mentor's use of the mobile device until she herself was more comfortable with her ST and placement. Study results included close observation of the system's use in the classroom and showed evidence of feedback, feed forward and dialogue, with the mobile device enabling accurate interpretation of its impact on preservice teachers' performance.

Using a Mobile Device in the Classroom

The device's mobility also appeared to influence use of the system, since it was able to be used in the classroom to collect data and allowed access to the data elsewhere and at other times. The data were primarily in the form of comments (n = 36), photographs (n = 11) and videos (n = 4). System analytics revealed particular times of the day when the mobile device was used most, with supervising teachers preferring to collect data between 10:00 am and 12:00 noon, frequently in key subject areas. They added their own comments during the lunch break and shared their views about the evidence of teacher candidates' performance on the App. Interview questions indicated there were more disruptions in the afternoons, and consequently mornings were preferred for focusing on teaching and learning activities with preservice teachers. Unlike the supervising teachers, the preservice teachers accessed the comments, videos and other evidence at home, demonstrating a preference for accessing the data when planning for the following day, often between 8:00 pm and midnight.

As the pairs gradually familiarized with the App, they began to plan specific lessons and aspects of lessons they wished to video, commencing a focused intent on the evidence and learning. PST1 said: "When I was having some management issues, [ST] would suggest that she video a particular lesson with the kids and then we could look at the behavior of the kids and how I went with it", illustrating a shared approach to using the system in the classroom. The PSTs also requested specific areas of performance be videoed or recorded as evidence in some way: "I knew I have been having trouble with questioning. I wanted to see what kinds of questions I asked and see if I was doing factual and open-ended [questions]" (PST2). Supervisors could add video, photos, comments, goals and photo video of their feedback. These data are presented below, and show the limited use of video and the introduction of goals in dialogic practices respectively.

Figure 2 shows the frequency of video collection over the practicum. One video was collected in each of weeks 1 and 2, while two videos were collected in week 3, featuring different criteria. Figure 3 shows the frequency of use of photographs. It shows an increased use of photographs up to week 3

Figure 2. Video frequency

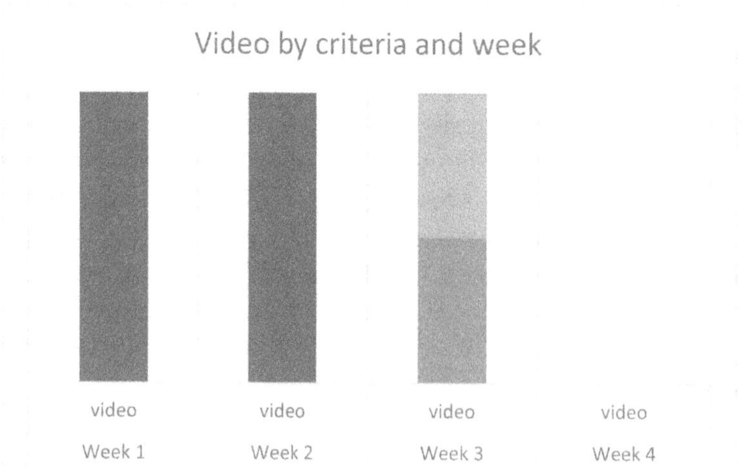

Figure 3. Photo frequency

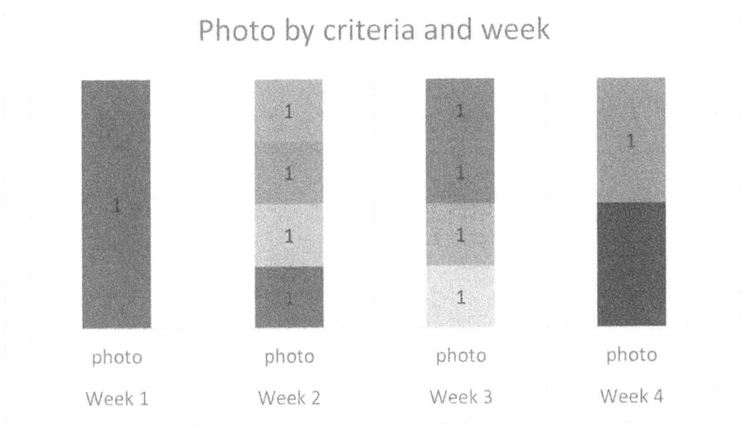

and a reduction in the final week of the practicum. Figure 4 shows the frequency of goals set during the practicum, illustrating a limited number, with only two goals set over the four weeks. Figure 5 depicts the increase in text comments and comments on criteria over the four weeks of the practicum.

Figure 4 illustrates the limited number of goals set during the practicum, with only two goals over the four weeks.

Feedback, Feed Forward and Dialogue

The video-enabled App was capable of storing data on usage, feedback and recordings. An analysis of the App's use identified random and sporadic use in the first week. This was anticipated while participants familiarized themselves and tested the system. From week two onwards, it became evident that the mobile App was being used to assist STs with evidence of alignment between their comments and

Figure 4. Goal frequency

Figure 5. Comment frequency

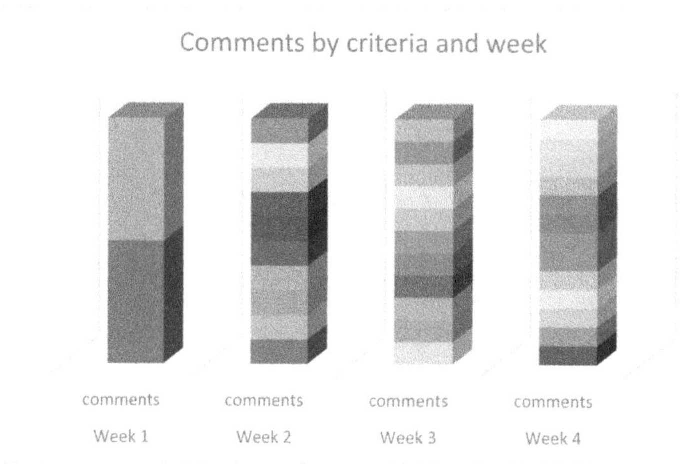

the assessment criteria. They were actively using the system to extend their feedback on the relevant learning expectations (see Table 2) that were accessible anywhere, anytime. This was highlighted by the meeting between PST1 and ST1 during their lunch break to view the video and engage in discussion on classroom management. Together, they entered the following criterion into the App: Reflects on observations of teaching practice across a range of learning areas to identify the strategies teachers use for fostering inclusive learning environments. Then they went on to discuss the teaching experience, taking note of areas of strengths and weaknesses, including a future strategy for the PST, namely, Think, Pair, Share. By incorporating a clear, shared goal and action plan into the feedback, the PST was able to focus her teaching on improving her practice. ST2 commented that the App provided "clear evidence and proof of practice", nevertheless choosing not to incorporate it into her practice.

In addition to the video examples of performance aligned with criteria, the comments aligned with criteria kept participants on track to reference them rather than random ideas. The PSTs were clearly able to identify and view, in real time, the assessment areas that needed to be addressed. One supervising teacher reported that it was not uncommon to focus on personal preferences: "The things that I really look for and value and I think I am really interested in and lead to longer term, better quality of practice are teaching and learning, assessment, management and professional learning" (ST1). Unfortunately, exercising this left other areas unassessed: "I could see that we hadn't done anything in the 6th and 7th standards (in reference to the National Teacher Standards), (PST1). So I could say something to my ST about that. Sometimes I find it hard to do reflections on standards that have nothing [no comments] with them" (PST1).

The mobile App also appeared to enhance the preservice teachers' reflective capacity during their school placements: "on evaluating what we actually taught and link it to the standards and how you are going to embed what you've learnt into your teaching" (PST2). It should be noted that reflection was a requirement, whether the App was used or not. Nevertheless, the reflective practice of PST1 was profoundly impacted by the feedback via the mobile/web technology by providing support through detailed criteria aligned and video to assist with reflecting. The ability to set goals, strategies and timelines for individual assessment criteria during the practice-based learning period, supported by visuals of performance, had changed her professional view of herself and her growth: "I've really found with this prac I've been able to critically reflect on that constructive criticism and put it into place next time I teach, which has been really useful. My ST looks at my lesson and then puts notes in the App under criteria. She also tells me stuff one-on-one, so if I say I need to work on this, then she would say 'I agree and I would extend by doing this'.

Furthermore, the App had a positive impact on the dialogue between mentor and student. For example, PST2 commented that the university used a paper-based approach which required her to reflect on certain topics in a booklet, such as adjustments for learners, while on placement. The supervising teacher was required to read it, provide comments and sign it off as satisfactory or unsatisfactory. The PST claimed the process was slow if feedback wasn't received the same day, adding "that is too late" for planning and improving the following lesson. On the other hand, PST1 favored the App because it "contained more space for comments" aligned to the assessment, "so I knew how I was going with the different criteria for the prac" (PST1).

The mobile system added a different dimension to the dialogue in practicums:

I am used to having short conversations throughout the day and longer conversations before and after school, and also written notes on my lesson plans from my mentor [supervising teacher], but using the App has been so different. I can go home and remind myself about our conversations because it is recorded in the App. Sometimes at night, I am so tired that I forget what my mentor [ST] said, but now it is there in the App. I really like that (PST1).

PST2 described the method of feedback from her ST as "one-way" and "not necessarily aligned to any of the assessment criteria". She added: "the school day is very busy and some days there is little time to have a good conversation with the [ST]". PST2 clearly did not recognize the App's ability to provide her with a real-life record of events for lesson plans at her convenience.

Impact on Preservice Teacher Performance

The ability of teacher candidates and supervising teachers to align their comments with video and photographic evidence served to contextualize the feedback against a framework of specific learning outcomes: "You can keep learning while you are not in the classroom environment. You don't often get to see yourself teach, so it was really great to have that recorded and notice things that you might do that you didn't know about, or things children were doing" (PST 1).

The use of video in placements was a new experience for all participants, but they readily accepted the opportunity once they realized the benefits, particularly since the video was only seen by the ST and PST and did not form part of a university assessment: "I think watching yourself teach in action and critiquing yourself, I find is really an important cycle to go through" (PST 2).

Contrary to the PSTs' positive opinions of the video evidence, one ST's personal feelings about being videoed deterred her from using the video component: "I would not want to be videoed, so when I videoed, I focused on short specific things like questioning, management, or a strategy, and not the whole lesson" (ST1). An added voice-to-text feature in the App reduced the time STs needed to spend at their keyboards. ST1 immediately recognized its value and added mastery of the system to her repertoire of skills, claiming: “it'll give a much richer perspective of where they are” (ST1). This kind of robust feedback can be viewed as feed forward, because it is timely and encourages improvement in students' practice. The preservice teachers explained that the App focused discussion on learning expectations and the actual data collected, and with the inclusion of video-enabled feedback and feed forward, had the potential to change practice in the classroom. The combination of video, photographs and comments, aligned with assessment criteria in the App, was able to concentrate the discussions between STs and PSTs on areas of relevance and successfully close the gaps in knowledge.

Sherin (2017) was of the view that for dialogue to lead to growth, supervising teachers must be able to articulate what they have “noticed” during a PST's performance and contextualize this within the professional expectations of teachers, as specified by the assessment criteria. The supervising teachers in this study provided written comments on lesson plans and had verbal discussions, as well as typing comments into the App:

Well, really it would have to be quite targeted with what you're videoing. So, it was more about when she's engaging children in that either shared experience, or guided kind of work, or cooperative tasks to take videos of what the kids were doing, how she was managing groups or giving some formative feedback, or things like that (ST1).

After taking the video I [ST1] would drop the video in [to the criteria] and then make a comment as well, so to sort of direct the learning, and I really felt that the video and the photographs, together with the comments gave her a good portfolio of [evidence]. The video became very powerful in that sense. I tend to be a fairly positive person so I'm looking for the good things, and I will offer some suggestions. [Like] one way to improve might be to... [And] here's evidence of what I've said here..., and this is what I've seen you do, and I think it's good practice. [So], here's the standard. This is what I'm supposed to be looking for. This is what I've seen and here's the evidence of it (ST1).

Table 2 provides examples of written comments alongside specific assessment criteria and video or photographic.

The findings confirmed that the supervising and preservice teachers who were inclined to use ICT were also in favor of using mobile devices for student assessments in school placements. Through trial and error, they came up with effective ways of incorporating the mobile device into their daily class-

Table 2. Alignment of teacher comments with assessment criteria

Assessment Criteria	Sample Written Comments
University reporting criteria	Plan and implement lessons and teaching sequences that show knowledge and understanding of curriculum content and focus on the core concepts to be taught.
Supervising teacher comments	Listening today, consolidating understanding around calendars was clear. Don't forget to remind students throughout the lesson what they are aiming to learn – you were repeating this yourself often, which is a great step and shows that you have considered our previous conversations and goals from last week. But you don't have to do all the talking. Remember that he who is talking is doing the learning, so for even greater impact and to be sure your students know what they are learning, have THEM articulate it – either to you or to one another. It was great that you kept making it explicit that what they were doing was helping them to achieve the learning intentions.

room routines. The feedback captured by the mobile device allowed for more finely-grained analyses of performance than is possible with paper-and-pen approaches. Key to the findings of this study was that mobile technologies particularly impact on the summative judgments of teacher candidates in a systematic, criteria-based testing environment that was focused on graduate standards, and where supervising teachers knew the standards expected by the university. Using the App also heightened participants' aptitude for noticing, in particular:

- Progression of learning and performance against the learning expectations.
- Personal areas successfully achieved.
- Personal areas not achieved.
- Areas that were overlooked and not addressed.

Supervising teachers reported these data were beneficial for completing final practicum reports: "I would like to see a combination of them all working together [written, photo and video] to reflect on [my teaching] in the best possible way" (ST 1).

DISCUSSION

This research provides insights into the role of a video-enabled App in assessment processes for practice-based learning by taking the first steps towards answering the question: How do preservice teachers and supervising teachers incorporate the use of mobile technologies for practicums, and how does this impact on pedagogy and assessment of the professional experiences of pre-service teachers? The study confirmed that video evidence of performance, captured on a video-enabled App and aligned with explicit criteria for preservice teachers as stipulated by accreditation authorities, heightened the awareness of PSTs and led to their improved performance.

It was essential to find an efficient way of capturing the performance data and feedback in the context of learning expectations in a busy classroom environment, so it was left to the supervising and preservice teachers to establish an efficient routine that suited them, rather than imposing one on them. Schools are notoriously busy and teachers have little time for planning and conferring with colleagues. It was therefore necessary for teachers to be allowed to incorporate the digital system into their classroom routines in their own ways, acknowledging that personal comfort with technology and classroom routines, to some extent, would dictate how they incorporated it. ST1 was more confident with ICT, and as a result, more flexible in her use of the system in the classroom. On the other hand, ST2 was lacking in confidence with ICT and needed more explicit information, frequently asking "what to do with it and when to use the system". This translated into her not using the system as much as ST1 and requiring more assistance from her preservice teacher. However, all the preservice and supervising teachers unanimously agreed that the system had benefited them and enhanced their roles by clarifying the learning objectives for the practicums. While the small number of participants in this case limited the research findings, it has opened up possibilities for further investigation with broader participant samples against any number of profiles to provide multimodal feedback linked to learning criteria.

Classroom observations of ST1 and PST1 demonstrated more interaction with the system and increased dialogue about expectations than observed previously. This was likely due to participants becoming more familiar with the technology and developing a positive attitude towards incorporating the system. Both ST and PST were eager to get involved and approach the practicum experience in a new way; in all likelihood this provided the fuel for these synergies and relational trust to further motivate their efforts and use of the system (Anderson et al., 2014).

It was evident that the supervising teachers had to adjust their usual routines while working with the preservice teachers and the video-enabled App in their classrooms. Despite their previous experience and past work with preservice teachers, they had established routines and habits for working with PSTs, built upon their personal belief systems (Weasmer & Woods, 2003). It is therefore possible that the use of digital devices to record evidence may have been at odds with their personal beliefs and professional habits. This was evident in the case of ST2, who appeared to have greater difficulty adapting to using the video-enabled App due to her self-confessed lack of confidence with ICT, and resultant unwillingness to engage.

In the study upon which this case was predicated, student teachers' interest in their performance against the learning expectations increased, and they began to drive more extensive use of the system with their supervising teachers in the classroom. This was also the case in the current study, for example, the PST pointing out areas on the App where the ST had made no comments or not provided any evidence. As they spent more time using the App, participants identified their own individual benefits and found effective ways of making the system work for them. This included making decisions about when to use video, which activities to record and/or comment on, and using the App discreetly in the classroom to avoid disruptions. In fact, this was easier than expected, since ICT had become a regular part of student learning and they had become familiar with tablets and laptops in the classroom.

Interestingly, the use of video also enhanced the ability of the preservice teachers to identify their own weaknesses once they realized the benefits of viewing their teaching. They especially liked the feature that allowed supervising teachers to add dialogue to the videos after recording, and the convenience of viewing them at home while preparing for the following teaching session. The student teachers reported that viewing the video allowed them to see aspects of themselves they had not noticed "in the moment" of teaching. This new mode of viewing the dialogue had enabled them to consider their planning and performance from the feed forward, outside of school hours, with the added bonus of repetition. The analytics confirmed the App information was primarily used in the evenings by the STs, and between 8:00 am and 12 noon by the PSTs to view the comments and videos. This was also the time that preservice teachers typically prepared for their teaching lessons.

The research confirmed that mobile technologies impact on the way practice-based learning assessments are undertaken, in particular, mobile devices positively influence both the reflective practices of teacher candidates and the formative feedback provided to them in the form of an itemized analysis of their performance aligned with appropriate assessment criteria.

The ubiquitous nature of mobile technologies has created valuable opportunities to enhance formative assessment feedback for teacher candidates in practice-based learning scenarios. Specifically, the ability to review visual recordings and commentary on their performance when their cognitive abilities are at a peak is clearly of benefit. In addition, the App's mobility gives easy access to performance data in learning environments where candidates are already familiar with mobile devices.

This research led to the conclusion that video feedback, with capabilities to incorporate comments, goals and images, does in fact have a positive impact on the learning of teacher candidates and the feedback provided by supervising teachers. The initial objective of the researcher was to reduce the variety and inconsistency in the data used to assess preservice teachers on school placements. It was subsequently determined that the data could be used more efficiently as feedback, feed forward and dialogue for improving learning, and the App was specifically designed to improve the sequence of learning, development (Boud & Molloy, 2013) and assessment. The systematic, criteria-based feedback process, using mobile-enhanced pedagogical approaches, was shown to improve practice-based learning assessments through the use of iPhone and tablet technology. It did so partly by focusing the feedback of supervising teachers on relevant performance criteria, rather than personal interpretations and preferences.

Implications

Due to the limitations of this case, the research has narrow implications for management and assessment practices of teacher candidates in practice-based learning. However, it does bear the promise of influencing areas outside of education that include field placements where students are required to meet specific standards of performance to qualify for certification. Furthermore, the move towards objective evidence of performance against criteria in the workplace persists, and the mobile device can be used as a powerful tool for guiding dialogue towards improvement, rather than emphasizing deficiencies. These implications can inform future directions in research.

Further examination is needed to understand the sustained impact of using video, and identify the pedagogical practices that can be used as a model for academics in order to systematize feedback and assessment processes linked to practical demonstrations. It would also be of interest to quantify the reduction in the knowledge gap achieved by PSTs through the use of video, as opposed to written and verbal feedback.

CONCLUSION

This paper set out to demonstrate how supervising and preservice teachers collaborate to incorporate a video-enabled App into classroom practice during a practicum experience. The findings showed that mobile devices with video collection capability can enhance the practical learning of higher education students, particularly when the supervising teacher has ICT experience and is interested in a new approach that offers evidence to enhance formative feedback for student teachers. More specifically, the use of the video-enabled App with assessment criteria embedded afforded opportunities for focused discussion and alignment to expectations that encouraged deeper engagement in dialogue and most importantly developed teaching and learning skills. However, lasting impact will require organizational commitment and systemic structure will have to form part of any attempt to achieve success for students and institutions.

ACKNOWLEDGMENT

The author would like to acknowledge the support of the University of Sunshine Coast through financial allocations from the school and internal grants from the university's C SALT department.

REFERENCES

Akkerman, S. F., & Meijer, P. C. (2011). A dialogical approach to conceptualizing teacher identity. *Teaching and Teacher Education*, *27*(2), 308–319. doi:10.1016/j.tate.2010.08.013

Ally, M., & Wark, N. (2018). *Online Student Use of Mobile Devices for Learning*. Chicago, IL: Concordia University Chicago. https://www.learntechlib.org/p/184917

Anderson, R., Feldman, S., & Minstrell, J. (2014). Understanding relationship: Maximizing the effects of science coaching. *Education Policy Analysis Archives*, *22*, 54.

Baecher, L., Kung, S.-C., Jewkes, A. M., & Rosalia, C. (2013). The role of video for self-evaluation in early field experiences. *Teaching and Teacher Education*, *36*, 189–197. doi:10.1016/j.tate.2013.08.001

Bennett, S., Dawson, P., Bearman, M., Molloy, E., & Boud, D. (2017). How technology shapes assessment design: Findings from a study of university teachers. *British Journal of Educational Technology*, *48*(2), 672–682. doi:10.1111/bjet.12439

Blomberg, G., Sherin, M. G., Renkl, A., Glogger, I., & Seidel, T. (2013). Understanding video as a tool for teacher education: Investigating instructional strategies to promote reflection. *Instructional Science*, *42*(3), 443–463. doi:10.100711251-013-9281-6

Boud, D., & Falchikov, N. (2005). Redesigning assessment for learning beyond higher education. *Research and Development in Higher Education*, *28*, 34–41.

Boud, D., & Falchikov, N. (2006). Aligning assessment with long-term learning. *Assessment & Evaluation in Higher Education*, *31*(4), 399–413. doi:10.1080/02602930600679050

Boud, D., & Molloy, E. (2013). Rethinking models of feedback for learning: The challenge of design. *Assessment & Evaluation in Higher Education*, *38*(6), 698–712. doi:10.1080/02602938.2012.691462

Brouwer, N., Besselink, E., & Oosterheert, I. (2017). The power of video feedback with structured viewing guides. *Teaching and Teacher Education*, *66*, 60–73. doi:10.1016/j.tate.2017.03.013

Carless, D. (2016). Feedback as dialogue. Encyclopedia of Educational Philosophy and Theory, 1-6. doi:10.1007/978-981-287-532-7_389-1

Carless, D., & Boud, D. (2018). The development of student feedback literacy: Enabling uptake of feedback. *Assessment & Evaluation in Higher Education*, *43*(8), 1315–1325. doi:10.1080/02602938.2018.1463354

Carver, M. (2017). Feedback, feedforward, or dialogue?: Defining a model for self-regulated learning. In E. Cano & G. Ion (Eds.), *Innovative practices for higher education assessment and measurement* (pp. 1–18). Hershey, PA: IGI Global. doi:10.4018/978-1-5225-0531-0.ch001

Christ, T., Arya, P., & Chiu, M. M. (2017). Video use in teacher education: An international survey of practices. *Teaching and Teacher Education, 63*, 22–35. doi:10.1016/j.tate.2016.12.005

Cochran-Smith, M., & Lytle, S. L. (2009). *Inquiry as stance: Practitioner research for the next generation*. New York: Teachers College Press.

Cochran-Smith, M., & Villegas, A. M. (2014). Framing Teacher Preparation Research. *Journal of Teacher Education, 66*(1), 7–20. doi:10.1177/0022487114549072

Cowie, B., Harrison, C., & Willis, J. (2018). Supporting teacher responsiveness in assessment for learning through disciplined noticing. *Curriculum Journal, 29*(4), 1–15. doi:10.1080/09585176.2018.1481442

Dann, C. (2014). Mobile video feedback: When 'seeing' informs reflection and learning. *International Journal for Cross-Disciplinary Subjects in Education, 5*(3), 1727–1734. doi:10.20533/ijcdse.2042.6364.2014.0241

Dann, C., & Allen, B. (2015). Mobile video collection in preservice teacher practicum placements. *Journal of Technology and Teacher Education, 23*(1), 5–27.

Dann, C., & Dann, B. (2018). Exemplifying formative assessment practices in the field through case study: concepts, issues, and challenges. In T. Richardson, B. Dann, C. Dann & S. O'Neill (Eds.), Formative assessment practices for pre-service teacher practicum feedback: Emerging research and opportunities (pp. 76-112). Hershey, PA: IGI Global. doi:10.4018/978-1-5225-2630-8.ch003

Dann, C., & O'Neill, S. (2018). Positioning preservice teacher formative assessment in the literature. In T. Richardson, B. Dann, C. Dann, & S. O'Neill (Eds.), *Formative assessment practices for pre-service teacher practicum feedback: Emerging research and opportunities* (pp. 1–48). Hershey, PA: IGI Global. doi:10.4018/978-1-5225-2630-8.ch001

Darling-Hammond, L. (2010). *Evaluating teacher effectiveness: How teacher performance assessments can measure and improve teaching*. Center for American Progress.

Deglau, D., Ward, P., O'Sullivan, M., & Bush, K. (2006). Chapter 5: Professional dialogue as professional development. *Journal of Teaching in Physical Education, 25*(4), 413–427. doi:10.1123/jtpe.25.4.413

Denton, P., & McIlroy, D. (2017). Response of students to statement bank feedback: The impact of assessment literacy on performances in summative tasks. *Assessment & Evaluation in Higher Education*, 1–10.

Dowden, T., Pittaway, S., Yost, H., & McCarthy, R. (2013). Students' perceptions of written feedback in teacher education: Ideally feedback is a continuing two-way communication that encourages progress. *Assessment & Evaluation in Higher Education, 38*(3), 349–362. doi:10.1080/02602938.2011.632676

Feilzer, Y. (2010). Doing mixed methods research pragmatically: Implications for the rediscovery of pragmatism as a research paradigm. *Journal of Mixed Methods Research, 4*(1), 6–16. doi:10.1177/1558689809349691

Grainger, P. R., Heck, D., & Carey, M. D. (2018). Are assessment exemplars perceived to support self-regulated learning in teacher education? *Frontiers in Education, 3*, 60. doi:10.3389/feduc.2018.00060

Hammerness, K., Darling-Hammond, L., Bransford, J., Berliner, D., Cochran-Smith, M., McDonald, M., & Zeichner, K. (2005). How teachers learn and develop. In *Teachers for a Changing World: What Teachers Should Learn and Be Able to Do* (p. 1). New York: John Wiley & Sons.

Hatch, T., Shuttleworth, J., Jaffee, A. T., & Marri, A. (2016). Videos, pairs, and peers: What connects theory and practice in teacher education? *Teaching and Teacher Education, 59*, 274–284. doi:10.1016/j.tate.2016.04.011

Hudson, P. B. (2010). Mentors report on their own mentoring practices. *Australian Journal of Teacher Education, 35*(7), 30–42. doi:10.14221/ajte.2010v35n7.3

Jacobs, V. R., Lamb, L. L., & Philipp, R. A. (2010). Professional noticing of children's mathematical thinking. *Journal for Research in Mathematics Education, 41*(2), 169–202.

Juwah, C., Macfarlane-Dick, D., Matthew, B., Nicol, D., Ross, D., & Smith, B. (2004). *Enhancing student learning through effective formative feedback.* The Higher Education Academy.

Kemmis, S., McTaggart, R., & Nixon, R. (2014). *The action research planner: Doing critical participatory action research.* Singapore: Springer; doi:10.1007/978-981-4560-67-2

Kleinknecht, M., & Schneider, J. (2013). What do teachers think and feel when analyzing videos of themselves and other teachers' teaching? *Teaching and Teacher Education, 33*, 13–23. doi:10.1016/j.tate.2013.02.002

Klenowski, V. (2013). Investigating the complexity of judgement practice. *Assessment in Education: Principles, Policy & Practice, 20*(1), 1–4. doi:10.1080/0969594X.2012.745702

Knight, P. T. (2002). Summative assessment in higher education: Practices in disarray. *Studies in Higher Education, 27*(3), 275–286. doi:10.1080/03075070220000662

Masats, D., & Dooly, M. (2011). Rethinking the use of video in teacher education: A holistic approach. *Teaching and Teacher Education, 27*(7), 1151–1162. doi:10.1016/j.tate.2011.04.004

Mena, J., García, M., Clarke, A., & Barkatsas, A. (2015). An analysis of three different approaches to student teacher mentoring and their impact on knowledge generation in practicum settings. *European Journal of Teacher Education, 39*(1), 53–76. doi:10.1080/02619768.2015.1011269

Molloy, E. K., & Boud, D. (2014). Feedback models for learning, teaching and performance. In J. M. Spector, M. D. Merrill, J. Elen, & M. J. Bishop (Eds.), *Handbook of research on educational communications and technology* (pp. 413–424). New York: Springer. doi:10.1007/978-1-4614-3185-5_33

Nicol, D., & Milligan, C. (2006). Rethinking technology-supported assessment practices in relation to the seven principles of good feedback practice. *Innovative Assessment in Higher Education*, 64-77.

Nicol, D. J., & Macfarlane-Dick, D. (2006). Formative assessment and self-regulated learning: A model and seven principles of good feedback practice. *Studies in Higher Education*, *31*(2), 199–218. doi:10.1080/03075070600572090

O-K, S., & Hong-Fa, H. (2014). Students' perceptions towards the use of dialogic feedback in mobile applications for students' writing: A qualitative case study. *Journal of e-Learning and Knowledge Society, 10*(3).

O'Neill, S., & Dann, C. (2018). Video feedback on practicum: Demonstrating effective pedagogy and and making children's progress explicit. In T. Richardson, B. Dann, C. Dann, & S. O'Neill (Eds.), *Formative assessment practices for pre-service teacher practicum feedback: Emerging research and opportunities* (pp. 113–139). IGI Global. doi:10.4018/978-1-5225-2630-8.ch004

Parker, M., Patton, K., & O'Sullivan, M. (2016). Signature pedagogies in support of teachers' professional learning. *Irish Educational Studies*, *35*(2), 137–153. doi:10.1080/03323315.2016.1141700

Plano Clark, V. L., & Creswell, J. W. (2008). *The mixed methods reader*. Thousand Oaks, CA: Sage.

Rushton, A. (2005). Formative assessment: A key to deep learning? *Medical Teacher*, *27*(6), 509–513. doi:10.1080/01421590500129159 PMID:16199357

Sherin, M. G. (2017). Exploring the boundaries of teacher noticing: Commentary. In E. Schack, M. Fisher, & J. Wilhelm (Eds.), *Teacher Noticing: Bridging and Broadening Perspectives, Contexts, and Frameworks* (pp. 401–408). Cham: Springer International Publishing. doi:10.1007/978-3-319-46753-5_23

Simon, S. E. (2013). Chaos of textures or 'tapisserie'? A model for creative teacher education curriculum design. *Australian Journal of Teacher Education*, *38*(11), 6. doi:10.14221/ajte.2013v38n11.2

Stellmack, M. A., Konheim-Kalkstein, Y. L., Manor, J. E., Massey, A. R., & Schmitz, J. A. P. (2009). An assessment of reliability and validity of a rubric for grading APA-style introductions. *Teaching of Psychology*, *36*(2), 102–107. doi:10.1080/00986280902739776

Tillema, H. H., Smith, K., & Leshem, S. (2011). Dual roles-conflicting purposes: A comparative study on perceptions of assessment in mentoring relations during practicum. *European Journal of Teacher Education*, *34*(2), 139–159. doi:10.1080/02619768.2010.543672

Weasmer, J., & Woods, A. (2003). The role of the host teacher in the student teaching experience. *The Clearing House: A Journal of Educational Strategies, Issues and Ideas*, *76*(4), 174–177. doi:10.1080/00098650309601998

Willis, M. J., Dann, C. E., Jones, C. M., Lowe, B. J., & Toohey, E. (2012, November). Designing mobile information systems to support WIL experiences. In *Proceedings of the 24th Australian Computer-Human Interaction Conference* (pp. 653-656). ACM. 10.1145/2414536.2414634

Zuber-Skerritt, O., Fletcher, M., & Kearney, J. (2015). *Professional learning in higher education and communities: Towards a new vision for action research*. London: Palgrave. doi:10.1057/9781137455185

Section 3

Approaches to Dialogue and Dialogic Practices

Chapter 11
Changing the LAB Experience in Undergraduate Engineering:
How an Online Approach Can Improve Formative Assessment Practices and Learning

Ali Mohamed Habibi
University of Southern Queensland, Australia

Ann Dashwood
University of Southern Queensland, Australia

ABSTRACT

The use of technology to enhance formative assessment in higher education continues to be a challenge regardless of advances in digital capabilities; yet research has shown its potential regardless of discipline. In undergraduate electrical and electronic engineering, which is the discipline focus in this chapter, lab work is an area that can be enhanced in this way but with such an enhancement comes a change in pedagogy from the conventional approach of in-lab physical practical work conducted by the individual student alone or in a group with limited support to one of working collaboratively in remote access laboratories scattered far and wide through an online learning systems that provides access to laboratory infrastructure and learning environments through the internet. In a collaborative learning environment, students work together to solve problems and need to become involved in dialogue to achieve a common goal where they depend on and are accountable to each other. This chapter explores students' experience of a collaborative approach to lab work regarding mastery of the voltage division rule and its relevance to formative assessment using remote access laboratories that depend on technology and internet access. The implications for task design and formative assessment are discussed based on the results of interviews with participating students. The nature of change in pedagogical practice is highlighted as are the implications for the design of formative assessment and the need to work at the level of "feedback markers" that are able to feed forward to progress learning.

DOI: 10.4018/978-1-7998-0426-0.ch011

INTRODUCTION

This chapter focuses on an exploration of undergraduate electrical and electronic engineering students' experience of learning through laboratory work (LAB work) that involves a collaborative approach 'on-line' as opposed to the conventional approach of individualised hands-on, in-LAB, learning alone. With the advent of the Internet the essential learning that requires students to be physically in a laboratory, as in conventional practice, is now able to be simulated online through remote laboratories where students are able to access from any place, subject to a computer, relevant program and stable Internet connection (Baladoh, Elgamal & Abas, 2017; Gadzhunov & Nafalski, 2010). Conventional practices in-LAB have typically required on campus presence of students to work with the specific equipment under the supervision of university staff, and as Schkoda, Schweisinger, and Wagner (2012) found, in their trials, the development of detailed manuals and training sessions improved student satisfaction. They also considered that such LAB work may be attached to a course or be provided separately to be successfully completed independently of standard courses or subjects of learning. In such a traditional pedagogical approach, this has meant the use of laboratories where students work with equipment in isolation receiving mostly only a summative assessment of their performance. Since contemporary approaches to teaching and learning are underpinned by social constructivist theory that advocate experiential learning ofproject- and problem- based learning types, where there is opportunity for critical reflection and dialogue, it would seem imperative that the vital knowledge and skills such as those acquired through the in-Lab learning experience, are taught in this way. In addition, this dialogic approach has implications for both the pedagogy and the assessment practices (Dann, Dann, & O'Neill, 2018), in that it supports the need for ongoing formative assessment in its acknowledgment of learning as inquiry with the co-constructing of knowledge during the experience (Gagnon & Collay, Hattie & Timperly, 2007; 2006; Pea, 2004).

The LAB learning experiencetherefore emerges as an important component of learning in Science, Technology, Engineering and Mathematics (STEM) undergraduate higher education (Dittmar & Kahlcke, 2016; Feisel & Rosa, 2005), and the choice of pedagogy, and assessment of learning continues to be under scrutiny to make improvements (Corter, Esche, Chassapis, Ma, & Nickerson, 2011). To explore this learning context in more depth with regards to engaging students in a more dialogic experience, the chapter reports case study research into undergraduate electrical and electronic engineering students' perceptions of working collaboratively in remote laboratories, simulated learning experiences. It deals with the underpinning pedagogical theory base for cooperative and collaborative learning and its application, and discusses its relevance to how these kinds of laboratory learning experiences are able to provide timely, useful feedback for the purposes of formative assessment. Besides providing a review of the related literature, it seeks to draw out the strengths of this formative approach from the perspective of students.

The chapter objectives are to explore how the conventional approach to LAB work and its assessment for undergraduate students in mechanical engineering may be enhanced through the use of technology by (1) engaging students in collaborative learning and (2) examining how this changed pedagogical practice relates to formative assessment and learning from the perspective of students. The chapter highlights students' perceptions of the pedagogical change involved that required them to interact with peers and use technology-supported techniques in their goal to master the application of the voltage division ruler (Boylestad, 2015).

BACKGROUND

The acquisition of knowledge and the mastery of specific skills to do with the use of specialised equipment and technology in higher education learning in STEM is essential for undergraduate students. In current times, as technology has advanced and instrumentation has become more sophisticated, and computers and software available to support online learning, the provision of quality learning experiences that are accessible by all students through remote laboratories is gaining momentum. Conventional LAB work has been geared to students being able to attend LAB sessions where equipment is available for them to individually apply their knowledge and demonstrate a particular skill. It is well established that this approach has limitations not only from the pedagogical perspective but from that of convenience and economics as well (Osen & Bye, 2017). Although in-LAB learning can be said to provide the 'real-life' 'hands on' authentic learning experience, because of the high number of students and the fact that physical laboratories are usually relatively few in number, students generally are required to attend at specific times in a 'shift-like' approach. Such scheduling becomes increasingly challenging as choice of timing competes with many other demands that today's students have in their busy lives, including the need to work. Moreover, when students are studying at a distance they are required to travel at preset times to the university campus to learn and be summatively assessed on the LAB components. In this conventional practice, time for learning is at a premium and tuition may largely depend on the provision of aninstructor? and minimal opportunity for discussion with a tutor or peers. In contrast to this traditional pedagogical approach, in this chapter, the authors take the stance that the provision of the online-LAB experience through remote laboratories is able to both counter these limitations as well as add value through facilitating a dialogic approach that involves collaboration with peers and critical reflective practice. In so doing, it also provides for the automatic building-in of formative assessment that "feeds forward" (Gonzalez, 2018) to enhance learning. Thus, technology enhanced approaches to LAB work that involve students in critical, reflective dialogue with their peers, through structured collaborative learning experiences, has been found to enhance learning outcomes, and provide students with increased satisfaction (Estriegana, Medina-Merodio, & Barchino, 2019).

Generally, a remote access laboratory (RAL) may be defined as a learning system that functions online by providing access to a specific infrastructure or software that creates a virtual learning environment through connecting students and staff via the Internet. Because of its ability to connect students into shared sessions where they are able to interact with each other in an online program, as well as tutors, the pedagogical approach lends itself to a collaborative model of learning. Dillenbourg, (1999, p. 12) cites Roschelle and Teasley's (1995) definition of collaboration as ". . . a coordinated, synchronous activity that is the result of a continued attempt to construct and maintain a shared conception of a problem (p. 70)". While Dillenbourg (1999) highlights the cognitive aspects of collaborative learning, he notes the importance of social practices as well. As a pedagogical approach, collaborative learning, is underpinned by social constructivist theory emphasising that learning occurs through the social construction and co-construction of knowledge through student engagement in social interactions. It is argued that the mobility provided by today's technology presents social computing tools for education that "increase students' opportunities for an active mind in multiple, contexts" (Dillenbourg, Järvelä, & Fischer, 2009), thus providing learners greater agency and independence in learning.

A collaborative problem-based learning pedagogical approach to LAB-learning as the focus of this chapter is both important in its exploration of pedagogical change in higher education for undergraduate students as for informing on the way technology used in RAL may contribute to learning and students' satisfaction through its provision of feedback. Collaboration is also viewed as an interactive process that engages two or more participants working together to achieve educational outcomes that they may not accomplish independently (Järvelä, Näykki, Laru, & Luokkanen, 2007). Thus, in a collaborative learning environment students are expected to participate, working together to achieve a joint outcome, such as solving a problem. Unlike traditional pedagogy or direct instruction, students engage in the task being dependent on one another, jointly accountable to achieve the goal Collaborative learning therefore warrants a change in tutors' approach to teaching and learning, and also students' expectations of how they should be learning at university. Osen and Bye (2017) note that many students are more concerned with achieving the requisite grade to pass the course, which reflects a surface view of learning rather than an appreciation of the possibility of deep learning. They cite Foldnes (2016) who emphasizes how learning environments that involve cooperation such as in collaborative learning are more effective in facilitating learning in depth.

Other research has found that when students collaborate with their peers to learn this experience can have a positive influence on outcomes (Lowe, Berry, Murray, & Lindsay, 2009). However, there continues to be the need for research into the effectiveness of collaborative learning in RALs. Although conventional LAB work may involve group work, which suggests a some form of cooperation is necessary, an authentic collaborative approach would require careful design of how the learning and the assessment would take place. According to Huang (2015) although currently RALs may not be maximizing the potential for a collaborative approach to learning, they are highly applicable to facilitating students' collaboration and helping them understandthe approach to learning.

In practice, learning in a RAL typically requires all students to learn about both the theoretical and practical aspects of the course content of their discipline as applied to the specific task or problem to be solved. Providing teaching through RALs allows a more flexible curriculum depending on the ways their use is organized and offered. As noted earlier RALs can provide more economical options and are attractive for teaching students who are at a distance, so long as they are supported through access to the Internet and computer or other mobile smart device. Once established as a practice, the same infrastructure may be applied to other LAB work. In terms of the research reported in this chapter, RALs provide a safer teaching and learning environment as otherwise students are exposed to electrical-related tasks, working with electrical currents that involve an element of danger.. RALs are able to address this issue since experiments are controlled, Output Current Level and Voltage Limit are preset, for instance, as exemplified in the Iordache, Pop, Samoila and Ursutia (2011) iLAB project. Nevertheless, although a safer experimental learning environment is created, students are still required to be aware of and adhere to safety regulations and instructions, though largely protected from being able to make errors that cause serious damage or harm. Other advantages of students being able to spend time in RALs is the convenience of being able to practise LAB work and prepare the specific tasks that may be part of the ongoing learning and formative assessment and also for summative assessment. In addition, students are not limited to a fixed set of functions or programmes whilst working in RALs, thus, their availability can add significant value to the way LAB learning is organized because of the adaptability of the software (Iordache, et al., 2011). By providing opportunities for students to learn more intensively because access can be made easier, RALs can enhance learning. The various elements of a task that students might need additional support for learning, including the basic concepts and discipline terminology and access to virtual instrumentation. These are all features that RALs can do well for teaching and improving students' knowledge and performance through collaboration. As noted by Zubía and Alves (2012) "regardless of the type of laboratory experience – hands-on, remote, or virtual – the instructional development process to plan laboratory instruction should not vary" (p. 13). Thus, there is much support for the enhancement of the LAB learning experience in undergraduate higher education through the use of RALs; and core to this, it is argued here, is the need for collaborative learning that is problem- or inquiry- based, where formative assessment plays a key role in sustaining students' engagement and learning.

MAIN FOCUS OF THE CHAPTER

Problem-solving by calculation in controlled laboratory settings has traditionally been an integral part of the undergraduate electrical and electronic engineering training across universities. Typically, LAB tasks are set and laboratory manuals provided from which students apply deductive reasoning to their encounters with the physics of tasks which, by extrapolation were applied to generate a series of events electronically. This system of training can be labour, equipment and time intensive for the instructor-supervisors and ultimately cost ineffective, since cost is tied to the provision and maintenance of LABs and instructor-monitored-on-campus presentations of diagrams and circuits and technical performance. In conventional practice, students are required to work on equipment in laboratories in relative isolation, in which approaching an instructor to interpret the written instructions may be confronting and the summative outcomes derived from a predominantly monologic context. This chapter proposes a more contemporary exploratory approach to the pedagogy of LAB tasks, and considers dialogic in its problem-solving stance. With the increase in online learning practices and assessment, RALs provide the opportunity to explore potentially more effective learning experiences beyond the traditional role of paper-based calculations for testing purposes.

Thus, the main focus of the Chapter is to highlight how technology-enhanced practices in higher education undergraduate electrical and electronic engineering programs are changing the conventional approaches to the in-LAB learning experience and enabling more focus on formative assessment. Through the reporting of case study research, that explored undergraduate higher education electrical and electronic engineering students' perceptions of online-LAB work learning using LabView, where the task was presented through RALs and students worked collaboratively in small groups, it explores what students perceived as advantages compared with conventional practice. In doing so, it reviews the literature in terms of pedagogical change, and the theoretical underpinnings of the need for a more collaborative approach to learning, and discusses the implications for those involved in teaching and learning LAB components. The introduction of working collaboratively in LAB work represents a major change in approach to both pedagogy and assessment, as does the use of technology through RALs, and therefore has implications particularly for the student experience. As such, the research sought to explore students' perceptions of this technology-enhanced approach. Not only did this approach cause students to have to work cooperatively by collaborating with their peers to successfully complete a set task, it also provided greater opportunity for learning through the collaborative process, building in opportunities for feedback to inform formative assessment.

ISSUES, CONTROVERSIES AND PROBLEMS

Traditional Approaches to Laboratory Learning and Assessment in Mechanical Engineering

The need for space (Balamuralithara & Woods, 2009) to accommodate students in training resulted in many sessions being scheduled to allow sufficient in-LAB time for tasks to be interpreted, experimented with, implemented and completed (Deniz, Bulancak, & Ozcan, 2003). Controlled in-LAB times required supervision to maintain safety standards (Deniz et al., 2003) yet at high cost (Elawady & Tolba, 2009). It was common for students to have relatively restricted interaction time with skilled technicians. Feedback from instructors was often visual or verbal only, causing retention issues for students in their sustaining their course of learning to achieve their qualification. The authenticity of the tasks set was also contestable, as Corter, Nickerson, Esche, and Chassapis (2004) argued. A change that ensured students were able to address more authentic problems would have greatly enhanced the learning experience. The need for change in this way has been reinforced by Chan and Fok (2009) who identified the need for the 'reality' of the created tasksto be sustained by the learners being on-task in theirinteractions. This was seen as ensuring opportunities to discuss elements of the task and develop comprehension skills associated with the problem to be solved, and the application of physics and mathematics to the whole task.

Feisel and Rosa (2005) note that there has been little focus on 'laboratory instruction' compared with curriculum and teaching methods, although it is a critical part of engineering education. They explain the role of the laboratory in undergraduate engineering as "an instructional laboratory to learn something that practicing engineers already know . . . [where] [T]that something needs to be better defined through carefully designed learning objectives if the considerable effort devoted to laboratories is to produce a concomitant benefit" (p. 121). Of note from this research is their recognition that the use of digital communication technology and systems of learning at a distance online via the Internet have been the disruptor to the conventional approach.. They argue the laboratory experience in engineering, while necessary for the learning essential skills, is also valuable for provding students with insights into their prospective job and in turn motivation to continue to learn (citing Leva, 2003; Moore & Voltmer, 2003; Wicker & Quintana, 2000). In addition, their work, in terms of identifying the objectives that are fundamental to LAB work, which they recommend to inform pedagogy and assessment as well as evaluating effectiveness in the context of online learning, are well justified in their applicability and exhaustive, although they may vary across disciplines. Taking the earlier work of Ernst (1983) into account, these objectives relate to (1) instrumentation, (2) models, (3) experiment, (4) data analysis, (5) design, (6) learn from failure, (7) creativity, (8) psychomotor, (9) safety, 10) communication, (11) teamwork, (12) ethics, and (13) sensory awareness.

It is important to consider these objectives in any LAB task design since as they specify, they deal with the cognitive domain, which has always been the 'province of engineering educators', the psychomotor domain and also the affective domain, and have relevance across disciplines.

So, the first five objectives dealing with cognition—Instrumentation, Models, Experiment, Data Analysis, and Design were expected. Then, two were specified that involved the psychomotor domain: Psychomotor (the ability to actually manipulate apparatus) and Sensory Awareness. Finally, the remaining objectives have a cognitive part but also include a significant component of the affective domain, that is behavior and attitudes: learning from failure, creativity, safety, communication, teamwork, and ethics in laboratory. Exposing students to all three of these domains is necessary to produce an effective engineer (Feisel & Rosa, 2005, p. 127).

The Rise of Opportunities for Online Learning and Assessment and the Potential of Remote Access Laboratory (RAL)

Formative Assessment and Learning

Drawing upon Wiliam (2011), formative assessment (FA) relates to how data may be provided during the learning process in such a form for use in facilitating and improving students' learning and teachers' teaching. This formative approach is in contrast to summative assessment where students' learning at the completion of a course of study is represented by a final score or benchmark or measure of performance or competency. Brookhart (2013) differentiates the two by describing formative assessment as an aid to learning and summative assessment being the goal of learning. Yet, Panadero, Andrade, and Brookhart (2018) highlight the link between self-regulated learning theory and the role of formative assessment, thus emphasizing the important role of learners in their own learning and the need for feedback or data on their learning progress that is able to 'feed forward' (citing the Council of Chief State School Officers, 2012). They emphasise how the impact of assessment in learning needs to be considered from the psychological perspective as well as the social:

We also need to examine the psychological and social effects of FA to better understand what happens with SRL. What types of mental and emotional processes do students activate during self- or peer assessing? How do those processes influence the activation of learning strategies? Interest in this type of research is growing, as demonstrated by the publication of a recent handbook on the human and social factors of assessment (Brown & Harris, 2016). Only by understanding internal cognitive and affective processes, we can truly understand the power of FA (Panadero, et al., 2018, n.p.)

Implementation of formative assessment involves the use of information retrieval cycles to assess and interpret the information collected and build on this information to improve the teaching and learning processes. In addition, Poth (2018) notes that formative assessment may also involve interaction between teachers, students and their peers, and allows instructions to be interpreted, applied and modified to guide students and support their learning outcomes. In advocating for technology enhanced formative assessment (TEFA) she acknowledges the complexity of learning environments and purposes for learning, stating that "we are only beginning to recognize the prominence of features present in the learning environment that explicitly check for understanding and then support learning through adjustments in instruction" (n.p.). Further, she cites Maier, Wolf and Randler (2016) as recognizing how TEFA has the potential to support and facilitate the provision of such features and ultimately contribute to improving the quality of higher education learning environments. Thus, the design of the learning experiences represents a challenge in ensuring their ability to respond to the need for appropriate feedback markers.

Panadero, et al. (2018) outline six attributes that are deemed critical to the design of effective formative assessment taken from the Council of Chief State School Officers (2012), as follows:

- *Learning progressions* Learning progressions should clearly articulate the subgoals of the ultimate learning goal.
- *Learning goals and criteria for success* Learning goals and criteria for success should be clearly identified and communicated to students.
- *Evidence of learning* Evidence of learning is elicited during instruction.
- *Descriptive feedback* Students should be provided with evidence-based feedback that is linked to the intended instructional outcomes and criteria for success.
- *Self- and peer assessments* Both self- and peer assessments are important for providing students an opportunity to think metacognitively about their learning.
- *Collaboration* A classroom culture in which teachers and students are partners in learning should be established.

These attributes are clearly in keeping with the Feisel and Rosa's (2005) objectives for the design of laboratory learning experiences, relating well to commmunication and teamwork in noting the need for collaboration and working with peers. Thus, the pedagogical approach becomes a major consideration when the conventional approach to LAB work, which has been typically students working in isolation from each other, is seen as needing to have students engaged in learning together.

Although the use of technology in online learning has been implemented in higher education for many years now, Poth (2018) notes that debate remains as to whether there are significant differences in the effectiveness of students learning asynchronously compared with those learning synchronously (able to communicate with each other in real time). She cites the research of Sánchez-Vera and Prendes-Espinosa (2015) in noting that:

on-demand access has led to the creation of masses of digitized lectures and learning activities and a marked increase in research that is assessing student experience and impacts on learning; for example, there has been an exploration of alternative means of assessment in massive open online courses, MOOCs (n.p.).

However, Poth's (2018) research that investigated online practice quizzes shows that technology enhanced formative assessment strategies can support students' motivation and self-regulated learning, successfully enabling instructional adjustments to be made for both immediate learning and in the longer term. She concludes that "technology enhanced formative assessment strategies are an important contributor towards creating continuous assessment systems where learning and assessment go hand in hand and students are provided the opportunity to drive their own learning". This analysis reiterates the need for pedagogical change towards a social constructivist view of learning where Feisel and Rosa's (2005) objectives could be more comprehensively addressed in the LAB learning experience. A collaborative approach to learning would be central to effecting this change. Thus, it needs to be noted that the approach to pedagogy and such formative assessment relies upon the implementation of a more student-centred approach where students have opportunity to discuss in a context of problem-based learning. Assessment may involve peer and self-assessments as well as group assessments that focus on collaborative tasks. Tasks such as planned in this research require minimal resources to support students to self- and peer- assess yet are able to ensure debriefing and assistance to students to encourage self-regulation (Boud, 2013).

THE COLLABORATIVE APPROACH TO LEARNING

Pedagogies that are underpinned by social constructivism are not new, yet the implementation of change from traditional approaches continues to be a challenge in many contexts (Mayer, 2012; Myllyäki, 2013; Quaye & Harper, 2015). Yadav, Subedi, Lundberg and Bunting (2011) explored the shift to learner-centred teaching through the introduction of problem-based learning in an electrical and electronic engineering undergraduate study. They pointed out that over the past decade or more research has shown that although the work of engineers requires highly effective communication skills and the ability to apply their specialist knowledge to solve problems on the job, these skills have been lacking in new graduates. They also noted that past statistics have shown an increase in jobs for engineers but a high attrition rate for undergraduate students and in turn lower graduation rates (citing National Science Board, 2008; see also Litzier & Young, 2012; Male & Bennett, 2014). Thus, the way courses have been taught traditionally has been called into question, highlighting the contrast between the information-transmission view of learning through the monologic lecture talk and in-LAB isolated activity. This is succinctly stated by Yadav et al. (2011, p. 254), "the main problem within engineering education is the gap between the active field and the passive classroom experience (Palmquist, 2007)". Therefore, despite such comprehensive advice as that contained in Feisel and Rosa's (2005) objectives for LAB work and Panadero, et al.'s (2018) extensive attributes applicable to the learning experience, there is a strong need to explore how a more social constructivist approach to learning in LAB work impacts on the learning and assessment experience of students.

While the desired change in pedagogy might be achieved through problem-based learning or project-based learning or inquiry-based learning, as they all require students to be involved in real-life, authentic learning experiences, as noted earlier in this research, collaborative learning was selected in the context of online learning and RALs. Given that at the core of social constructivist approaches to learning are the social practices associated with the way teachers and students interact in the co-construction of knowledge and the scaffolding of learning, a collaborative approach, which may also be problem-based, is ideal (Odeh, Mckenna, & Abu-Mulaweh, 2017, O'Neill, 2016). It is able to facilitate learning experience design that can reflect the skills required by practicing engineers that require effective communication to successfully work in a team environment and involves the cognitive, psychomotor and affective skills in keeping with Feisel and Rosa (2005).

In additon, what is not so much discussed in relation to how social constructivist pedagogical approaches to learning is the contrast with the traditional approach, which can be conceptualised as information transmission, is the impact on the interactive classroom talk (Edwards-Groves, Anstey, & Bull, 2014; O'Neill, 2018; Van Es & Sherin, 2002) and the relationship to formative assessments need to focus on features of practice that allow the checking for understanding (Maier, et al., 2016). Dann et al. (2018) differentiate formative assessment from summative by its capacity to 'feed forward' rather than 'back' and thus progress a student's learning and the teacher's teaching through two-way communication, so dialogic, compared with the one-way monologic nature of only 'information transmission'. Dialogic pedagogies involve critical reflective practices and require teachers to be aware of the quality of dialogue that classroom interactions/learning experiences create. O'Neill, (2017, p. 75) states they involve "teachers being aware of their metacognitive processes in being able to formulate the most useful cognitive moves in the scaffolding of learning in response to the emergent dialogue with students". Similarly, in a team problem-solving situation the design of the learning experiences would need to stimulate constructive, critical conversation. Much research has idemonstrated how the analysis of classroom dialogue can provide insights into the scaffolding of learning and features of practice that allow checking for understanding and thus inform formative assessment (Kollöffel & Jong, 2013; Leininger & Robinson, 1995; Walsh, 2006). As a guide to developing collaborative learning environments Doolittle's (1995) work on cooperative learning applies, as it relates well to the needs of formative assessment through consideration of Vygotsky's Zone of Proximal Development (ZDP) in his formulation of eight constructivist principles to create engaging learning experiences. However, given that Herrington, Reeves and Oliver (2010) used these as a basis to develop guidelines on how to design and teach authentic learning experiences, the project drew upon these guidelines to ensure the design of the RALs' tasks reflected authentic learning experiences in the context of collaborative learning. The following section lists these guides and provides an overview of the relevance of each principle to the design of LAB work.

Guides to LAB Learning Experience Design

Guide one, *Teach using whole and authentic activities* highlights the need for the learning experience to directly relate to the real world activity of the work of the engineer. In the context of RALs the pedagogy and practices rely on the technology that is accessed via the Internet. This means instrumentation is accessed and controlled remotely through the browser-based interface such that students do need to learn how to function in the online learning space and the protocols involved, particularly in the collaborative learning environment where clarity of roles and responsibilities is essential to practice. In this case protocols that facilitate collaboration, should ensure effective practice, and therefore some preparation is necessary, as would be in using 'high-end' scientific simulation tools such as iLab and Nano HUB Gateway (Purdue). Importantly, as Gibbings (2014) discusses, the quality of experience in RALs and the way it depends on a variety of factors, including the efficiency of the technology, and the equipment as in any LAB, but also the design and set up of the learning experience, the linking to authentic professional work and facilitation of the collaboration and opportunity for critical reflection and deeper learning.

Guide two, *Create a 'need' for what is to be learned* relates to the importance of students being able to have agency within their own learning. Williams and Williams (2011) note the importance of the student learning experience being positive such that they are motivated to participate. The LAB experience does have the advantage of allowing students to directly put theory into practice although in RALS it is mediated through technology. Again, the adoption of problem- or project based learning is applicable to meet this guide.

Similarly, Guide three, *Create classroom exercises that require social interaction with peers, parents, teachers, or professionals,* can be achieved through learning experiences being designed to facilitate peer interactions and collaboration to address the task and its goal. It is in keeping with constructivist views of learning, supporting social interaction in the course of ideas exchange and their internalizstion. In turn learning experiences can be designed to foster critical reflective practices that encourage behavioral change and cognitive change, and independence in learning (Doolittle, 1995; Gagné, 2009). Related to this and examining this pedagogy in more depth Guide four, *Encourage self-talk or egocentric speech,* acknowledges the role of the learner and the metacognitive processes that effectively designed collaborative processes should facilitate to develop problem-solving skills. Doolittle (1995) recognised that talking to oneself or thinking out aloud can be an effective tool for students in problem-solving. This behavior also supports students' personal cognition since as Vygotsky believed when expressing thoughts aloud, and in the case of engineering students, internalizion becomes part of the problem-solving strategies they use. Adopting this strategy is also seen as contributing to developing students' self-control in keeping with a professional approach to solving problems on the job. Guide five, *Provide opportunities for verbal interactions,* also notes the importance of creating opportunities for verbal interactions, in LAB work, because it allows students and tutors to question and discuss the task or problem to be solved. As Pais Marden and Herrington (2011) emphasise, use of language in interactive situations involves patterns of behavior, and in the process of facilitating cognitive processing/thinking, mental functions restructure to accommodate new knowledge and skills, in turn facilitating opportunities for learners to move forward through their Zone of Proximal Development (ZPD). This leads into Guide six, that considers the importance of formative assessment during a course of learning: *Closely monitor student progress in order to avoid assigning tasks that are not within a student's zone of proximal development.*

This guide is an essential prompt to ensure tasks are set at the appropriate level to progress students' learning. Therefore, task design needs to support the close monitoring each student's learning as noted above with respect to the Maier, Wolf and Randler' (2016) requirement to identify feedback markers. However, these markers need to be formative and so feedforward to progress the learning at hand (rather than summative). Further to this, Guide seven which notes that *Instruction or activities must precede a student's development,* is crucial to facilitating learning, since learning experiences that present as too difficult for students to achieve the goal would be clearly offputting. Thus, the task designer needs to have a thorough awareness of both students' existing knowledge and skills and the new demands of the task to be mastered. Learning experiences should be designed such that students' capabilities evolve in order to control the task. Since the learning experience design as advocated by Herrington, et al. (2006; 2010) uses the concept of the ZDP to underpin how learning develops in relation to task design, Guide eight is important in pointing out the need, at the start, to *Present tasks that students can perform successfully with assistance,.* It is important for the student to be aware of the availability of assistance while the anticipated outcomes of the course of learning are to become independent. Guide nine, *Provide sufficient support to enable the student to perform challenging tasks successfully, and then gradually withdraw support as the student becomes more skilled,* expands on this. Guide nine recognizes that while assistance may be needed initially, as learning progresses, this can be gradually withdrawn, not withstanding some element of challenge being necessary to maintain interest and motivation. The fact that learning experiences should be linked to the real world work experiences of engineers (Guide one) the LAB experience design would be expected to have students take responsibility for implementing tasks independently and in the collaborative learning process assist others.

An additional important aspect of the learning experience is Guide ten, *Students must be given the opportunity to demonstrate learning independent of others.* This guide is a reminder to avoid losing sight of the fact that each team member in collaborative learning needs to acquire and demonstrate the planned knowledge, skills and understanding independently. To achieve this independence, it is vital for students to have the opportunity personally to carry out the activities to build their confidence and self-efficacy as engineers. Thus, the nature of the learning experience is highlighted again as crucial in the learning process but needing to link closely to formative assessment to support the design of ongoing learning. Aligning with Guide ten is the final Guide eleven- *Construct activities that are designed to stimulate both behavioral changes and meet cognitive changes,* emphasizing Osberg's (1997) stance that learning experiences should be presented within an organized structure that links to the various behaviors that students need to acquire as a result of the learning. Thus, designers/tutors also need to have the skills to plan and organize to be able to ensure students have the opportunity to learn and demonstrate the behaviors. In this way, it is expected that the student learning experience will allow students to construct and build knowledge independently as well as developing practical skills and capabilities. All-in-all, Herrington et al. (2010) provide comprehensive guidelines that focus on learning and the essentials of authentic task design. In the context of this research it is important to gain insights into the task that was used for the present research that explored students' collaborative learning with respect to the voltage divider task in electrical and electronic engineering.

THE LAB VIEW PROGRAM: HOW THE VOLTAGE CURCUIT DIVIDER TASK IS IMPLEMENTED

Figure 1 and Figure 2 provide insights into how students experience the use of the LabVIEW software, which is accessed online via the computer, and which allows students to dialogue from where ever they are situated through the use of Zoom video conferencing software. The voltage divider circuit is shown in Figure 1. It represents the basic understanding for students that the voltage divider is a simple circuit where the application of two or more series resistors and an input voltage can produce a reduced output voltage that is part of the input in direct proportion to the resistors' values.

As fundamental circuits in electronics, understanding how volatage dividers work is core to undergraduate electrical and electronic engineering students' learning, which requires a practical component. It has application in cases where the voltage involved in an electrical issue may need to be reduced in order to investigate the problem through the use of an instrument that can only be used with lower volatages e.g. using a five volt microprocessor to monitor a 12 volt battery.

Voltage divider rule: $$V_{out} = V_{R2} = \frac{V_{in} R2}{R1 + R2 + R3} \qquad (1)$$

When using LabVIEW students are shown the instrumentation to be able to manipulate the variables and measure voltage changes. Figure 2 depicts how the voltage divider code appears via LabVIEW. The left hand side of the figure shows the 'front panel' which is the user interface, where the running of the program and so manipulation of switches occurs. The right hand side presents the block diagram, where the coding occurs to design and create the program, and where the processing occurs that is not seen by the user. Students follow the instructions given to manipulate the various knobs and are guided through the task sheet to consider the changes that occur as a result of their actions. A more technical explanation is provided that reflects the level of mathematics required. Students are also supported in their course

Figure 1. Voltage divider circuit

Where:

V out = Output voltage. VR1=voltage across Resistor 1, VR2= voltage across Resistor 2, VR3= voltage across Resistor 3, V in= Input voltage, R1 and R2 = Resistor values

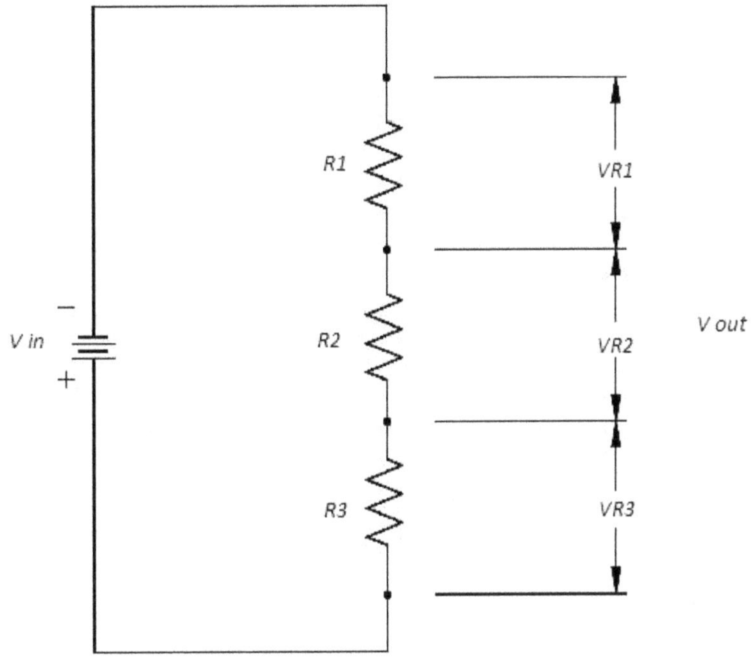

Figure 2. Voltage divider code in LabVIEW

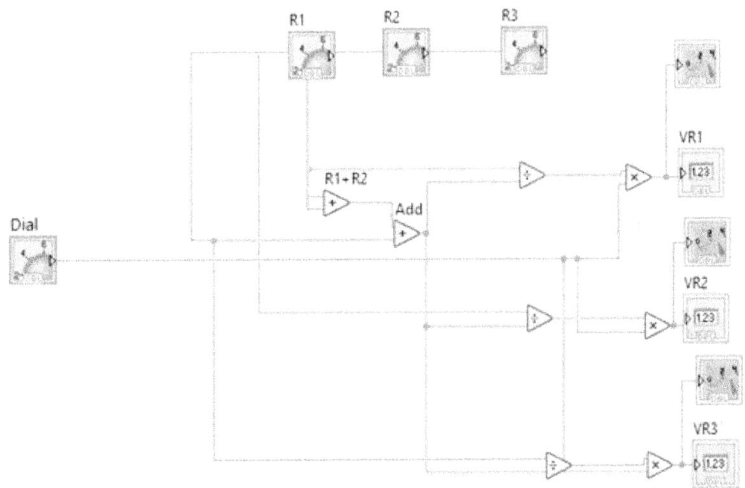

through the use of a text book and as noted below are provided with links to online learning materials. To work on the voltage divider circuit, the student enters data into the front panel, which presents the controls for the resistors R1, R2 and R3. The required basic mathematical functions need to be entered which support the necessary calculations as shown below to produce new values that register in the indicator terminals., For example, the front panel indicators displays the following equation.

(R1+R2+R3) and (V in * R2) and division $\left(V_{out} = V_{R2} = \frac{V_{in} R2}{R1 + R2 + R3} \right)$.

Team members are able to view and control this Virtual Instruments (VI) front panel remotely. This can be from within LabVIEW, which is the approach taken by the team leader or from within a Web browser since LabVIEW is built-in to the the Web Server. There is much flexibility for the users in their ability to interact with the program, publish front panel content and control the learning experience. Control may be shared and moved from team member to team member, with the facility to request control. Should there be queuing then the user at the server computer has the control of turn-taking. The circuit is typically already written by the tutor/team leader or depending on the pedagogical approach students may participate in preparatory work prior to be able to build the circuit themselves in LabVIEW.

The LabVIEW program can be accessed by any participant anywhere in the world where there is an Internet connection. There are many resources to support its use (e.g. see National Instruments, 2019) Steps for the task considered here include the following:

- Sweep the Potentiometer (resisters) continuously up and down over a range which includes the entire range during which maximum and minimum voltage to each resistor is achieved.
- The OUTPUT voltage cross each resistor and should look similar to the calculation result on the screen display.
- Using the voltmeter as shown on the front panel measure VR1, VR2 and VR3 and compare them with the theory shown in the calculations method.

CASE STUDY OF THE TRIAL OF RAL

This section reports the results of the trial of RALS through a case study of electrical and electronic engineering undergraduate students learning of the Voltage Divider Curcuit Task. As part of a larger project, this chapter's focus is confined to the results of interviews to identify students' perceptions of the opportunity to experience learning through the collaborative model in the context of a technology enhanced LAB learning experiences using LabVIEW.

Participants

A total of 12 students participated with 3 to 5 per group, with 10 participating in follow up semi-structured interviews. Convenience sampling was applied owing to the fact that students typically attended on campus for Lab work and needed to travel at a particular time in the program. The time taken for each RAL learning experience was between 1 to 2 hours depending on the students completion of activities and the duration of follow up interviews was approximately one hour. The students were placed under the same conditions, but in different remote locations. Manual instructions were prepared and trialled to support the conduct of the Labwork and the role of the team leader including a guide to support the collaborative approach in the online environment.

Data Analysis

The data collected from interviews were qualitatively analysed by Leximancer software. Unlike NVivo, ATLAS.ti, and CATPAC, Leximancer does not apply word frequency, or coding of terms and phrases. Leximancer, applies its own algorithms, and is used for analysing the meanings within passages of text by extracting the main concepts and ideas (Tseng, Wu, Morrison, Zhang, & Chen, 2015). In this analysis, Leximancer software was applied to analyse the qualitative interview data in order to identify the main themes from written transcripts to explore students views of their experience in RALs and the change to collaborative learning online.

FINDINGS

Following the thematic analysis, data showed that students' perceptions were positive and they appreciated the change in approach to Lab work bearing in mind that the students were participating in the research project and would still be required to complete the standard in-Lab work on campus later. In this sense the students found the experience valuable as an additional opportunity for learning. Overall, they found the collaborative approach one that although they were not very familiar with, it was valuable as was the use of RALs in supportingtheir learning compared with the conventional in-LAB experience.

Key Themes and Connectivity Rate in the RALs Experience

Analysis of the interview data using the Leximancer software identified eight key themes and a concept map of their connectivity rates. Figure 3 shows the key themes and their connectivity rates in order of relative importance. These were Team (100%), Task (47%), Collaboration (38%), Time (33%), Work (13%), Language (02%) and People (02%). There relative importance is shown in the concept map displayed in Figure 4. These also were the main themes representing collaboration on remote access laboratory experiences for students.

Figure 3. Key themes and connectivity rates based on analysis of student interviews

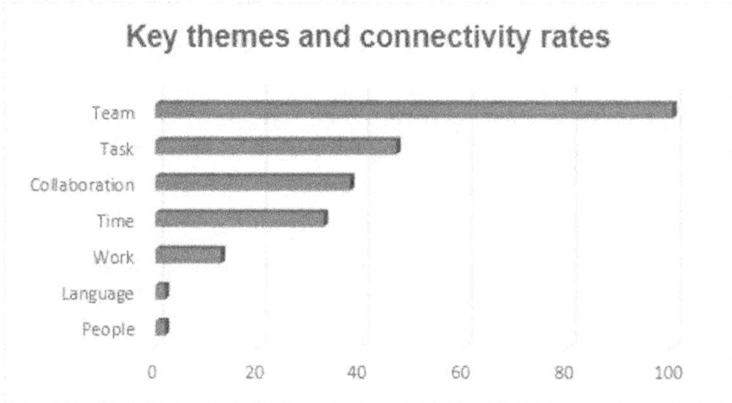

Figure 4. Concept map of themes and connectivity links derived from student interviews on their RALs collaborative learning experience

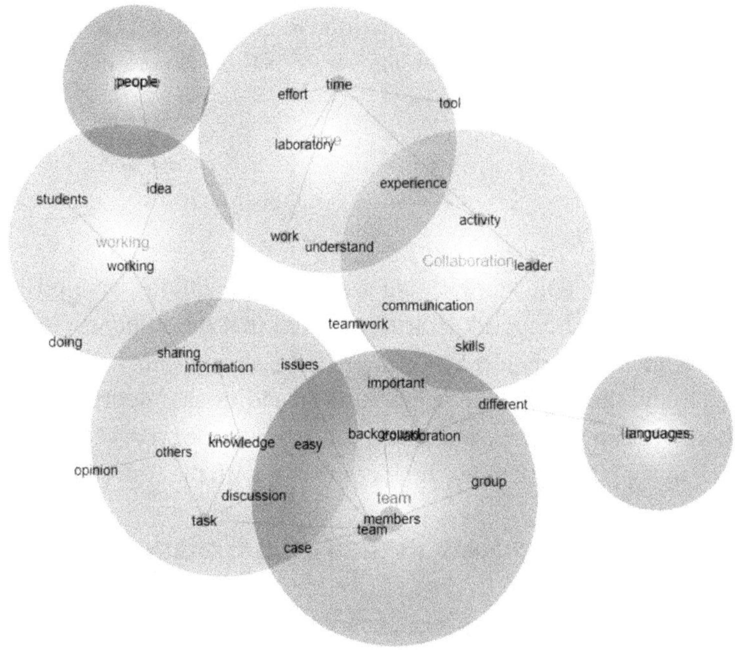

Overall, thirty-four concepts emerged from the student interviews, with the key themes showing a strong focus on learning and the fact that a collaborative approach was adopted as opposed to the students learning in-Lab work in the conventional way. Therefore, these findings, on the one hand, may not be surprising but at the same time the learning experience raised the issues of ‘time’ and ‘work’, in addition to the obvious focus on teamwork, and the task and the collaborative approach. Time and work in linking to the collaborative approach also showed students were considering the Lab experience from the perspective of working and doing the task and also in terms of time the amount of time and effort involved as well as the tool involved and their understanding.

Although the importance of 'people' and 'language' had minimal connectivity these ideas relate to the way the collaborative learning approach depends on the members of the group working cooperatively, and where the ability to communicate and use language appropriately underpins the successful grasp of the task requirements, the ability to discuss and its completion. The notion of formative assessment, although not explicitly emerging, is evident in students' consideration of task difficulty and amount of effort involved with regard to needing time to understand, discuss related issues involved, thus identifying potential feedback markers, and offer their opinions.

Students' interview reponses highlighted their awareness of the pedagogical change and the issues that it raised in the collaborative learning context. This was demonstrated by Interviewee 1 who noted that on occasions the team needed to deal with a peer who gave his/her opinion even though it was wrong, which meant that "each team member needed to become more transparent than ever before". Similarly, Interviewee 5 noted that it was important to be able to "accept the arguing, debate and disagreement when another other member of the team was incorrect". The need to be able to clarify task-related issues and questions was also raised by Interviewee 1 who explained that the collaborative approach to learning encouraged some team members to clarify and explain the theoretical background and willingly answer questions of peers. As Interviewee 8 also pointed out, it was possible that in the conventional Lab experience, in the absence of a strong team, an error may accrue without anyone recognizing it, such that the problem-solving task would go ahead without this knowledge and as a consequence time and effort would be wasted and failure occur. Whereas, in the collaborative approach it is highly likely that one or more team members would be able to identify a concern and work towards a solution. It was also pointed out that the group was able to work more effectively when each team member was clear about the task and activity as this prevented participants interrupting the group process because of lack of knowledge.

Interviewees also identified how their participation in the collaborative Lab experience highlighted the need for pedagogical change in LAB work, since it highlighted the different expectations of those who were more familiar with working in isolation. For instance, Interviewee 3 noted that the strategy of "brainstorming ideas as a group avoids old viewpoints that often come out of working solo" and that "the opportunity to combine unique perspectives from each team member creates more effective solutions". In addition, the students found that they were more motivated in this context for learning because of the opportunity to ask questions of each other and also gain different perspectives from across the group. Interviewee 7 viewed the changed approach as allowing team members to discuss "new fresh ideas" through "freely ask questions of all members" such that participants gained insights into "their talents". He noted that "working as a team is a new idea and the knowledge helped me to discover more and more". Needing to cooperate with peers was also highlighted in the interviews, with interviewees noting that their peers were cooperative and helpful. Interviewee 5 noted, everyone participated and was attentive, giving a chance to others to express their opinion and talk, and also no-one took excess time over other team members. It was specifically noted that participants were respectful in waiting to offer their idea or give their opinion until the appropriate moment. Generally, it was the belief that participants needed to be conscious of ensuring all team members were able to participate in the collaborative process. However, it was pointed out that when a technical problem occurred then this meant the person dealing with it would have more time than peers but unavoidable. Being able to collaborate in the problem solving process was seen as advantageous because of the opportunity to discuss and clarify ideas to solve the issue in a timely manner.

Students also recognized the importance of the role of the team leader, with Interviewee 8 noting their extra responsibility as "a big mission" in ensuring all members participated. Interviewee 5. reported "in our team the leader was guide and managed the procedure", which included allocating other roles in controlling and monitoring the experiment, which supported the collaborative activity. The session tended to be structured around the leader's initial responsibility to organize and explain the task followed by the team members carrying out their roles which took the majority of the time due to the number of question and the increasing difficulty level as particpants delved deeper into their learning.

In terms of difficulties, Interviewee 5 raised concerns regarding the need for participants to adhere to cooperative, communicative protocols to avoid the intrusion of personalization, which could be interpreted as negative. Interviewee 5 also emphasized that "all participants should have prior knowledge of the computer setup of LabView and the Zoom software, and know how to communicate with people and have the patience with other team members to accept all opinions, arguments, disagreements, and build on other members ideas rather than belittle them". The need for pior knowledge and preparation for the Lab learning was raised in a variety of ways but clearly linked to the importance of Herrington et al.'s (2010) guides for practice. Not only did participants highlight the need for participants entering the Lab work with identified prior skills knowledge, they also raised the issue of formative assessment all be it implicitly. This was evident in Interviewee 8's advice to increase the team members skills at the start and then alluding to the need for feedback markers, he made the point that "team members cannot go ahead without understanding the previous step, but unfortunately it was left to the team working together to ensure understanding in order to achieve success for the team". As a learning experience reliant upon collaborative learning through problem-solving this observation raises the issue as to what extent members of the team are able and willing to contribute and how can the relevant feedback markers be built in to ensure the gap in learning is not too great as might have been in Interviewee 8's experience. Thus, the design of the learning experience and the preliminary support and follow up are all elements of the learning experience that need to be thoroughly considered. In addition, in spite of the detailed planning of the sessions, the research shows that a much deeper analysis of the task needs to occur at the start so that feedback markers are identified relative to Guides 6, 7, 8, 9 and 10:

Guide six: *Closely monitor student progress in order to avoid assigning tasks that are not within a student's zone of proximal development.*
Guide seven: *Instruction or activities must precede a student's development.*
Guide eight: *Present tasks that students can perform successfully with assistance.*
Guide nine: *Provide sufficient support to enable the student to perform challenging tasks successfully and then gradually withdraw support as the student becomes more skilled,* expands
Guide ten: *Students must be given the opportunity to demonstrate learning independent of others.*

These Guides indicate implications for the way the collaborative approach operates, since to achieve a smooth operation, where team members are working within their ZDP and are all able to grow in their learning the importance of task and skills, analyses as part of preparation must be paramount, as is identifying the feedback markers to strengthen formative assessment as drivers of the learning process.

SOLUTIONS AND RECOMMENDATIONS

While the students preferred the collaborative learning experience over the conventional approach, they raised several issues that need to be taken into account before making a decision to move Lab work online through the use of technology such as LabView. This is not to say that the LabView program is the reason, as this works very well. Rather it is the design of the learning and assessment experiences that need careful consideration to better address Herrington et al's (2010) guides that are based on the Doolittle (1995) principles. Although the design of the learning experiences for the purposes of the research was detailed and trialled, the research results imply that a much more comprehensive and in-depth analysis of the task is necessary. This includes the knowledge and skills that students need at the start of their course of learning and identificatiom of the way they are expected to develop to eventually be able to address more sophisticated problems associated with the Lab work to enable their engagement with the expected range of tasks and challenges they will meet when they join the profession. Within this the need for the explicit building of formative assessment into the design featured as core to practice in being able to move students smoothly through their ZPD, beginning with identification of the skills to start.

A further outcome of interest was the students' awareness of how the collaborative learning process had the potential to impact both positively and negatively on the social relations, depending on whether a team member required extra assistance or not. Their responses suggested that there might be an optimum level of skill required to participate and contribute but still be able to learn with peers as opposed to a member possibly appearing too dependent. Thus, it is recommended that in line with Herrington et al's (2010) guides and the six attributes advocated by Panadero, et al. (2018) that are deemed critical to the design of effective formative assessment (Council of Chief State School Officers, 2012), much greater attention is paid to identifying the feedback markers that 'mark' the student's learning journey for Lab work requirements. By clearly articulating learning progressions with regards to sub-goals and identifying success criteria and evidence of learning both tutor and students will be clearer about what a task entails. In addition, the incorporation of descriptive feedback that 'feeds forward' to assist students to grow in their learning would improve practice. Ensuring greater clarity with regards to the progressive skills development and associated feedback markers would also support the effectiveness of the collaborative approach in helping to ensure that all students are able to participate at a level that contributes to constructive discussions and a cooperative learning culture.

FUTURE RESEARCH DIRECTIONS

Although there were limitations with respect to the present study of a small sample size and also only one kind of LAB task being addressed, there are implications for future research. Since the research suggests that RALs can provide distinct advantages for both learners and their tutors and they can facilitate students acquisition of collaborative skills that are also required by their profession, it is recommended that a more in-depth study be conducted that focuses very closely on planning for 'feedback markers', their exemplification, and tracking more closely students' progression via their ZDP. At the same time, there needs to be closer examination of the dialogue involved in during collaboration to clarify how learning might be scaffolded in the collaborative approach. This focus should include an investigation of how the student-peer communicative interactions during collaborative learning sessions impacts on learning and development of interpersonal skills relevant to work needs. Both of these recommendation would assist in improving pedagogy, including the identification of formative assessment design that optimizes students' growth in learning and preparation for the workplace.

CONCLUSION

This Chapter explored how the conventional approach to LAB work and its assessment for undergraduate students in electrical and electronic engineering studies may be enhanced through their use of technology in RALs using a collaborative approach to learning that represented a pedagogical change to conventional practice and provided opportunities for improving formative assessment. Thus, the purpose of the exploratory case study was to investigate the students' perceptions of their RALs learning experience during an intensive collaborative task that required them to master the voltage divider rule. While acknowledging the limitations of the study in terms of the small number of groups the results were favourable regarding the practicality, changed pedagogical approach and opportunity for learning in RALs using the LabView program. The study highlighted the importance of the design of the Lab work experience and the need to address how technology enhanced formative assessment (TEFA) practices can be incorporated. In considering the Lab learning experience on the basis of Ernst's (1983) objectives and the Doolittle (1995) principles, the research found that Herrington et al.'s (2010) guidelines, which draw upon the latter, provided a sound basis to address how formative assessment can be designed to identify feedback markers as recommened by Poth (2018). Without having teased out such markers that are able to 'feed forward' in practice to support students' growth in learning and build them in to task design the learning experience may be less beneficial or even frustrating for students. Thus, task design may also be enhanced by ensuring the subgoals of learning are analysed and explicit, and how students' progression is evidenced or made visible. In the context of collaborative learning the research showed the internal dynamics of the group work can influence the way students perceive the usefulness of their interactions, which also, in moving from a conventional model of in Lab learning, shows it is important for students to be included in the discussion of the purposes of the changes to pedagogy and their potential value for learning. This collaborative learning trial also showed that the need for students to socially interact in the problem-solving process in the Lab experience highlighted the importance of taking advantage of Lab work to develop interpersonal skills because of their relevance to their eventual work in the profession.

Thus, this research supports the idea that technology enhanced formative assessment through the Internet in the higher education sector provides a contemporary yet manageable and pro-active challenge to traditional in-Lab physical, practical pedagogy in undergraduate electrical and electronic engineering studies. This student-experience study of a dialogic collaborative approach to LAB work tasks on the Voltage Division Rule using Remote Access Laboratories, offers evidence of the need for innovation in task design preparation for deeper thinking collaborative approaches to assessment incorporating co-constructed feedback. Many skills are required of a 'communicative engineer' in addition to the expected cognitive skills of instrumentation, modelling, problem-solving, experimenting, analyzing data and designing. The psychomotor ability to manipulate instruments, learning from failure, adjusting creatively, by communicating and through teamwork, along with enhanced sensory awareness for safety and ethics in the laboratory, were additional challenges that a cooperative approach with peers enabled students to face in terms of their attitude and behavior. This chapter highlighted student perceptions of the benefits, challenges and need for a pedagogical change towards more social constructivist approaches towards mastering the use of authentic RALs applications in engineering tasks. Collaborative interaction with peers and tutors in the RALs technology-supported online environment contrasted positively against conventional isolation of completing summative assessment tasks.

ACKNOWLEDGMENT

We acknowledge and thank the participants for their willingness to participate and contribute to the research.

REFERENCES

Baladoh, S. M., Elgamal, A. F., & Abas, H. A. (2017). Virtual lab to develop achievement in electronic circuits for hearing-impaired students. *Education and Information Technologies*, *22*(5), 2071–2085. doi:10.100710639-016-9532-7

Balamuralithara, B., & Woods, P. C. (2009). Virtual laboratories in engineering education: The simulation lab and remote lab. *Computer Applications in Engineering Education*, *17*(1), 108–118. doi:10.1002/cae.20186

Bennett, R. E. (2011). Formative assessment: A critical review. *Assessment in Education: Principles, Policy & Practice*, *18*(1), 5–25. doi:10.1080/0969594X.2010.513678

Boud, D. (2013). *Enhancing learning through self-assessment*. Routledge. doi:10.4324/9781315041520

Boylestad, R. L. (2015). Introductory circuit analysis (13th ed.). Boston: Pearson.

Chan, C., & Fok, W. (2009). Evaluating learning experiences in virtual laboratory training through student perceptions: A case study in Electrical and Electronic Engineering at the University of Hong Kong. *Engineering Education*, *4*(2), 70–75. doi:10.11120/ened.2009.04020070

Corter, J. E., Esche, S. K., Chassapis, C., Ma, J., & Nickerson, J. V. (2011). Process and learning outcomes from remotely-operated, simulated, and hands-on student laboratories. *Computers & Education*, *57*(3), 2054–2067. doi:10.1016/j.compedu.2011.04.009

Council of Chief State School Officers. (2012). *Distinguishing formative assessment from other educational assessment labels*. Washington, DC: Author. Retrieved from http://www.ccsso.org/Documents/FASTLabels.pdf

Dann, B., Dann, C., & O'Neill, S. (2018). Formative assessment via video feedback on practicum: Implications for higher education and professional teacher accreditation bodies. In T. Richardson, B. Dann, C. Dann, & S. O'Neill (Eds.), *Formative assessment practices for pre-service teacher practicum feedback: Emerging research and opportunities* (pp. 158–183). Hershey, PA: IGI Global. doi:10.4018/978-1-5225-2630-8.ch006

Deniz, D. Z., Bulancak, A., & Ozcan, G. (2003, November). A novel approach to remote laboratories. In 33rd Annual Frontiers in Education, 2003. FIE 2003 (Vol. 1, pp. T3E-T3E). IEEE. doi:10.1109/FIE.2003.1315161

Dillenbourg, P. (1999). What do you mean by collaborative learning? In P. Dillenbourg (Ed.), *Collaborative-learning: Cognitive and computational approaches* (pp. 1–19). Oxford, UK: Elselvier Publishing.

Dillenbourg, P., Järvelä, S., & Fischer, F. (2009). The evolution of research on computer-supported collaborative learning: from design to orchestration. In N. Balacheff, S. Ludvigsen, T. de John, T. Lazonder, & S. Barnes (Eds.), *Technology-enhanced learning: Principles and products* (pp. 3–19). Heidelberg, Germany: Springer. doi:10.1007/978-1-4020-9827-7_1

Dittmar, R., & Kahlcke, T. (2016). A lab for undergraduate control engineering education equipped with industrial distributed control systems. *Computer Applications in Engineering Education, 24*(2), 288–296. doi:10.1002/cae.21708

Doolittle, P. E. (1995). *Understanding cooperative learning through Vygotsky's Zone of Proximal Development.* Paper presented at Lilly Conference on Excellence in College Teaching, Columbia, SC.

Edwards-Groves, C., Anstey, M., & Bull, G. (2014). Classroom talk: Understanding dialogue, pedagogy and practice. Newtown, Australia: Primary English Teaching Association Australia (PETAA).

Elawady, Y., & Tolba, A. S. (2009). *Educational objectives of different laboratory types: A comparative study.* arXiv preprint arXiv:0912.0932

Ernst, E. W. (1883). A new role for the undergraduate engineering laboratory. *IEEE Transactions on Education, 26*(2), 49-51.

Estriegana, R., Medina-Merodio, J.-A., & Barchino, R. (2019). Student acceptance of virtual laboratory and practical work: An extension of the technology acceptance model. *Computers & Education, 135*, 1–14. doi:10.1016/j.compedu.2019.02.010

Feisel, L. D., & Rosa, A. J. (2005). The role of the laboratory in undergraduate engineering education. *Journal of Engineering Education, 94*(1), 121–130. doi:10.1002/j.2168-9830.2005.tb00833.x

Foldnes, N. (2016). The flipped classroom and cooperative learning: Evidence from a random experiment. *Active Learning in Higher Education, 17*(1), 39–49. doi:10.1177/1469787415616726

Gadzhunov, S., & Nafalski, A. (2010). Pedagogical effectiveness of remote laboratories for measurement and control. *World Transactions on Engineering and Technology Education, 8*(2), 162–167.

Gagné, N. (2009). *Cooperative learning and oral interaction activities in an intensive Grade 6 ESL program in Quebec* (Unpublished master's thesis). Université Laval, Quebec City, Canada.

Gagnon, G. W., & Collay, M. (2006). *Constructivist learning design.* Corwin Press.

Gibbings, P. (2014). Qualitatively different ways students experience remote access laboratories. *International Journal of Engineering Education, 30*(5), 1120–1129.

Gonzalez, J. (2018). *Moving from feedback to feedforward: Cult of pedagogy*. Retrieved from https://www.cultofpedagogy.com/feedforward/

Herrington, J., Reeves, T. C., & Oliver, R. (2006). Authentic tasks online: A synergy among learner, task, and technology. *Distance Education*, *27*(2), 233–247. doi:10.1080/01587910600789639

Herrington, J., Reeves, T. C., & Oliver, R. (2010). *A guide to authentic e-learning*. New York: Routledge.

Huang, Y.-M. (2015). Exploring the factors that affect the intention to use collaborative technologies: The differing perspectives of sequential/global learners. *Australasian Journal of Educational Technology*, *31*(3), 278–292. doi:10.14742/ajet.1868

Iordache, D., Pop, D., Samoila, C., & Ursutia, D. (2011). Multifunctional laboratories based on Agilent USB technology LabView generated web interface. In J. G. Zubía & G. R. Alves (Eds.), *Using remote labs in education: Two little ducks in remote experimentation* (pp. 135–156). Bilbao, Spain: University of Deusto.

Järvelä, S., Näykki, P., Laru, J., & Luokkanen, T. (2007). Structuring and regulating collaborative learning in higher education with wireless networks and mobile tools. *Journal of Educational Technology & Society*, *10*(4), 71–79.

Kennepohl, D. K., & Shaw, L. (2010). *Accessible elements: teaching science online and at a distance*. Athabasca University Press.

Kollöffel, B., & Jong, T. (2013). Conceptual understanding of electrical circuits in secondary vocational engineering education: Combining traditional instruction with inquiry learning in a virtual lab. *Journal of Engineering Education*, *102*(3), 375–393. doi:10.1002/jee.20022

Leininger, J., & Robinson, J. (1995, November). Combining groupware and total quality management principles to improve engineering students' teamwork skills. In *Proceedings Frontiers in Education 1995 25th Annual Conference. Engineering Education for the 21st Century* (Vol. 2, pp. 4c3-10). IEEE.

Litzier, E., & Young, J. (2012). Understanding the risk of attrition in undergraduate Engineering: Results from the project to assess climate in engineering. *Journal of Engineering Education*, *101*(2), 319–345. doi:10.1002/j.2168-9830.2012.tb00052.x

Lowe, D. B., Berry, C., Murray, S., & Lindsay, E. (2009). Adapting a remote laboratory architecture to support collaboration and supervision. *International Journal of Engineering*, *5*(S1), 51–56.

Maier, W., Wolf, N., & Randler, C. (2016). Effects of a computer-assisted formative assessment intervention based on multiple-tier diagnostic items and different feedback types. *Computers & Education*, *95*, 85–98. doi:10.1016/j.compedu.2015.12.002

Male, S. A., & Bennett, D. (2014). Don't know what we'll be doing yet: Enhancing career preview and engagement among undergraduate engineering students. In *25th Annual Conference of the Australasian Association for Engineering Education: Engineering the Knowledge Economy: Collaboration, Engagement & Employability* (pp.1387-1394). Academic Press.

Mayer, S. J. (2012). *Classroom discourse and democracy: Making meanings together*. New York, NY: Peter Lang.

Moore, D. J., & Voltmer, D. R. (2003). Curriculum for an engineering renaissance. *IEEE Transactions on Education*, *46*(4), 452–455. doi:10.1109/TE.2003.818754

Myllyäki, S. (2013). Incorporation of continuous assessment into lectures in engineering education. *European Journal of Engineering Education*, *1*(1), 1–9.

National Instruments. (2019). *Learn LabView*. Retrieved from http://www.ni.com/academic/students/learn-labview/

O'Neill, S. (2016). Educational growth and the semiotics of the tree: Exploring the OAK as a framework for deep learning. In F. V. Tochon (Ed.), *Deep education in the disciplines and beyond* (pp. 43–69). Blue Mounds, WI: Deep University Press.

O'Neill, S. (2018). Building students' capacity to write English for academic purposes: Pedagogy and the demands of writing persuasively. In L. T. Wong & W. L. Wong (Eds.), *Teaching and learning English for Academic Purposes: Current research and practices* (pp. 69–96). New York: Nova Science Publishers, Inc.

Odeh, S., Mckenna, S., & Abu-Mulaweh, H. (2017). A unified first-year engineering design-based learning course. *International Journal of Mechanical Engineering Education*, *45*(1), 47–58. doi:10.1177/0306419016674508

Osberg, K. M. (1997). *Constructivism in practice: The case for meaning-making in the virtual world* (Unpublished Doctoral Dissertation). University of Washington, Washington, DC.

Osen, O. L., & Bye, R. T. (2017). Reflections on teaching electrical and computer engineering courses at the bachelor level. In *Proceedings of the 9th International Conference on Computer Supported Education (CSEDU 2017)* (vol. 2, pp. 57-68). Academic Press. 10.5220/0006359000570068

Pais Marden, M., & Herrington, J. (2011). Supporting interaction and collaboration in the language classroom through computer mediated communication. In T. Theo Bastiaens & M. Ebner (Eds.), *ED-MEDIA, World Conference on Educational Multimedia, Hypermedia, and Telecommunications* (pp. 1161-1168). AACE.

Palmquist, S. (2007). *Active project-based learning in structural analysis: Field inspection of steel truss bridge*. Paper presented at the American Society for Engineering Education, Honolulu, HI.

Panadero, P., Andrade, H., & Brookhart, S. (2018). Fusing self-regulated learning and formative assessment: A roadmap of where we are, how we got here, and where we are going. *Australian Educational Researcher*, *45*(1), 13–31. doi:10.100713384-018-0258-y

Pea, R. D. (2004). The social and technological dimensions of scaffolding and related theoretical concepts for learning, education, and human activity. *Journal of the Learning Sciences*, *13*(3), 423–451. doi:10.120715327809jls1303_6

Poth, C.-A. (2018). The contributions of mixed insights to advancing technology-enhanced formative assessments within higher education learning environments: An illustrative example. *International Journal of Educational Technology in Higher Education, 15*(9). doi:10.118641239-018-0090-5

Quaye, S. J., & Harper, S. R. (Eds.). (2015). *Student engagement in higher education: Theoretical perspectives and practical approaches for diverse populations* (2nd ed.). New York, NY: Routledge.

Roschelle, J., & Teasley, S. D. (1995). The construction of shared knowledge in collaborative problem solving. In C. E. O'Malley (Ed.), *Computer-supported collaborative learning* (pp. 69–197). Berlin: Springer-Verlag. doi:10.1007/978-3-642-85098-1_5

Sánchez-Vera, M. M., & Prendes-Espinosa, M. P. (2015). Beyond objective testing and peer assessment: Alternative ways of assessment in MOOCs. RUSC. *Universities and Knowledge Society Journal, 12*(1), 119–130. doi:10.7238/rusc.v12i1.2262

Schkoda, R. F., Schweisinger, T. A., & Wagner, J. R. (2012). An improved undergraduate mechanical engineering laboratory structure and curriculum: Design and assessment. *International Journal of Mechanical Engineering*, *40*(3), 182–195.

Tseng, C., Wu, B., Morrison, A. M., Zhang, J., & Chen, Y. (2015). Travel blogs on China as a destination image formation agent: A qualitative analysis using Leximancer. *Tourism Management*, *46*, 347–358. doi:10.1016/j.tourman.2014.07.012

Walsh, S. (2006). Talking the talk of the TESOL classroom. *ELT Journal*, *60*(2), 133–141. doi:10.1093/elt/cci100

Wicker, R. B., & Quintana, R. (2000). An innovation-based fluid mechanics design and fabrication laboratory. *Journal of Engineering Eucation*, *39*(3), 361–367.

Wiliam, D. (2011). What is assessment for learning? *Studies in Educational Evaluation*, *37*(1), 3–14. doi:10.1016/j.stueduc.2011.03.001 PMID:22114905

Williams, K. C., & Williams, C. C. (2011). Five key ingredients for improving student motivation. *Research in Higher Education*, *12*, 1–23.

Yadav, A., Subedi, D., Lundeberg, M. A., & Bunting, C. F. (2011). Problem-based learning: Influence on students' learning in an electrical engineering course. *Journal of Engineering Education*, *100*(2), 253–280. doi:10.1002/j.2168-9830.2011.tb00013.x

Zubía, J. G., & Alves, G. R. (Eds.). (2012). *Using remote labs in education: two little ducks in remote experimentation* (Vol. 8). Universidad de Deusto.

Chapter 12
An Exploratory Study to Understand the Phenomena of Eye-Tracking Technology:
A Case of the Education Environment

Ahmed Syed
University of Southern Queensland, Australia

Subrata Chakraborty
University of Technology Sydney, Australia

Abdul Hafeez-Baig
University of Southern Queensland, Australia

Anuradha Mandal
University of Southern Queensland, Australia

ABSTRACT

Technology has played a pivotal role in revolutionizing the formative aspects of learning and teaching in the current digital age. Due to technology, there is an expectation of having customized medicine, customized interaction, and customized formative communication instead of traditional mass reporting approaches. Formative assessment within higher education teaching and learning environments are no exception to such an approach in the 21st century digital environment. Eye-tracking technology in recent years has provided an insight to understand the human eye movements and concentration patterns, which has application in education. Eye-tracking can be used to examine the processes of individuals in their learning to establish how learning contents are delivered and perceived by all involved (e.g., teaching staff, students, and markers). This chapter proposes that critical and specific information from eye-tracking software can lead to tailored educational content to accommodate, customize, and optimize the unique learning methods for an individual student as per their learning habits. This chapter describes the available eye-tracking technologies and their application in educational processes.

DOI: 10.4018/978-1-7998-0426-0.ch012

INTRODUCTION

A new era of physiological assessment has arrived. How do our eyes react to various sorts of visual stimulation? What do students look at and where, and for how long? What cognitive workload does their gaze create and how can learning designers use this understanding to optimize the educational process? The answer to such queries is within the realms of possibilities if revolutionary eye-tracking technology and its derivatives can be harnessed. Eye-tracking technology has made it possible to measure cognitive workload objectively and inobtrusively in real time (Palinko, Kun, Shyrokov, & Heeman, 2010). Eye-tracking describes the eyes position and movement recorded in an environment based on the optical monitoring of corneal reflections. While the impression of eye-tracking is quite straightforward, the technology behind it is evolving every day (Farnsworth, 2018a). Eye-Tracking has numerous potential usages, and it can make a significant contribution to formative information provided to learners in the educational process. The innovation is utilized to quantify the effect of visual stimuli and gives understanding into subjective and cognitive engagement (Wang, Yang, Liu, Cao, & Ma, 2014). It discloses what individuals see, to what extent they take notice and what they see or do not take notice of. Eye-tracking is a tool to study visual focus and social interaction in all education settings but the focus of this chapter is its application in the formative learning space and in reducing assessment bias. By researching how different features of the knowledge process shape educational results, researchers can effectively design, evaluate, and improve education practices (Pro, 2019). However, before we ponder into the effectiveness of Eye-tracking technology in the field of formative assessment, let's first understand the basic principles this technology is based upon and its evolution over the period.

WHAT IS EYE-TRACKING?

Eye-tracking is the way of defining the motion of an eye relative to the head (Doherty-Sneddon, 2008). They are the interplay of saccades and fixations movements (Hoffman & Subramaniam, 1995) as shown in Figure 1. Although there are other types of eye movements like pursuit eye movements, where eyes follow a moving target and the action is much slower than saccades. Alternatively, there are the vergence eye movements where eyes are moving inward to fixate a close object or the vestibular eye movements

Figure 1. Eye-tracking trajectories consist of fixations and saccades. When a human focuses on something of interest the eye jitters around this area
(Krueger, Koch, & Ertl, 2016)

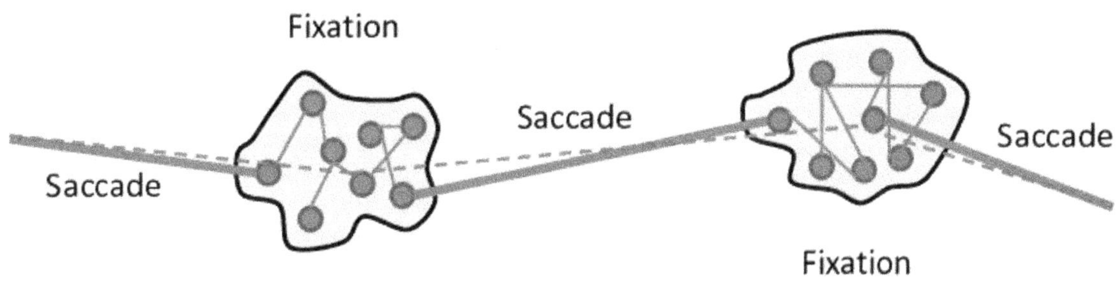

where eyes rotate to compensate for head and body movements (Rashbass & Westheimer, 1961). For eye-tracking information processing in the education environment as a learner or educator, we are interested in the interplay of saccades and fixations i.e. where the eyes move to and how long it stays focused on the item. We can understand this phenomenon better if we try to reveal how students distribute their attention to sections of academic material. It may show, for example, that students do not spend enough time looking at topic sentences or fail to interpret diagrams with text properly. In such cases, by studying the interplay of saccades and fixations, we can gain more insight. Such data analysis could help improving how information is presented for achieving better student focus to significant areas.

When we talk about eye-tracking, fixations and gaze points are the two necessary output measures of our interests and are often the most used terms (Farnsworth, 2018b). Gaze points demonstrate what the eyes are looking at (Farnsworth, 2018b). The eye movements between fixations are usually stated as saccades (Salvucci & Goldberg, 2000). When we read, for example, our eyes do not travel smoothly. We lock our eyes to every 5-7 letters. The term "visual span" refers to how many words we can read before and after the currently fixated word. Trained readers have a higher visual span, allowing them to cover more text with fewer fixations (Farnsworth, 2018b). Another term of importance here is the visual field for learner perspectives. Visual environment represented on the retina when fixating one point. As described below in Figure 2, from the fixated point, the visual field can be divided into three areas: Fovea, Parafovea, Peripheries (Rayner, Inhoff, Morrison, Slowiaczek, & Bertera, 1981). Foveal vision refers to vision in the center of the field of vision, where the highest visual acuity is. Peripheral vision, or indirect vision, is a vision as it occurs outside the fixation point, i.e. outside the gaze center. The Parafovea is the circular area between the Fovea and the Periphery.

Figure 2. The three regions of visual perception: fovea, parafovea and peripheral vision (not to scale) *(Norling, 2009)*

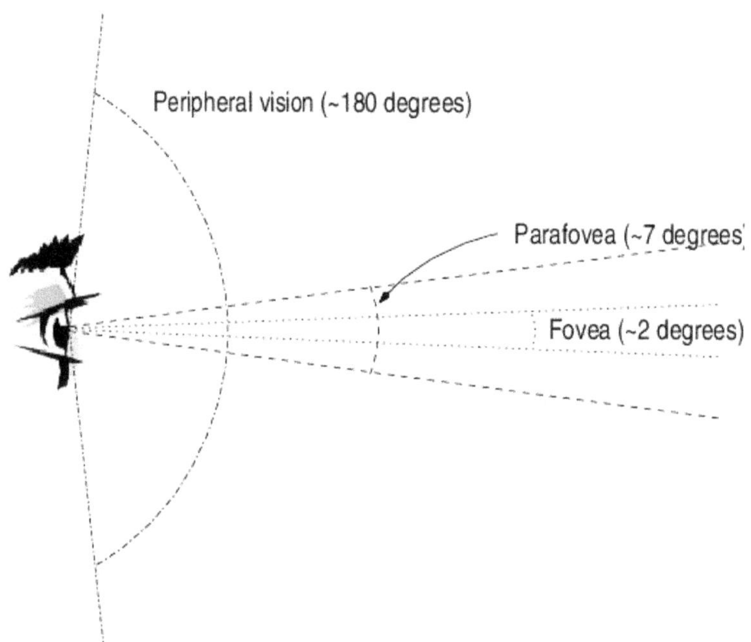

We need to make eye movements to perceive all the details of a visual scene. Our brain uses Saccadic suppression, i.e. reduced sensitivity for optical input during a saccade to "put together the picture" and gives us one coherent visual scene (Ibbotson & Krekelberg, 2011). Pinpoint accuracy in terms of tracking eye pupil movements during foveal and parafoveal focus is critical for accurate eye-tracking. The measure of fixations or gaze points that are coordinated towards a specific area of an image (in respect to different parts) demonstrates that increasingly visual consideration has been organized there. This provides a starting point for understanding which parts of a scene best catch and support our focus. One way to represent this concentration of gaze points is through heatmaps. As Figure 3 below explains, heatmaps are Visualizations that show the overall gaze point distribution. They are typically displayed on the image or stimulus presented as a color gradient overlay. In descending order, the red, yellow, and green colors represent the number of gaze points directed to parts of the image (Bojko, 2009).

Using a heatmaps is an easy way to visualize which elements attract more attention than others quickly. Heatmaps can be compared across individual respondents as well as participant groups, which can be useful in understanding how different people might view a stimulus in alternative ways (Farnsworth, 2018b). Another essential primary term is an Area of Interest, also referred to as an AOI, is a tool for selecting regions of a displayed stimulus and extracting metrics specifically for those regions (Hessels, Kemner, van den Boomen, & Hooge, 2016). Although not strictly a parameter by itself, the area by which other metrics are calculated is defined. For instance, if someone shows a person's picture, separate AOIs can be drawn around the body and face. This phenomenon is also measured as The Time to First Fixation (TTFF) which shows how long it requires a respondent to look at a certain stimulus-initiated AOI. TTFF may indicate both bottom-up stimulus-driven searches and top-down attention-driven searches (Noland, Weiner, Gao, Cook, & Nelessen, 2017).

Figure 3. Example of eye-tracking used in the pre-testing of a questionnaire, heatmaps of reading patterns (Koller, Salzberger, Brenner, & Walla, 2012)

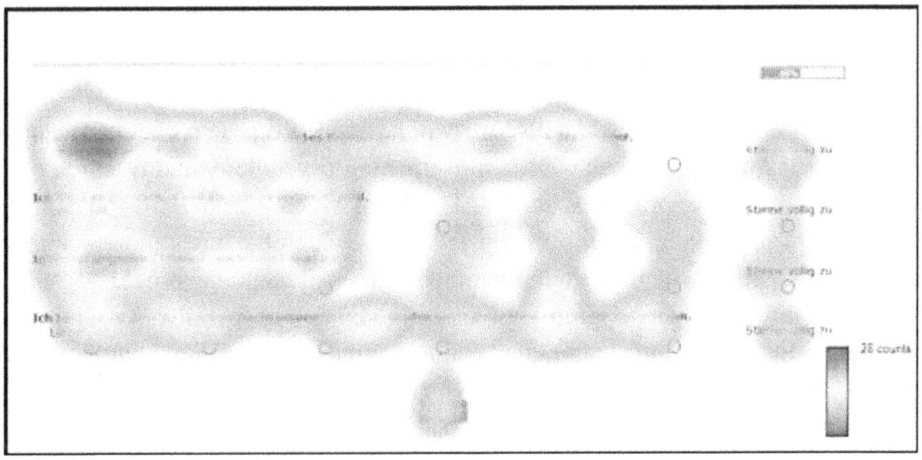

In some cases, a relative increment in time spent on a particular part of an image could be related to motivation (Farnsworth, 2018b). A long period of viewing a particular region may indicate a high level of interest, while shorter periods of viewing may suggest that other areas on the screen or in the environment may be more interesting. Sequences of fixations are also important which are based on information both spatial and temporal, i.e. when and where a participant looked. This makes it possible to create a picture of what a participant prioritizes when they see a visual scene. This often starts with the central fixation bias in the middle of the image, but the following components will be representative of what is most motivating to look for (Tsang, Tory, & Swindells, 2010). As we explore with our eyes a visual scene, we move through saccades before fixing on the part of the image. The first duration of fixation provides information on how long the first fixation lasted. This can be especially valuable when used in conjunction with the Time to First Fixation (TTFF) as an indicator of how much attention an aspect of the scene attracted initially. This is also useful as obtained from an AOI, as it provides information on how long it lasted for the first fixation in a particular region (Farnsworth, 2018b), which can be compared with other regions. This may be useful in determining AOI's first impressions.

Consequently, the average fixation duration tells us how long the average fixation lasted for individuals or groups. This may be useful as a baseline measure and may be useful in understanding the perception of the stimuli among the participants to understand the AOIs (Nyström & Holmqvist, 2010). This section has explained the key terms and concepts of eye-tracking technology. Now to comprehend the advances of the eye-tracking technology, we should give a glimpse of how the technology has evolved.

HISTORY OF EYE-TRACKING

Movements of the eyes have been monitored since the late-1800s. Émile Javal (1839-1907) was a French eye specialist and politician and Director of the ophthalmological laboratory at Sorbonne University (1878-1900), is widely given credit for the discovery of the interplay of saccades and fixations during reading (Lupu & Ungureanu, 2013). Ewald Hering (1834-1918) a German physiologist, described jerky eye movements when eyes go over a line of text (Alpern, 1972). We can divide the evolution of eye-tracking technology in three eras of eye movement research. The first Era was 1879-1920 when first observations on eye movements during reading and discovery of basic facts on eye movements were found. The second Era is 1920-1950, its time of behaviorist movement in psychology and research on the relation of eye movements and cognitive processes and the Third Era which is mid-1970-today. In this era, lots of technical improvements around eye movement recording which becomes more accurate and more accessible to conduct together with the development of theories of language processing which becomes a tool for understanding underlying cognitive processes (Mohamed, Da Silva, & Courboulay, 2007). This understanding will help us understand how technology has evolved and what the latest trends are while we will discuss the specific use cases of eye-tracking technology. In the past decade, eye-tracking technology has branched into many derivatives and has impacted multiple industries.

While eye-tracking technology has been widely used in cognitive psychology for many decades now, it has only recently been used to conduct educational research. In particular, eye tracking was used to study how students solve physics and mathematics problems (Barmby, Andrà, Arzarello, Gomez, & Shvarts, 2014; Hegarty & Just, 1993). Reading and understanding, and learning and interaction in multimedia (Rayner, Chace, Slattery, & Ashby, 2006; van Gog & Jarodzka, 2013). Eye-movement modeling potentials have remained unexplored in computer programming education until a study has done focusing on Source Code Comprehension (Bednarik, Schulte, Budde, Heinemann, & Vrzakova, 2018). Also, eye movements were recorded to investigate possible differences in attention while reading in a classroom setting (Panse, Alcock, & Inglis, 2018).

EYE-TRACKING TODAY AND BEYOND

The eye-tracking technology – how exactly does eye-tracking work today? While early devices were highly intrusive, involving particularly complicated setup procedures, in recent years, modern eye trackers have undergone quite a technological evolution. The rigid experimental configurations and seating arrangements that you might think of are long gone. Current eye trackers are hardly larger than smartphones and provide the respondents with an incredibly natural experience. Figure 4 below is showcasing model for head-mounted eye trackers comprising two main building blocks: Pupil Position Mapping and Display Detection and Gaze Mapping. Remote, non-intrusive methods have made eye-tracking an easy-to-use and accessible tool in human behavior research that enables objective real-time measurement of eye movements (Duchowski, 2007). Most modern eye trackers, together with a high-resolution camera (or other optical sensors), use near-infrared technology to track the direction of the gaze.

Figure 4. Gaze estimation error model for head-mounted eye trackers comprising two main building blocks: Pupil Position Mapping and Display Detection and Gaze Mapping. Model inputs include parameters for Calibration, Eye Tracker and Display
(Barz, Daiber, & Bulling, 2015)

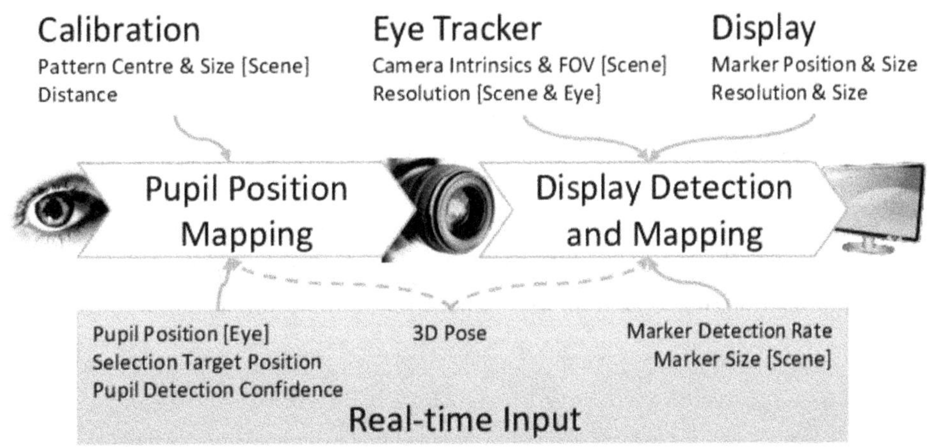

The underlying concept is quite simple, commonly referred to as the Pupil Center Corneal Reflection (PCCR) (Zhu & Ji, 2007). The basic idea is to use a light source to illuminate the eye that causes highly visible reflections, and a camera to monitor the eye image that shows these reflections. The camera captured image is then used to identify on the cornea (glint) and in the pupil the reflection of the light source. Then we can calculate a vector formed by the angle between the reflections of the cornea and the pupil — the direction of this vector is then used to calculate the direction of the gaze, combined with other geometric characteristics of the reflections (Kar & Corcoran, 2017). The visible spectrum is likely to produce uncontrolled reflections, while illuminating the eye with infrared light (Farnsworth, 2018a), which the human eye cannot perceive, makes pupil and iris demarcation an easy task, i.e. while the light enters the pupil directly, it only reflects from the iris. Does the question arise that why we use infrared spectrum imaging? The accuracy of measuring eye movement is heavily based on clear pupil demarcation and corneal reflection detection. This means a clear contrast (with little noise) is generated and therefore algorithms (running inside the eye tracker) can be quickly followed (Kar & Corcoran, 2017). Based on these technologies we have three main eye tracker types: screen-based, wearable glasses and VR headset eye-tracking. Eye-tracking based on webcam was seen as an option, but this technology is inherently inferior to eye trackers based on infrared. To summarize, we have seen remarkable improvements in the technology of eye-tracking. Studies of eye movements may have started more than 100 years ago, but it is safe to say that eye-tracking has come a long way. Psychologists, neuroscientists, marketers, designers, architects are currently employing eye-tracking (Farnsworth, 2018a). You name it; it is happening. In the pages below, we will go through some of the most common eye-tracking application areas and see how it helps guide discoveries and insights in each.

Eye-Tracking Equipment Set Up

Which is the right eye tracker for you? Will your respondents be sitting in front of a session computer? Go for an eye tracker based on a screen. In a natural setting, do your respondents need to move freely? Alternatively, if you are in a formative assessment environment like a classroom, you may need to choose a head-mounted system that enables mobility of the head and body (Evans, Jacobs, Tarduno, & Pelz, 2012). Formative assessment refers to a wide range of methods used by teachers during a lesson, unit, or course to conduct in-process evaluations of student understanding, learning needs, and academic progress. The overall purpose of a formative evaluation is to collect detailed information that can be used to improve instruction and student learning as it occurs.

In addition to the hardware, we certainly need to consider about which recording and data analysis software to use before starting eye-tracking based research. For data acquisition and data processing, separate software is usually required. Although some manufacturers offer integrated solutions, for data inspection and further processing, we will most likely have to export the raw data to a dedicated analysis software (Krassanakis, Filippakopoulou, & Nakos, 2014). In the following sections, we discuss current best practices, available system and general experiment set up to perform an eye-tracking experiment.

Best Practices in Eye-Tracking Experiments

Conducting an eye-tracking study in a formative assessment setting requires a more precise setup as many factors can have an impact on the outcome. For example, a complex experimental design, new respondents, various technologies, various pieces of hardware, various operators all can impact the results of a study. In the educational setting, this is confounded by the inherent complexities of learners. Study failures or complications most often occur because of minor errors. Below we are explaining some of the key practical aspects which should be considered while setting up for a study such as capturing learners reading patterns to improve learning content quality.

Firstly, we need to keep the experimental setup for all participants of the specific study as consistent as possible. Lighting conditions are essential for eye-tracking. Avoid sunlight as it contains infrared light which can affect the measurements of eye-tracking. Use of ambient light is ideally recommended. Maintaining the lighting levels consistent when measuring eye pupil dilation levels is particularly important. However, it is not possible all the time, especially if the experiment requires natural settings like studying the eye movement while driving. Figure 5 is an example of such a setup where driver's eye movements and fixations are monitored to understand driver focus and concentration patterns. In the later sections, we discuss in details how eye tracking can be useful in education settings and provide an example case study showing how marking bias could be reduced with such technology.

Secondly, ensure that all involved participants (such as students) are properly trained. It is essential to train the participants involved in data collection on the systems used in order to have a level of knowledge that allows them to run a study smoothly.

Thirdly, simplifying the technology setup is equally important. Each biometric sensor can reveal a specific human cognition, emotion, and behavioral aspect. Consider combining eye tracking with two or more additional biosensors to gain meaningful insights into the dynamics of attention, emotion, and motivation, depending on your individual research question in formative assessment and other educa-

Figure 5. Experimental setup inside the vehicle
(Kun, Paek, Medenica, Memarovic, & Palinko, 2009)

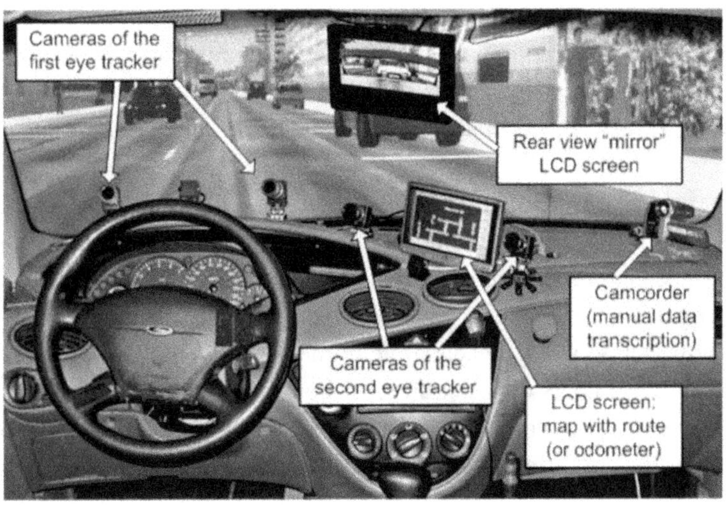

tional studies. We should use as few vendors as possible for both hardware and software to ensure that they interact well and are integrated well with each other.

Choosing the right eye-tracking equipment for the planned study is of high importance. In a classroom setting, we can have screen-based eye trackers which require respondents to sit before a screen or near the stimulus used in the experiment. Although screen-based systems only track the eyes within certain limits, they can move a limited amount as long as the limits of the eye tracker's range; this range is called the head box range. Another option would be to use Eye tracking glasses. Eye Tracking glasses are fitted close to your eyes and thus allow respondents to move as freely as they want – certainly a plus if the study design requires respondents to be present in different areas. Some of the key specifications to look for to get the ideal eye tracker for formative assessment is explained here:

- **Accuracy:** The accuracy of measurement is measured in degrees. Standard is approximately 0.5 degree. Low - end hardware starts at about 1.0 degree, medium 0.5 degrees, and high - end at or below 0.1 degrees (Imotions, 2017). Better accuracy will help identify areas of interests (AOI) for a student correctly while he is reading a document such as marking rubrics or assignment specifications.
- **Trackability:** Another criterion to look for is the ability to track. How many participants can be tracked? The best systems track about 95% of the population, with fewer low-end systems (Imotions, 2017). This will be particularly useful if the study involves tracking multiple students at the same time.
- **Sampling:** The sampling rate is a major factor in hardware selection. The sampling rate is how many times per second equipment measures the eye position? The typical range of values are between 30 and 60 Hz for low end systems. Research projects usually require equipment with a sampling rate of 120 - 1000 + Hz (Imotions, 2017). Higher sampling rate provides more accurate tracking.
- **Size:** Headbox size is also important. Does it mean how far it is allowed for the respondent to move to the eye tracker in the relative distance? A good system typically allows movement in each direction around 11 inches (28 cm) (Imotions, 2017). This aspect needs to be considered while running study with different screen sizes.
- **Recapture Rate:** How quickly the eye tracker detects the position of the eye after the eyes are out of sight for a moment (Imotions, 2017). We need equipment with higher recapture rates for research studies. For studies in education space, having a system with better recapture rate would help with reducing errors related to large head movements.

Aside from the hardware, it is also vital to select the right software for eye-tracking studies. An appropriate software should be scalable to meet the requirements of the studies such as the tracking in formative assessment. Considerations should be given whether there is a need to add other biometric sensors that capture cognitive, emotional or physiological processes conveniently. Each of these are important pieces of information to assist an academic in the assessment of student progress towards their learning goals or the learning goals of the course. It should be able to adapt data recording as well as data analysis. Therefore, it can be synthesized that integration of both hardware and software is critical reliability and validity of the data captured through eye-tracking technology in a learning environment.

Leading Eye-Tracking Systems

Over the past few years, eye-tracking technology has improved dramatically, leading to both more advanced and more accessible hardware development. As with most products, there is a more significant difference between the market's low-end and high-end hardware and the firms that produce it. Table 1 shows the comparison of different eye-tracking hardware producers and products available in the market.

Generic Setup for Eye-Tracking System

In order to conduct studies including formative assessment tests with an Eye-tracker, the system must be set up first to be able to capture and analyze data. Although there may be some variations, the key set up process can be described as a two-stage process discussed below.

Stage 1 (equipment set up): In this stage the equipment is installed and consists of two steps.

Table 1. Comparison of several market leading eye-tracking products (Farnsworth, 2017)

Producers	Number of Research Publications	Products	Key Information
Tobii Pro	9230	• Tobii Pro Spectrum • Tobii Pro X3 • Tobi Pro X2 • Tobii Pro Nano • Glasses 2 • Vive VR Headset	• Wide product range (30 – 1200 Hz) • Excellent suite of software available • The low-end products mainly work with Tobii Gaming SDKs, while mid-range and high-end products are more suitable for research studies. • Can integrate with biometric data such as EEG, ECG, EMG, GSR and NIRS. • https://www.tobiipro.com/
Gazepoint	39	• GP3 HD • GP3 • Biometrics	• Eye tracking and Biometric equipment (60/150 Hz) • Capture and analytics software available • Biometric integration supported • https://www.gazept.com/
SensoMotoric Instruments (SMI)	6040	• SMI RED • SMI RED-M • SMI Glasses	• Both glass and screen-based systems (60-500 Hz) • Software package available • Can integrate biometrics • Product currently may be unavailable due to the company acquisition by Apple Inc. • https://www.smivision.com/
Eyetech Digital Systems	441	• VT3 Mini • VT3 XL	• Screen-based systems (60 Hz) • Capture and analysis software available • https://www.eyetechds.com
EyeLink	5530	• EyeLink 1000 Plus • EyeLink Portable Duo • EyeLink II	• Screen-based and head mounted systems (up to 2000 Hz) • Software available • Can be used with biometric systems • https://www.sr-research.com/
ISCAN Inc.	2650	• RK-446 • RK-447	• Head-mounted and remote eye trackers (up to 1000Hz) • Software available • Can track multiple targets • http://www.iscaninc.com
LC Technologies	1130	• Eyegaze Edge	• Small screen-based system (60 Hz) • Software available • https://eyegaze.com/

Figure 6. Heat map is showing visual attention aggregated via an eye-tracking camera. Warmer colors (e.g., red) on the heat map represent higher concentrations of visual attention
(Scherr et al., 2017)

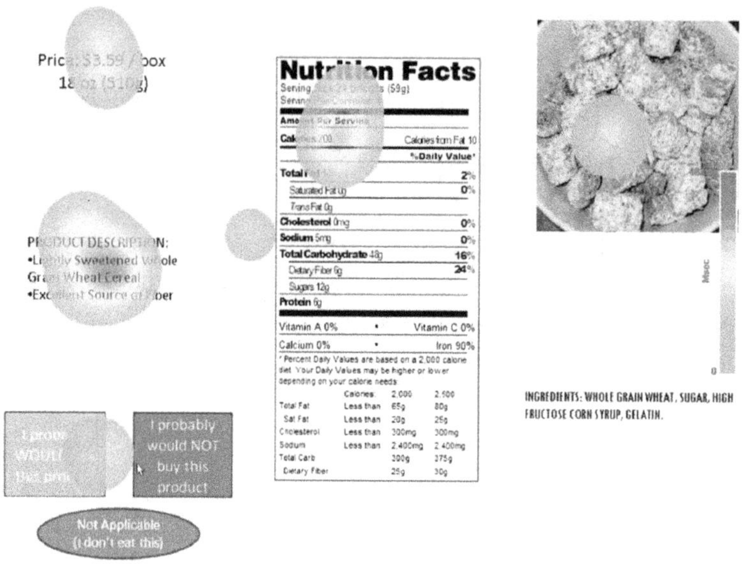

Step 1 *(hardware):* Install the hardware by making the connections between computer systems, screens, peripherals and any other supporting systems.

Step 2 *(software):* Install the software necessary for data capture and data analysis.

Stage 2 (Study set up): In this stage the equipment is used for studies in four steps.

Step 1 *(stimuli content preparation):* The resources for the visual stimulating contents are selected and related files or links are prepared. These resources could be text, graphic, video, websites etc.

Step 2 *(calibration):* The tracking equipment is calibrated using a nine or five-point calibration process. It ensures that the participant is sitting at a correct distance in front of the display screen and the head movement is within the acceptable range.

Step 3 *(data capture):* Run the stimuli contents on the screen and record the eye movement using the software. The capture process overlays the eye-movements and gaze on top of the contents of the stimuli.

Step 4 *(data analysis):* Use software to analyze the captured data to produce results such as heatmaps. If other data such as audio and biometric is captured, they are compared with eye tracker data based on timestamps.

APPLICATION OF EYE TRACKING

Think about all the manners in which you will utilise your eyes today. For example, you may drive a vehicle you may read a magazine, you may surf the web or search the planogram of a supermarket, play a computer game, watch a motion picture, or take pictures on your cell phone. These are the use cases of eye-tracking. Without many exemptions, anything with a visual part can be eye-tracked. We utilise our eyes always and seeing how we utilise them has turned into a critical thought leading towards cutting edge research and further innovation.

For example, healthcare, transport, and defence & security industries have used eye-tracking innovations to make us more secure. The fields like marketing, entertainment, and website architecture have all benefited fundamentally from studying the visual conduct of the user. Research has produced exciting leaps forward in cognitive and physiology study. Consistently, as eye-tracking is utilised in innovative new ways, the use-cases are growing. An American airline is using eye-tracking advancements for assessment of the carrier's email correspondences, internet publicising and self-check-in terminal kiosks. So also, a large chain of retail discount stores in North America is utilising a similar innovation for assessing the convenience and permeability of their website, intended to help consumers. Another global security organisation is utilising eye-tracking advances for the Index of Cognitive Activity (ICA) to assess two renditions of a mission-critical system intended to monitor air, surface and subsurface traffic for the Space and Naval Warfare Command in the USA. The objective was to evaluate the psychological workload of operators of both systems.

Neuroscience and psychology use eye-tracking to analyze the gaze pattern sequence to gain deeper insight into cognitive processes that underlie attention, learning, and memory. Eye-tracking in combination with conventional research methods or other biosensors can help (Farnsworth, 2018a) to evaluate and potentially diagnose conditions such as ADHD, Autism Spectrum Disorder (ASD), Obsessive Compulsive Disorder (OCD), schizophrenia, Parkinson's disease, and Alzheimer's disease (Vidal, Turner, Bulling, & Gellersen, 2012). Besides, eye-tracking technology can be used to detect drowsiness or support multiple other medical, quality assurance or monitoring fields (Vidal et al., 2012).

Why do some products give customers an impression while others just do not get it right? In market research, eye-tracking has become a popular and increasingly vital tool. Many leading brands are actively using eye-tracking to evaluate customer attention to key messages and advertising, as well as evaluating product performance, product and package design, and overall customer experience (Wedel & Pieters, 2008). Figure 6 as an example of label design and use of heatmap. Similar application can be brought into education and learning such as students reading feedbacks.

When applied to in-store testing, eye-tracking provides information about in-store navigation (Farnsworth, 2018b) ease and difficulty, search behaviour, and buying choices (Wedel & Pieters, 2008). Figure 7, below, is an example of how head-mounted eye-tracking is used to observe the search behaviour and buying choices.

Driving Simulation often uses eye-tracking in combination with several other sensors to gain a better understanding of human behaviour in hazardous situations. When facing obstacles on the street, where do drivers look? How does it affect driving behaviour by talking on the phone (Kun et al., 2009)? How exactly does visual attention compromise speed (Palinko et al., 2010)? How exactly does visual attention compromise speed? Such insights can help improve the awareness of hazards and be applied to enhance the safety of future drivers (Palinko et al., 2010). For a long time, automotive research has embraced eye-tracking glasses to assess the visual attention of drivers – both in terms of navigation and dashboard layout (Manary, Flannagan, Reed, & Schneider, 1998). Cars may even be able to respond to the eye gaze, eye movements, or pupil dilation of the drivers in the near future (Lawson, Salanitri, & Waterfield, 2016). Within the education space students can be monitored while learning using latest technology assisted learning space such as virtual reality based learning contents.

Figure 7. User is wearing eye-tracking glasses in a supermarket shopping task (Blascheck, Burch, Raschke, & Weiskopf, 2015)

Human-Computer Interaction (HCI) research focuses on how computers are used and designed, and how people use them. It is possible to evaluate the use of technology from Laptops, Tablets, Smartphones and beyond by measuring our visual attention to the devices we use (Bruneau, Sasse, & McCarthy, 2002). Usability and user experience testing is a rapidly growing field that uses eye-tracking as a methodology for evaluation. A frequently used approach is eye-tracking for website testing, giving insights into how websites are viewed and experienced. How do people take care of advertisements, communications? Eye-tracking data can provide valuable insights into the website visitor's gaze patterns–how long does it take them to find a particular product on your website, what kind of visual information do they ignore (Bergstrom & Schall, 2014)? Even mobile apps on tablets and smartphones can use the same investigations. Figure 8 below is an example of the use of eye-tracking heatmap for webpage designing.

Eye-tracking has been introduced in the gaming industry recently and has since become an increasingly prominent tool as designers are now able to evaluate and quantify measures such as visual attention and reactions to key moments during gameplay to improve the overall gaming experience (Bernhaupt, Eckschlager, & Tscheligi, 2007). The data can be used by designers to measure emotional and cognitive responses to gaming when combined with other biometric sensors. Based on pupil dilation and eye movements, new trends and developments may soon make it possible to control the game (Jönsson, 2005).

Figure 8. Heat map for Kalahari.net home page
(Pretorius & Calitz, 2010)

Figure 9. Using mobile eye-tracking to count read words. Left: A user reads while wearing the SMI iView X eye tracker. Lower centre: The eye tracker uses infrared light and stereo camera views to track eye gaze. Right: The Wordometer application maps.
(Kunze, Iwamura, Kise, Uchida, & Omachi, 2013)

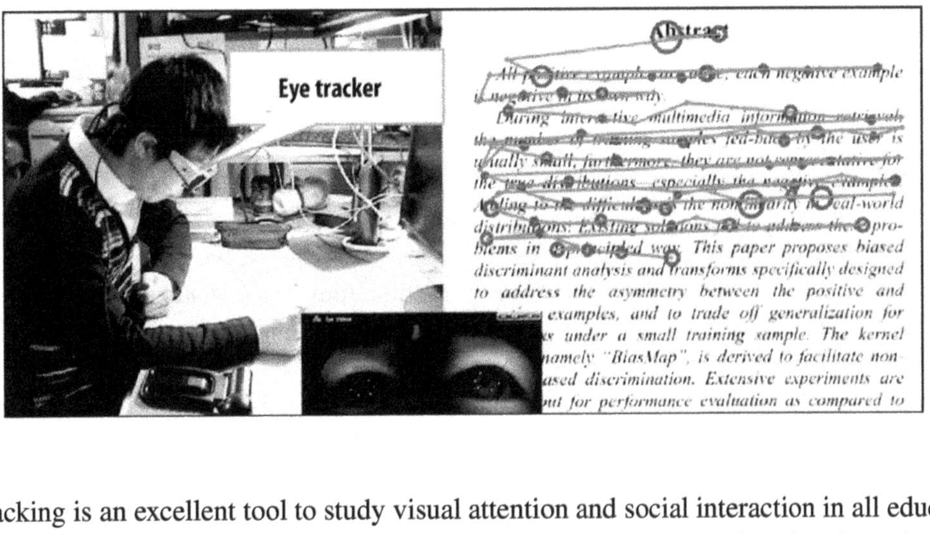

Eye-tracking is an excellent tool to study visual attention and social interaction in all education settings. By knowing how different aspects of the knowledge process shape educational results, researchers can effectively design, evaluate, and improve education. One of the goals of biometric eye-tracking in education is that it can help teachers understand student behaviour such as how course material is absorbed (Hooijdonk, 2016). A group of researchers in Open University of the Netherlands and Lund University, Sweden performed a study on three major research areas in Educational Science that have already successfully used eye-tracking (Jarodzka, Holmqvist, & Gruber, 2017). In the following sections, the way eye-tracking technologies can be used in the education sector are explored.

EYE-TRACKING IN EDUCATION

What if learning for all of us could be an equally satisfying experience? What exactly is required to make learning a success? Over the past few years, eye-tracking technology has made its way into educational science to help gain insights into learning behaviour in a variety of settings ranging from traditional "chalk and talk" approaches to teaching through digital learning. Eye-tracking is an extraordinary tool to understand the visual focus and social association in all learning settings. By seeing how unique parts of the learning procedure shape teaching results, analysts can successfully configure, assess, and improve teaching contents and approaches. Figure 9 below shows one such study where the user was asked to read a paragraph with a head-based eye-tracking device. One of the advantages of biometric eye-tracking in education is that it can enable educators to comprehend student's conduct, for example, how course material is consumed. A gathering of analysts in Open University of the Netherlands and Lund University, Sweden performed an investigation on three research zones in Educational Science that have as of now effectively utilised eye-tracking. (Jarodzka et al., 2017). First, eye-tracking has been used to enhance the instructional design of computer-based learning and testing environments, often using hypermedia or multimedia. Second, eye-tracking has given light on expertise and its development in visual domains. Third, eye-tracking has also been used to focus on visual expertise using eye movement modelling examples (Jarodzka et al., 2017). Although the study proposes further research on the topic of the role of eye-tracking to further corroborate, improve, and enrich educational theories on human cognitive processing, it does explain how eye-tracking can help us to understand how learners process instructional material.

Also, with eye-tracking analysis students' visual attention during class can be determined. Such analyses provide valuable information about which elements they are interested in and which distract or go unseen.

In another study, researchers have analyzed relevant works of 13 years (2000- 2012) to review empirical studies that have employed the eye-tracking technology to probe the cognitive processes during learning (Lai et al., 2013). The fundamental objective for this study was to distinguish how the eye movement approach has been applied to the researches of various learning topics, and what eye movement measures have been utilised to build a framework uncovering how the eye-tracking methods can teach us about cognitive exercises identified with learning. The study results demonstrate that the utilisation of eye-tracking technologies in contemplating learning-related ways gives a promising result to associate learning results in relevant cognitive processes.

Eye-tracking can also be used in assessing teacher's situational awareness while supervising a classroom. Maintaining few of the main variables of the classroom low limits is one of the most crucial goals of every teacher; this activity is performed by continuous visual information processing. A study was done to investigate which cues teachers detect and process from their students during instruction

(Dessus, Cosnefroy, & Luengo, 2016). Teachers' attentional capacity is limited, so they cannot equally draw their attention to every event occurring during instruction, or on every learner. A well-established fact is that every teacher has to keep an overall awareness of the instructional situation (Kounin, 1970). In any case, the sort of psychological procedures embraced to keep up this mindfulness has been concentrated so far generally from verbalisation methods, which are known to offer deficient access to the activity and choice subjective procedures because of their partly inherent nature (Ericsson & Simon, 1980). Eye-tracking technologies have turned into a dependable method to beat this issue (Salvucci & Anderson, 2001). They empower the eye fixation and saccades with the goal that two snippets of data can be surmised (Mele, Federici, & Technology, 2012).

Marker Bias Analysis as a Case Study

It is not only students, who could be analyzed with eye-tracking capabilities, but they can also expect more accurate marking of most subject's assessment scripts. Research studies investigating the relationship between examiners' background and marking performances suggested that different subjective standards affect marking behaviours (Meadows, 2006). These subjective norms differ among individuals and change in the same assessment to different junctures of marking performed.

Debnath et al. (2016) conducted a study to understand how different markers read exam scripts while marking. The study utilised eye-tracker to track markers' reading and focus patterns while marking. The study utilised a Tobii Pro TX300 Eye tracker with a desktop computer having two 23" monitors with screen resolution of 1920x1080.

The study followed the two-stage setting up process discussed earlier. The TX300 Eye tracker was connected to the desktop computer and was set underneath one of the screens set for the participant markers. The necessary software package from Tobii was installed on the desktop which is capable of both capturing and analysing the data. The experiment stage first collected the stimuli content for the study. In this study exam answer scripts from 12 students were scanned to be used as the stimuli content. Calibration was performed using nine-point built in calibration system of the Tobii software. The exam scripts for each student were shown on screen while each marker viewed them separately and marked them. Eyes of the markers were tracked on each occasion. The collected data was then analyzed to understand each marker's focus areas and reading patterns. Results obtained through use of eye-tracking technologies with multiple assessors indicate that different evaluators marked the same response script differently and their visual pattern is also varied, although they were given the same instructions (Debnath et al., 2016).

Figure 10, below demonstrates the resulting gaze plots and heat maps for one such experiment in the study. This evidence indicates that the similarity of the visual contour with the benchmarked is a good indicator of quality marking. The heat maps provide information about the time spent by the individual marker in various areas. The heat map for marker 2 is vastly different from other markers including the subject coordinator (Debnath et al., 2016). The case study shows a practical example of how eye-tracking technology could be adopted in education practices. Similar usage could be to improve content quality presented to the students. For example, information provided on university websites could be tested with students using eye-trackers to find out if students are finding key information easily, thus providing design

Figure 10. Gazemap and heatmap for 3 markers for the same assessment item (Script)

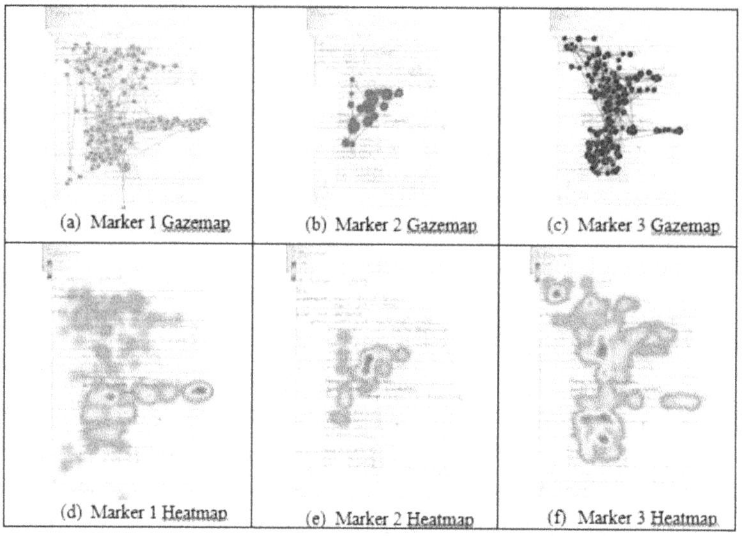

and content improvement feedbacks. Further, course contents and assignment requirement specifications can be tested with students utilising eye-trackers. This will help academics to understand if students are looking at the right areas, providing visual feedbacks for reorganising key information presented. Eye-tracking can be highly useful for visual content-based teaching where photograph, art, videos are used.

CONCLUSION

A key pillar in education research is helping students to learn. In this endeavour, eye-tracking can help by providing an additional piece of data that allows researchers to understand better students and the impact of what teachers are doing during lectures with students. A key aspect is to investigate student attentiveness patterns in the classroom environment to gain a better understanding of what keeps students focused during the lecture and generate new insights into effective teaching techniques and strategies to keep students engaged and motivated to learn during the lecture. Eye-tracking can provide useful insights to be able to develop effective learning and teaching strategies. It is not new to measure and respond to human eye movement or eye-tracking, but the recent advances in technology are unlocking the potential of eye-tracking technology. Soon eye-tracking will be a standard feature of a new generation of smartphones, laptops and desktop monitors, setting the stage for a massive re-evaluation of how we communicate with devices and/or how they interact with us. Eye-tracking can also be a breakthrough tool in all learning settings to study visual attention and social interaction. By understanding how various aspects of the knowledge process shape educational outcomes, researchers can design, evaluate and improve educational environments effectively.

REFERENCES

Alpern, M. (1972). Eye movements. In D. Jameson & L. M. Hurvich (Eds.), Handbook of sensory physiology VII/4: Visual psychophysics (pp. 303-330). Berlin: Springer. doi:10.1007/978-3-642-88658-4_12

Barmby, P., Andrà, C., Arzarello, F., Gomez, D. O. A., & Shvarts, A. (2014). The use of eye-tracking technology in mathematics education research. In P. Liljedahl, C. Nicol, S. Oesterle, & D. Allan (Eds.), *Proceedings of the Joint Meeting of PME 38 and PME-NA 36* (Vol. 1, pp. 253-371). Vancouver: Psychology of Mathematics Education.

Barz, M., Daiber, F., & Bulling, A. (2015). *Computational modelling and prediction of gaze estimation error for head-mounted eye trackers*. Retrieved from http://citeseerx.ist.psu.edu/viewdoc/download?doi=10.1.1.717.2872&rep=rep1&type=pdf

Bednarik, R., Schulte, C., Budde, L., Heinemann, B., & Vrzakova, H. (2018). Eye-movement modeling examples in source code comprehension: A classroom study. *Proceeding. Koli Calling'18 Proceedingsof the 18th Koli Calling International Conference on Computing Education Research, Article No. 12*. New York: ACM. doi10.1145/3279720.3279722

Bergstrom, J. R., & Schall, A. (2014). *Eye tracking in user experience design*. Amsterdam: Elsevier.

Bernhaupt, R., Eckschlager, M., & Tscheligi, M. (2007). Methods for evaluating games: how to measure usability and user experience in games? *Proceedings of the International Conference on Advances in Computer Entertainment Technology*. 10.1145/1255047.1255142

Blascheck, T., Burch, M., Raschke, M., & Weiskopf, D. (2015). Challenges and perspectives in big eye-movement data visual analytics. In *Proceedings of the 1st International Symposium on Big Data Visual Analytics*. IEEE. 10.1109/BDVA.2015.7314288

Bojko, A. A. (2009). Informative or misleading? Heatmaps deconstructed. In J. A. Jacko (Ed.), Lecture Notes in Computer Science: Vol. 5610. *Human-computer interaction: New trends. HCI 2009* (pp. 30–39). Berlin: Springer. doi:10.1007/978-3-642-02574-7_4

Bruneau, D., Sasse, M. A., & McCarthy, J. (2002). The eyes never lie: The use of eyetracking data in HCI Research. In *Proceedings at Conference on Human Factors in Computing Systems*. ACM.

Debnath, T., Paul, M., Chakraborty, S., Podder, P. K., Gururajan, R., & Hafeez-Baig, A. (2016). *An experimental analysis of assessor specific bias in a programming assessment in multi-assessor scenarios utilizing an eye tracker*. Paper presented at the London International Conference on Education: Proceedings (LICE-2016).

Dessus, P., Cosnefroy, O., & Luengo, V. (2016). "Keep your eyes on'em all!": A mobile eye-tracking analysis of teachers' sensitivity to students. In K. Verbert M. Sharples & T. Klobučar (Eds.), *Adaptive and adaptable learning. Proc. 11th European Conf. on Technology Enhanced Learning (EC-TEL 2016)* (pp. 72–84). New York: Springer. 10.1007/978-3-319-45153-4_6

Doherty-Sneddon, G. (2008). Eyes: A new window on mental disorders. *Scientific American*. Retrieved from https://www.scientificamerican.com/article/eyes-mental-disorders/

Duchowski, A. T. (2007). Eye tracking methodology: Theory and Practice (2nd ed.). Dordrecht: Springer.

Ericsson, K. A., & Simon, H. A. J. P. (1980). Verbal reports as data. *Psychological Review, 87*(3), 215–251. doi:10.1037/0033-295X.87.3.215

Evans, K. M., Jacobs, R. A., Tarduno, J. A., & Pelz, J. B. (2012). Collecting and analyzing eye tracking data in outdoor environments. *Journal of Eye Movement Research, 5*(2), 6.

FarnsworthB. (2017). Retrieved from https://imotions.com/blog/top-eye-tracking-hardware-companies/

Farnsworth, B. (2018a). *Eye tracking: The complete pocket guide.* Available at https://imotions.com/blog/eye-tracking/

Farnsworth, B. (2018b). *7 most used eye tracking metrics and terms.* Retrieved from https://imotions.com/blog/7-terms-metrics-eye-tracking/

Hegarty, M., & Just, M.-A. (1993). Constructing mental models of machines from text and diagrams. *Journal of Memory and Language, 32*(6), 717–742. doi:10.1006/jmla.1993.1036

Hessels, R. S., Kemner, C., van den Boomen, C., & Hooge, I. T. (2016). The area-of-interest problem in eyetracking research: A noise-robust solution for face and sparse stimuli. *Behavior Research Methods, 48*(4), 1694–1712. doi:10.375813428-015-0676-y PMID:26563395

Hoffman, J. E., & Subramaniam, B. (1995). The role of visual attention in saccadic eye movements. *Perception & Psychophysics, 57*(6), 787–795. doi:10.3758/BF03206794 PMID:7651803

Hooijdonk, R. v. (2016). *The magic of biometric eye-tracking in education.* Retrieved from https://www.emergingedtech.com/2016/06/the-magic-of-biometric-eye-tracking-in-education/

Ibbotson, M., & Krekelberg, B. (2011). Visual perception and saccadic eye movements. *Current Opinion in Neurobiology, 21*(4), 553–558. doi:10.1016/j.conb.2011.05.012 PMID:21646014

Imotions. (2017). Retrieved from https://imotions.com/blog/eye-tracking/

Jarodzka, H., Holmqvist, K., & Gruber, H. (2017). Eye tracking in educational science: Theoretical frameworks and research agendas. *Journal of Eye Movement Research, 10*(1), 1–18.

Jönsson, E. (2005). *If looks could kill–an evaluation of eye tracking in computer games* (Unpublished Master's Thesis). Royal Institute of Technology (KTH), Stockholm, Sweden.

Kar, A., & Corcoran, P. (2017). A review and analysis of eye-gaze estimation systems, algorithms and performance evaluation methods in consumer platforms. *IEEE Access: Practical Innovations, Open Solutions, 5*, 16495–16519. doi:10.1109/ACCESS.2017.2735633

Koller, M., Salzberger, T., Brenner, G., & Walla, P. (2012). Broadening the range of applications of eye-tracking in business research. *Proto Alegre, 23*(1), 71–77.

Kounin, J. S. (1970). *Discipline and group management in classrooms.* New York: Holt, Rinehard & Winston.

Krassanakis, V., Filippakopoulou, V., & Nakos, B. (2014). EyeMMV toolbox: An eye movement post-analysis tool based on a two-step spatial dispersion threshold for fixation identification. *Journal of Eye Movement Research, 7*(1), 1–10.

Krueger, R., Koch, S., & Ertl, T. (2016). SaccadeLenses: Interactive exploratory fltering of eye tracking trajectories. Presentation at the *2016 Conference at IEEE Second Workshop on Eye Tracking and Visualisation*, Baltimore, MD. 10.1109/ETVIS.2016.7851162

Kun, A., Paek, T., Medenica, Z., Memarovic, N., & Palinko, O. (2009). Glancing at personal navigation devices can affect driving: Experimental results and design implications. In *Proceedings of the 1st International Conference on Automotive User Interfaces and Interactive Vehicular Applications* (pp. 129-136). Academic Press. 10.1145/1620509.1620534

Kunze, K., Iwamura, M., Kise, K., Uchida, S., & Omachi, S. (2013). Activity recognition for the mind: Toward a cognitive "quantified self". *Computer, 46*(10), 105–108. doi:10.1109/MC.2013.339

Lai, M.-L., Tsai, M.-J., Yang, F.-Y., Hsu, C.-Y., Liu, T.-C., Lee, S. W.-Y., ... Tsai, C.-C. J. E. r. (2013). A review of using eye-tracking technology in exploring learning from 2000 to 2012. *Educational Research, 10*, 90–115.

Lawson, G., Salanitri, D., & Waterfield, B. (2016). Future directions for the development of virtual reality within an automotive manufacturer. *Applied Ergonomics, 53*, 323–330. doi:10.1016/j.apergo.2015.06.024 PMID:26164106

Lupu, R. G., & Ungureanu, F. (2013). A survey of eye tracking methods and applications. *Buletinul Institutului Politehnic din Iasi. Automatic Control and Computer Science Section, 3*, 72–86.

Manary, M. A., Flannagan, C. A. C., Reed, M. P., & Schneider, L. W. (1998). Development of an improved driver eye position model. Technical Paper 980856. *SAE Transactions: Journal of Passenger Cars*, 107.

Meadows, M. (2006). *Can we predict who will be a reliable marker.* AQA Centre for Education Research and Policy.

Mele, M. L., Federici, S. J. D., & Technology, R. A. (2012). A psychotechnological review on eye-tracking systems: Towards user experience. *Disability and Rehabilitation. Assistive Technology, 7*(4), 261–281. doi:10.3109/17483107.2011.635326 PMID:22117604

Mohamed, A. O., Da Silva, M. P., & Courboulay, V. (2007). A history of eye gaze tracking. *Laboratoire Informatique, Image et Interaction.* Rochelle: Universite de La Rochelle.

Noland, R. B., Weiner, M. D., Gao, D., Cook, M. P., & Nelessen, A. (2017). Eye-tracking technology, visual preference surveys, and urban design: Preliminary evidence of an effective methodology. *Journal of Urbanism, 10*(1), 98–110. doi:10.1080/17549175.2016.1187197

Norling, E. (2009). *Modelling Human Behaviour with BDI Agents* (PhD thesis). Engineering - Computer Science and Software Engineering, The University of Melbourne.

Nyström, M., & Holmqvist, K. (2010). An adaptive algorithm for fixation, saccade, and glissade detection in eyetracking data. *Behavior Research Methods*, *42*(1), 188–204. doi:10.3758/BRM.42.1.188 PMID:20160299

Palinko, O., Kun, A. L., Shyrokov, A., & Heeman, P. (2010). Estimating cognitive load using remote eye tracking in a driving simulator. *Proceedings of the 2010 symposium on eye-tracking research & applications*. 10.1145/1743666.1743701

Panse, A., Alcock, L., & Inglis, M. (2018). Reading proofs for validation and comprehension: An expert-novice eye-movement study. *International Journal of Research in Undergraduate Mathematics Education*, *4*(3), 357–375. doi:10.100740753-018-0077-6

Pretorius, M., & Calitz, A. (2010). *The Development of a Usability Methodology Incorporating Eye Tracking for Developing Countries*. Fourth International Business Conference, Victoria Waterfalls, Zambia.

Rashbass, C., & Westheimer, G. (1961). Disjunctive eye movements. *The Journal of Physiology*, *159*(2), 339–360. doi:10.1113/jphysiol.1961.sp006812 PMID:14490420

Rayner, K., Chace, K. H., Slattery, T. J., & Ashby, J. (2006). Eye movements as reflections of comprehension processes in reading. *Scientific Studies of Reading*, *10*(3), 241–255. doi:10.12071532799xssr1003_3

Rayner, K., Inhoff, A. W., Morrison, R. E., Slowiaczek, M. L., & Bertera, J. H. (1981). Masking of foveal and parafoveal vision during eye fixations in reading. *Journal of Experimental Psychology. Human Perception and Performance*, *7*(1), 167–179. doi:10.1037/0096-1523.7.1.167 PMID:6452494

Salvucci, D. D., & Anderson, J. R. J. H.-C. I. (2001). Automated eye-movement protocol analysis. *Human-Computer Interaction*, *16*(1), 39–86. doi:10.1207/S15327051HCI1601_2

Salvucci, D. D., & Goldberg, J. H. (2000). Identifying fixations and saccades in eye-tracking protocols. *Proceedings of the 2000 symposium on Eye tracking research & applications*. 10.1145/355017.355028

Scherr, R. E., Laugero, K. D., Graham, D. J., Cunningham, B. T., Jahns, L., Lora, K. R., ... Mobley, A. R. (2017). Innovative techniques for evaluating behavioral nutrition interventions. *Advances in Nutrition*, *8*(1), 113–125. doi:10.3945/an.116.013862 PMID:28096132

Tsang, H. Y., Tory, M., & Swindells, C. (2010). eSeeTrack: Visualizing sequential fixation patterns. *IEEE Transactions on Visualization and Computer Graphics*, *16*(6), 953–962. doi:10.1109/TVCG.2010.149 PMID:20975132

van Gog, T., & Jarodzka, H. (2013). Eye tracking as a tool to study and enhance cognitive and metacognitive processes in computer-based learning environments. In *International Handbook of Metacognition and Learning Technologies* (pp. 143–156). Sydney: Springer. doi:10.1007/978-1-4419-5546-3_10

Vidal, M., Turner, J., Bulling, A., & Gellersen, H. (2012). Wearable eye tracking for mental health monitoring. *Computer Communications*, *35*(11), 1306–1311. doi:10.1016/j.comcom.2011.11.002

Wang, Q., Yang, S., Liu, M., Cao, Z., & Ma, Q. (2014). An eye-tracking study of website complexity from cognitive load perspective. *Decision Support Systems*, *62*, 1–10. doi:10.1016/j.dss.2014.02.007

Wedel, M., & Pieters, R. (2008). A review of eye-tracking research in marketing. *Review of Marketing Research.*, *4*, 123–147. doi:10.1108/S1548-6435(2008)0000004009

Zhu, Z., & Ji, Q. (2007). Novel eye gaze tracking techniques under natural head movement. *IEEE Transactions on Biomedical Engineering*, *54*(12), 2246–2260. doi:10.1109/TBME.2007.895750 PMID:18075041

Chapter 13
A Cross-Professional Collaborative Educational Approach to Building Student Feedback Literacy

Anbarasu Thangavelu
University of Southern Queensland, Australia

Tyler Cawthray
University of Southern Queensland, Australia

Ron Pauley
University of Southern Queensland, Australia

ABSTRACT

Understanding and engaging with assessment feedback is a crucial step in contributing to a student's development of academic skills and content knowledge. However, in order for feedback to be effective, students need to be aware of its importance and how to engage with it. Preliminary benchmarking of the tertiary education sector in Australia demonstrates that publicly and openly accessible student resources on feedback are not available. Rather, most student resources at universities focus on common academic and study skills. At the University of Southern Queensland, a student resource was developed to inform students on the value of feedback and how to engage with it through a three-step process. This resource was embedded in first-year courses as part of a suite of academic and study skills resources. This chapter explores how student feedback resources have the potential to positively contribute to student feedback literacy, learning and development.

DOI: 10.4018/978-1-7998-0426-0.ch013

INTRODUCTION

Students' engagement with and understanding of feedback represents an important aspect of their learning journey at a tertiary institution. However, it has been suggested that the usefulness of feedback is limited by factors such as students' lack of engagement with feedback (Hounsell, 1987; Orsmond, Merry & Callaghan, 2004; Price, Handley, Millar, & O'Donovan, 2010) or not understanding how to interpret or utilize it (Gibbs & Simpson, 2004; Lea & Street, 1998; McCune, 2004; Orsmond, Merry & Reiling, 2002; Price, Handley, Millar & O'Donovan, 2010). This highlights the need for academic staff to utilize educational strategies that will teach students the value and importance of feedback, and how to engage with it effectively. This chapter explores a cross-professional collaborative educational approach employed at the University of Southern Queensland (USQ) and reasons for the development of a student resource that focused exclusively on feedback. Effective embedding and teaching strategies that target the wider student cohort will also be discussed. In addition, this chapter will examine the potential impact of this approach to educating students about feedback, building capacity and implications for teaching practices at tertiary institutions in terms of student learning and development.

STUDENT ENGAGEMENT WITH FEEDBACK IN THE HIGHER EDUCATION CONTEXT

This section explores the importance of student engagement and how a lack of engagement can diminish the effectiveness of feedback. Factors that impact or constrain engagement, from the perspective of students and teaching staff are discussed. The aim of this section is to provide the higher education context and understanding of feedback engagement, or lack thereof, thus raising important points for consideration.

The importance of feedback to student learning and development has been widely acknowledged within the literature (Carless, Salter, Yang, & Lam, 2011; Hattie & Timperley, 2007; Hounsell, 2003), and has drawn much attention regarding its functionality, and its importance in contributing to the effectiveness of learning and teaching practices (Ackerman & Gross, 2010; Carless et al. 2011; Hattie & Timperley, 2007; Hounsell, 2003; Mulliner & Tucker, 2017). However, feedback is not as effective when students do not understand, engage and utilize it, and are instead only interested in their results (Gibbs & Simpson, 2004; Price et al. 2010; Mulliner & Tucker, 2017). Essentially, students have the choice on whether to engage with any feedback they receive (Price et al. 2010). The reasons for why students do or fail to act on feedback could be a result of reasoned objection to it following extensive consideration, or not trusting the feedback provider (Price et al. 2010). Negative feedback, while meant to be constructive, can lead to a negative reaction from students as it is critiquing their performance and potentially their individual self (Ackerman & Gross, 2010). It could also be due to when feedback is provided and if it can be applied (Crisp, 2007; Price et al. 2010) or students might be receiving inconsistent feedback from teaching staff in different courses (Crisp, 2007). Consequently, lack of engagement could result in students committing the same mistakes in subsequent assignments, which can be frustrating for markers who had spent considerable amount of time and energy in providing the initial feedback (Crisp, 2007).

In attempting to understand the reasons why students may not engage with their assessment feedback, it is important to explore the perceptions about feedback between students and academics. In a study focusing on student engagement with assessment feedback, Price et al. (2010), note that the application of feedback to ensuing assessments was important from the perspective of teaching staff, whereas students focused on the timing and applicability of feedback to content. However, in terms of applicability, it also depends on the quality and type of feedback that is provided (Price et al. 2010). Therefore, there should not be an underlying assumption amongst markers that students will act on written feedback as intended by the marker (Crisp, 2007). In contrast, Mulliner and Tucker (2017) challenge the notion that students do not act on feedback as their study demonstrates that the majority of students do engage with feedback by reading and applying it, thus the assumption amongst teaching staff that students do not engage with feedback is not entirely valid. Yet, this does not necessarily mean that students know how to use the feedback to maximize its benefits (Burke, 2009; Mulliner & Tucker, 2017), as they have not often been taught how to effectively use it (Carless et al. 2011; Weaver, 2006).

A study on students' perceptions of feedback found that when it came to effective feedback and acting on it, students did not have a homogenous view, which has implications for teaching students about feedback (Poulos & Mahony, 2008). A more proactive approach to educating students about feedback should therefore be used and as stated by Crisp (2007, p. 578), "for academics, a less comfortable option than blaming students for their apparent ignoring of feedback is to critically reflect on their own practices". The approach should be centered on enabling students to understand what feedback is, and its purpose (Crisp, 2007). It should also highlight the importance of feedback to learning and success, as students will then consider it more seriously (Poulos & Mahony, 2008). Taking into account the various reasons and perceptions, an innovative, applied and extensive approach to teaching students about feedback has the potential to positively impact their learning and development.

In exploring student engagement, the complexity of this topic is clearly evident with the identification of varying factors contributing to students reading, understanding and acting on feedback effectively. The literature demonstrates, from the perspectives of teaching staff and students, that there needs to be further consideration of building student capacity to engage with feedback. Therefore, innovative approaches need to be utilized by teaching staff to enabling students with capacity building and increasing feedback literacy. Similarly, teaching staff need to reflect on their own feedback practices to ensure that constructive, clear, understandable feedback is provided to students.

BUILDING FEEDBACK LITERACY: THE LAYERS

Given the complexity of this topic, it is crucial for a consideration of the concepts of feedback and academic literacy. This section aims to briefly deconstruct the layers associated with feedback and academic literacy, in relation to generic skills and graduate attributes. In doing so, the importance of building feedback literacy and the impetus for teaching staff to dedicate greater attention to teaching students about feedback will be emphasized.

The importance of building student capacity to engage and act on feedback has been widely acknowledged. However, feedback literacy is a crucial subset of academic literacy and this needs to be explored further. The idea of feedback literacy as an important aspect of academic literacy, was proposed by Sutton (2012) and its focal point is about student's "ability to read, interpret and use written feedback" (Carless & Boud, 2018, p. 1316). According to, Carless and Boud (2018, p. 1316), feedback literacy also encompasses "the understandings, capacities and dispositions needed to make sense of information and use it to enhance work or learning strategies", which clearly articulates the multidimensional aspect of feedback literacy. Importantly, feedback literacy extends beyond university as it is an important lifelong skill for learning and the professional work environment (Carless & Boud, 2018). This idea raises an important point about how feedback literacy is part of, or feeds into, 'generic' skills and by extension graduate attributes.

The notion of generic skills has been defined as a core set of attributes or skills that can be applied in varying contexts, that is non-discipline specific skills and transferrable (Badcock, Pattison & Harris, 2010; Barrie 2006; Drummond, Nikon & Wiltshire, 1998). These set of skills may in some cases be referred to as lifelong skills. The common list of generic skills comprises of interpersonal skills, problem solving, intellectual curiosity and rigor, communication and information management skills, creativity, integrity and tolerance, ethical awareness and practice, capacity for logical and independent thought, and critical thinking (Badcock et al., 2010; Bath, Smith, Stein, & Swann, 2004). In general, these generic skills are, in certain instances, seen more favorably than specific discipline knowledge by employers as graduates who are able to think critically, start and respond to change, problem solve and communicate confidently are required (Badcock et al., 2010; Harvey, Geall, & Moon, 1997; Hesketh, 2000; Holmes, 2001).

Considering the value and importance of generic skills, it might be useful to examine the following definition of feedback by Carless and Boud (2018, p. 1315), "feedback is defined as a process through which learners make sense of information from various sources and use it to enhance their work or learning strategies". This definition suggests the role of the student in trying to understand feedback and acting on it to improve subsequent work (Carless & Boud, 2018). Inherent in this definition and a crucial element of making sense of feedback, reflecting and acting on it, is the use of critical thinking. Therefore, feedback literacy and the process of engagement should be viewed as an applied example of critical thinking through critical reflection, thus representing an important generic skill and graduate attribute.

In deconstructing the layers around building feedback literacy, there seems to be multiple layers as feedback literacy should not only be considered as an important skill on its own for learning content, but it is also a vital generic skill and graduate attribute that is applicable in different contexts, and it develops critical thinking and reflective skills. Being able to read, engage and act on feedback to make improvements is also a valuable lifelong skill. Therefore, given the focus among tertiary institutions for the development of generic skills, it is recommended that strategies specific to building student feedback literacy are incorporated by academic staff at an institution.

DEVELOPING RESOURCES ON FEEDBACK: AN INSTITUTIONAL CONTEXT

Following on from exploring student engagement and feedback literacy, this section discusses the relevance and need for a feedback resource for students in the higher education sector. The recent development of a feedback resource at USQ and its holistic design are explored. The aim of this section is to discuss the approach of developing a specific resource on feedback, as an additional method for building student feedback literacy.

Considering the issues surrounding student engagement, understanding and utilization of feedback, tertiary institutions need to explore new ways of addressing this area of concern. While teaching staff may attempt to communicate with their students about this topic, developing and disseminating student-centric resources that focus specifically on feedback may be beneficial. Within the higher education sector, workshops and resources on academic skills, such as essay and report writing or referencing, are commonly available to students through academic libraries or learning support services. However, workshops or resources that focus specifically on feedback are not common. Evidently, benchmarking of academic skills resources provided by members of the Group of Eight Universities', and the Regional Universities Network whose institutions are comparable to USQ in Australia, revealed that detailed resources on feedback are not available for students. This does not necessarily suggest that student resources on feedback are not available at these institutions as they may only be accessed internally by staff and students of those universities. Nevertheless, as students do access academic skills resources, especially in their first-year, it would be valuable to develop a student resource or workshop on feedback. Importantly, as feedback at university level may be different to school, students might not know what represents feedback at this level (Mulliner & Tucker, 2017). For first-year students, 'effective feedback' assists them with their transition to tertiary education as evident in a study by Poulos and Mahony (2008). Likewise, international students originating from countries where assessments are exam based, such as India, may not understand what assignment feedback is.

Recognizing this gap, a student-centric feedback resource was developed by academic staff from the Library at USQ in 2016. The resource titled Understanding Assessment Feedback, was initially developed as an hour long on-campus workshop for students. Subsequently, a shorter recording was developed as a self-help student resource and to also cater to students studying online. Prezi, an innovative Web 2.0 digital presentation tool, was used to develop the resource instead of PowerPoint. The major difference between Prezi and PowerPoint is with how information is presented as Prezi has a nonlinear presentation style (Chou, Chang, & Lu, 2015). While there was anecdotal evidence for increased student engagement in previous workshops at USQ that used Prezi, and students had provided unsolicited feedback on its benefits to their learning, research also suggests that Prezi is a more effective presentation tool for knowledge acquisition than traditional instruction (Chou et al., 2015).

The feedback resource was designed following three main learning objectives for students:

1. Increase students' understanding of why and how feedback is given on assignments, and its significance.
2. Develop students' ability to find and interpret feedback in a positive and constructive manner.
3. Build students' capacity to engage with markers to grow and improve academically.

To address the learning objectives, the resource was structured to include the following:

1. Definition of feedback – explores what feedback actually is with the overall message that feedback is constructive, timely, and meaningful advice given to help students improve their skillset.
2. Importance of feedback - examines the importance of feedback with the use of a fun activity to encourage student engagement. Emphasis placed on the importance of feedback as an opportunity to grow and improve, rather than stagnate and decline.
3. Types of written feedback - states the purpose of and difference between written summative and formative feedback. Examples of each are provided and explained to students. However, it is clearly expressed that feedback is only valuable when it is received, understood, and acted upon. This transitions to the following sections which outline a 3-step process to engaging with feedback successfully.

Step 1: Preparing to receiving feedback - discusses the importance of creating an open and positive attitude to receiving feedback. Explores the relevance of students' attitude by comparing the difference between a 'fixed mindset' to a 'growth mindset'. An activity was included for students to reflect about how they can best prepare themselves for feedback.

Step 2: Engaging with received feedback - examines the notion that feedback is a discussion and the actions to take when feedback is provided. States the importance of having a positive and objective mindset, and how to positive conversations even with critical feedback. Examples of formative and summative feedback are provided.

Step 3: Acting on feedback to improve - explores how utilizing feedback may improve their work and the idea that repeating their actions would not lead to different results. Differences in the approach of two students to feedback are used as examples.

Step 4: Points to remember – provides a reminder of the main points relating to feedback.

This resource generally aligns with the features of student feedback literacy model proposed by Carless and Boud (2018), comprising of appreciating feedback, making judgements, managing effect, and taking action. Overall, this resource explores feedback holistically in order to raise awareness amongst students and build their capacity to self-regulate their learning with the potential to make a positive contribution to their study experience.

After benchmarking academic skills resources available publically amongst the Australian Group of Eight and Regional University Network (RUN) member institutions, it was evident that a resource dedicated to feedback was not available. Furthermore, when the search term feedback or assignment feedback was used, the results, in some cases, presented information relating to where feedback could be found. In line with the approach taken at USQ, this gap can, and should, be addressed by intuitions through the provision of a dedicated feedback resource for students, in the form of self-help information on a website such as a guide or tip sheet, or ideally as a short recording or workshop for students to engage with. Essentially, providing a dedicated resource on feedback provides an additional measure that may assist in building feedback literacy.

EMBEDDING RESOURCES ON FEEDBACK: MODELLING PRACTICE

The availability of skills resources for students on academic libraries or learning support websites, while beneficial, may not necessarily assist in their learning and development over the course of their education. Instead, a strategic approach is required for the effective delivery of materials that results in greater use of resources by students to develop their skills. This section discusses the benefits of a cross-professional collaborative educational approach to teaching academic and generic skills, with a particular focus on the strategy that was utilized to build student feedback literacy. Discussion of this approach will provide teaching staff with an additional strategy that they could use to inform their practice.

Cross-Professional Collaboration

While resources or workshops on academic skills can be useful, the lack of a strategic approach to disseminating information to students can negate their reach and effectiveness. For example, academic skills resources hosted on a library's website might not be regularly accessed by students or they might not even be aware of the available self-help support. Similarly, one-off sessions on academic skills, even if they are recorded, may not be accessible or looked at by students, especially if the majority of the cohort are online. Instead, a more effective strategy, is for a cross-professional collaborative approach to teaching academic skills in a course with teaching staff whereby generic skills are embedded and taught within the module (Masiss, 2012; Mavodza, 2011; McInnes Bowers, Chew, Bowers, Ford, Smith, & Herrington, 2009; Rushton & Lahlafi, 2013; Shumaker, 2010; Tucci, 2011). Research also demonstrates that students prefer generic skills to be taught as part of the module instead of an additional workshop (Hall, Nix, & Baker, 2012).

In addition to the positive impact on students' learning and development, cross-professional collaboration has potential benefits for staff. Following their experience, Rushton and Lahlafi (2013) discuss the value and impact of cross-professional collaboration between an academic, a librarian, an academic skills tutor, and an e-learning expert by highlighting several benefits. The collaboration not only validated the role of the librarian and academic skills tutor in the classroom, but it also increased students' understanding of the importance of academic and information literacy skills to their academic studies and lifelong learning from embedding these skills in the modules (Rushton & Lahlafi, 2013). Furthermore, students took skills development more seriously, and the use of a reflective approach allowed students to think more about the development of their own research skills (Rushton & Lahlafi, 2013). For the academic, the collaboration with staff from other sections of the university assisted with the development of course modules that focused on content and provided students with the necessary generic skills to be effective learners (Rushton & Lahlafi, 2013). In addition, the academic valued and understood the need for a holistic approach to enhance student learning and development (Rushton & Lahlafi, 2013). By bringing in other specialist staff, new perspectives were also introduced to the modules, which led to better student engagement with improvements in the use of academic sources and critical writing amongst students (Rushton & Lahlafi, 2013).

Embedding the Feedback Resource

Similarly, a cross-professional collaborative approach was used at USQ to embed information literacy and academic skills into the modules of a first-year Education course. In 2017, the discipline support team (DST) for Education, comprising of a liaison librarian, academic language skills and mathematics learning advisors, and an educational designer were approached by a course examiner about teaching and embedding generic information and academic literacy skills into the module. This provided the DST with an opportunity for cross-professional collaboration on the design of the course, choice of topics to be covered and the sequence of delivery over the course of a semester. This collaboration also allowed the course examiner to leverage of the individual specializations of staff in the DST to potentially produce better outcomes. Specifically, each individual staff in the DST was able to provide the following specialist advice in accordance with their role to address several aspects of student literacy and numeracy skills development:

- **Liaison Librarian:** Enhancing students' information and digital literacy skills.
- **Academic Skills and Mathematics Learning Advisors:** Enhancing student learning and development in the areas of academic, generic, and mathematics skills.
- **Educational Designer:** Enhancing course learning design and delivery, assessment and resources.

During the subsequent planning phase and in consultation with the course examiner, academic skills topics that would be valuable and important for students in terms of their learning and development were agreed upon and mapped out. Topics were also selected from the point of view of a first-year student, and what would be most beneficial for them in terms of their transition to university study. Examples of topics included reading and notetaking, academic writing, communication, and referencing. As part of this suite, it was decided that the Understanding Assessment Feedback resource should be included as it provided an opportunity to educate and upskill students about feedback. This aligns with the idea that students should be informed about feedback when they commence their studies (Carless & Boud, 2018). Importantly, by teaching and embedding information and academic literacy skills into a core first-year course, it allowed the DST and course academic to ensure that these important skills are covered right at the beginning of the student's university learning journey which they can apply to other courses in the future. This approach is in contrast to providing a one-off session on academic skills or hosting a resource on the library's website, as both strategies may not necessarily be as impactful to student learning and development.

Some of the topics were taught in person but recordings of all topics were embedded as the majority of students in the course were online. The resources were also sequenced strategically to coincide with start of the semester, assessments and semester end. For example, the resource on feedback was embedded at a later point within the course to correspond with the release of results for the first assessment, to encourage students to engage with feedback provided by markers. An additional engagement strategy was for students to review the weekly subject content and academic skills recordings, with online students posting the review on the online course forum while on-campus students discussed the topic with their tutors. However, as this weekly exercise was not part of their assessment, students' engagement and usage of certain resources was not as high as expected, especially for online students. In contrast, the first three recordings on communication, reading and notetaking, and time management, were accessed by the majority of the students as the first assessment, an online quiz, was based on the content in these recordings. While certain students provided positive unsolicited feedback on the feedback resource, data shows that 17% of the students in the course accessed the resource in Semester 1, 2018. In comparison, there was greater engagement with this resource in Semester 2, 2018 with 39.3% of students accessing it. Although the data suggests that this embedding strategy might be beneficial, the aim is for the majority of the student cohort to access all of the resources. To achieve this in future offerings of this course, an end of semester exam drawing on the skills resources will be included as part of the assessment. It is envisioned that the inclusion of an exam will not only increase student engagement with available resources, but also have a positive impact on student learning and development of generic academic and literacy skills, and build their feedback literacy.

Analysis of this impact will be ongoing in future semesters. Importantly, by embedding this resource, it assisted students in understanding the new academic learning environment they are part of, which can be a challenge for first-year students, especially in relation to feedback practices, assessments and interacting with teaching staff (Blair, 2017). Apart from this course, the feedback recording, along with other resources were made available to other course examiners as a suite of academic skills resources that can be embedded in their respective courses. Resources were also embedded in the assessment task sheet in collaboration with a course examiner and this approach was disseminated to the rest of the academic staff in the discipline. By utilizing a strategic cross-professional collaborative approach, there are multiple points where students will come across these resources, thereby maximizing their reach by disseminating to the wider student cohort and increasing its potential effectiveness.

Steps for Successful Cross-Professional Collaboration

Achieving a successful outcome with cross-professional collaboration requires effective communication, planning and understanding of the learning needs of the respective student cohort (e.g. needs may vary between 1st and 3rd year students) or the issues highlighted by a course examiner (e.g. students lack of familiarity with writing critical reflections). For example, it was identified by the course examiner and the DST during the collaboration that 1st year students transitioning to university needed to understand and develop a number of generic and academic skills, including feedback literacy.

The five main steps to effective cross-professional collaboration are:

1. Discussion and identification of needs.
2. Engagement in design of course materials and assessment, and delivery of material.
3. Leveraging individual specializations of roles to produce the best outcomes.
4. Data collection and analysis, including student feedback.
5. Iterative development at the end of each teaching cycle.

The first step to engage in a discussion and identify students' learning needs which is essential as it informs further actions and strategies that may need to be implemented. Specialist staff continue the collaboration with the course examiner by engaging in course design, resource development, and devise strategies for effective delivery of material. Leveraging individual specializations allows for the development of material specific to the role (e.g. academic skills learning advisor developing the feedback resource) and the specialists staff collaborate with the course examiner on scaffolding and embedding material into the curriculum to achieve best outcomes. Throughout the semester, feedback is solicited from students, and other forms of data are collected and analyzed (e.g. end of semester student survey data). An evaluation of the collaboration and the effectiveness of the course design and resources then provides an opportunity for further development, if required, before the next teaching cycle. Depending on the area needing further development, other specialist staff may be required. It is envisioned that the steps, discussed above, will provide a useful guide for teaching staff that may add to successful course delivery and student learning outcomes.

Even though resources are available for students online via a library or learning support website, not all students might be aware or use these materials. By scaffolding and embedding resources within a course, students are more likely to use the material. To achieve this, teaching staff should collaborate with other specialist staff for course development and/or redesign, and the required materials that not only teach subject specific content, but also build important generic skills. Essentially, in this example, it is the intersection of the cross-professional collaboration and the dialogical approach to feedback in a course that has provided an avenue for students to build their feedback literacy. Especially for on-campus students who were able to discuss this topic with their tutors. Therefore, meaningful impact could potentially be achieved through collaboration by melding together complex and two multilayered approaches to student-centered learning design.

DISCUSSION

Student engagement with and effective use of feedback is an area of concern for teaching staff. However, academics should explore innovative approaches to teaching students about feedback, and other academic skills. The cross-professional collaborative educational approach taken here, demonstrates an innovative way of teaching students about feedback. Importantly, just as how resources and classes are available for academic writing or referencing, feedback should be considered as an area of its own with specific resources or classes to educate and encourage students to engage with feedback and build their feedback literacy. This approach not only teaches students about feedback, it also helps them to become self-regulatory learners and build capacity as they understand and take ownership of their learning.

While teaching staff may not view specifically focusing on feedback instead of content favorably, the value and importance of feedback to students actually learning subject content needs to be acknowledged. To become self-regulated learners who engage with course content, specific resources or communications about feedback are required. As Carless et al. (2011, p. 406) state,

Students may need to be pushed to involve themselves in developing self-regulatory practices consistent with sustainable feedback. This reinforces the need for more to be done in terms of communicating with students the purposes of feedback and their central role in the process. It would be necessary to articulate consistently, on a program-wide basis, the benefits for students of being able to self-regulate their own learning.

In line with this view, the development and delivery of the feedback resource communicates specifically to students and teaches them how to use it effectively. Therefore, formal training would likely be beneficial for students to prepare them to get the most out any feedback, and the three-step process communicated to students in the feedback resource, prepare, engage, improve aimed at providing students with an effective strategy to approach their feedback. There are however limitations with this resource, as its focus is on written assignment feedback instead of emphasizing that feedback is as an ongoing dialogue between academic staff and students over the course of the semester and their university study. To address this limitation, it is likely that further refinements will be made to this resource to clearly articulate the importance of a feedforward and dialogic approach to students, with the aim of embedding the updated resource to coincide with the start of semester in order to facilitate the ongoing dialogue with teaching staff throughout the semester.

A further point, is that teaching staff should reflect on their practices, and collaborate with other specialist staff across the university to leverage on their individual specializations to enhance their course and embed the necessary resources to develop students' skills, and not assume that students have the adequate skills or that their assessments will do this. A proactive cross-professional collaborative educational approach, will be beneficial for students in terms of their learning and development as demonstrated in this chapter, and conversely assist the academic as students are likely to be more prepared, and be able to engage with their course content and feedback. Successful cross-professional collaboration to enhance course focus on student-centered feedback literacy development requires a dialogical feedback loop between teaching staff who act as a conduit for student views or required areas of skills development, and other specialist professionals to adapt the necessary resources and deliver the content. This strategic collaborative approach is essentially a method for building students' feedback literacy, generic and academic skills.

CONCLUSION

While student engagement with feedback is considered an issue, developing and teaching students specifically about feedback can assist to address this. To achieve this, student-centric feedback resources need to be developed and taught. Equally important is the need for a strategic approach to embedding resources into courses, or dissemination to students as lack of engagement reduces the effectiveness of the resource. Academic staff should not assume that students will read and act on feedback, instead they need to be educating students about it. Using this approach, and in collaboration with other specialist staff has the potential to contribute to developing feedback literacy amongst students, and to facilitate and/or enhance overall student development and learning outcomes.

FUNDING AGENCY

This research received no specific grant from any funding agency in the public, commercial, or not-for-profit sectors.

ACKNOWLEDGMENT

The authors acknowledge Mark Emmerson for developing the initial version of the feedback resource, and Stephen Smith (University of Southern Queensland) for his support with the cross-professional collaboration and for providing permission to access the data on students' engagement with the resource. We also thank Melissa Fanshawe (University of Southern Queensland) for the opportunity to collaborate and embed resources in an assessment task sheet.

REFERENCES

Ackerman, D. S., & Gross, B. L. (2010). Instructor feedback: How much do students really want? *Journal of Marketing Education*, *32*(2), 172–181. doi:10.1177/0273475309360159

Badcock, P. B., Pattison, P. E., & Harris, K. L. (2010). Developing generic skills through university study: A study of arts, science and engineering in Australia. *Higher Education*, *60*(4), 441–458. doi:10.100710734-010-9308-8

Barrie, S. C. (2006). Understanding what we mean by the generic attributes of graduates. *Higher Education*, *51*(2), 215–241. doi:10.100710734-004-6384-7

Bath, D., Smith, C., Stein, S., & Swann, R. (2004). Beyond mapping and embedding graduate attributes: Bringing together quality assurance and action learning to create a validated and living curriculum. *Higher Education Research & Development*, *23*(3), 313–328. doi:10.1080/0729436042000235427

Blair, A. (2017). Understanding first-year students' transition to university: A pilot study with implications for student engagement, assessment, and feedback. *Politics*, *37*(2), 215–228. doi:10.1177/0263395716633904

Burke, D. (2009). Strategies for using feedback students bring to higher education. *Assessment & Evaluation in Higher Education, 34*(1), 41–50. doi:10.1080/02602930801895711

Carless, D., & Boud, D. (2018). The development of student feedback literacy: Enabling uptake of feedback. *Assessment & Evaluation in Higher Education, 43*(8), 1315–1325. doi:10.1080/02602938.2018.1463354

Carless, D., Salter, D., Yang, M., & Lam, J. (2011). Developing sustainable feedback practices. *Studies in Higher Education, 36*(4), 395–407. doi:10.1080/03075071003642449

Chou, P. N., Chang, C. C., & Lu, P. F. (2015). Prezi versus PowerPoint: The effects of varied digital presentation tools on students' learning performance. *Computers & Education, 91*, 73–82. doi:10.1016/j.compedu.2015.10.020

Crisp, B. R. (2007). Is it worth the effort? How feedback influences students' subsequent submission of assessable work. *Assessment & Evaluation in Higher Education, 32*(5), 571–581. doi:10.1080/02602930601116912

Drummond, I., Nixon, I., & Wiltshire, J. (1998). Personal transferable skills in higher education: The problems of implementing good practice. *Quality Assurance in Education, 6*(1), 19–27. doi:10.1108/09684889810200359

Gibbs, G., & Simpson, C. (2004). Conditions under which assessment supports students' learning. *Learning and Teaching in Higher Education, 1*, 1–31. Retrieved from http://eprints.glos.ac.uk/3609/1/LATHE%201.%20Conditions%20Under%20Which%20Assessment%20Supports%20Students'%20Learning%20Gibbs_Simpson.pdf

Hall, M., Nix, I., & Baker, K. (2012). "Why should I?" Engaging learners in digital literacy skills development. *Proceedings of the 11th European Conference on e-Learning,* 220-229. Retrieved from http://oro.open.ac.uk/35574/1/ProceedingsECEL2012-MHall.pdf

Harvey, L., Geall, V., & Moon, S. (1997). Graduates' work: Implications of organizational change for the development of student attributes. *Industry and Higher Education, 11*(5), 287–296. doi:10.1177/095042229701100504

Hattie, J., & Timperley, H. (2007). The power of feedback. *Review of Educational Research, 77*(1), 81–112. doi:10.3102/003465430298487

Hesketh, A. J. (2000). Recruiting an elite? Employers' perceptions of graduate education and training. *Journal of Education and Work, 13*(3), 245–271. doi:10.1080/713676992

Holmes, L. (2001). Reconsidering graduate employability: The 'graduate identity' approach. *Quality in Higher Education, 7*(2), 111–119. doi:10.1080/13538320120060006

Hounsell, D. (1987). Essay writing and the quality of feedback. In d. J.T.E. Richardson, M.W. Eysenck, & D. Warren-Piper (Eds.), Student learning: Research in education and cognitive psychology (pp. 109–19). Milton Keynes, UK: Open University Press and Society for Research into Higher Education.

Hounsell, D. (2003). Student feedback, learning and development. In M. Slowey & D. Watson (Eds.), *Higher education and the life course* (pp. 67–78). Maidenhead, UK: Open University Press.

Lea, M., & Street, B. (1998). Student writing in higher education: An academic literacies approach. *Studies in Higher Education, 23*(2), 157–172. doi:10.1080/03075079812331380364

Massis, B. E. (2012). Librarians and faculty collaboration – partners in student success. *New Library World, 113*(1/2), 90–93. doi:10.1108/03074801211199077

Mavodza, J. (2011). The academic librarian and the academe. *New Library World, 112*(9/10), 446–451. doi:10.1108/03074801111182030

McCune, V. (2004). Development of first-year students' conceptions of essay writing. *Higher Education, 47*(3), 257–282. doi:10.1023/B:HIGH.0000016419.61481.f9

McInnes Bowers, C. V., Chew, B., Bowers, M. R., Ford, C. E., Smith, C., & Herrington, C. (2009). Interdisciplinary synergy: A partnership between business and library faculty and its effects on students' information literacy. *Journal of Business & Finance Librarianship, 14*(2), 110–127. doi:10.1080/08963560802362179

Mulliner, E., & Tucker, M. (2017). Feedback on feedback practice: Perceptions of students and academics. *Assessment & Evaluation in Higher Education, 42*(2), 266–288. doi:10.1080/02602938.2015.1103365

Orsmond, P., Merry, S., & Callaghan, A. C. (2004). Implementation of a formative assessment model incorporating peer and self assessment. *Innovations in Education & Training International, 41*(3), 273–290. doi:10.1080/14703290410001733294

Orsmond, P., Merry, S., & Reiling, K. (2002). The use of exemplars and formative feedback when using student derived marking criteria in peer and self assessment. *Assessment & Evaluation in Higher Education, 27*(4), 309–323. doi:10.1080/0260293022000001337

Poulos, A., & Mahony, M. J. (2008). Effectiveness of feedback: The students' perspective. *Assessment & Evaluation in Higher Education, 33*(2), 143–154. doi:10.1080/02602930601127869

Price, M., Handley, K., Millar, J., & O'Donovan, B. (2010). Feedback: All that effort, but what is the effect? *Assessment & Evaluation in Higher Education, 35*(3), 277–289. doi:10.1080/02602930903541007

Rushton, D., & Lahlafi, A. (2013). The value and impact of cross professional collaborations in developing student information and academic literacy skills at Sheffield Hallam University, UK. *Nordic Journal of Information Literacy in Higher Education, 5*(1), 38–43. doi:10.15845/noril.v5i1.178

Shumaker, D. (2010). A wide range of approaches. *Information Outlook, 14*(1), 10–11.

Sutton, P. (2012). Conceptualizing feedback literacy: Knowing, being, and acting. *Innovations in Education and Teaching International, 49*(1), 31–40. doi:10.1080/14703297.2012.647781

Tucci, V. K. (2011). Faculty/librarian collaboration: Catalyst for student learning and librarian growth. *Science & Technology Libraries, 30*(3), 292–305. doi:10.1080/0194262X.2011.596792

Weaver, M. R. (2006). Do students value feedback? Student perceptions of tutors' written responses. *Assessment & Evaluation in Higher Education, 31*(3), 379–394. doi:10.1080/02602930500353061

Chapter 14
Are You Feeding Back or Is It Taking Students Forward?
Changing the Traditional Narrative to Ensure a Dialogic Approach in Formative Assessment

Christopher Dann
University of Southern Queensland, Australia

Shirley O'Neill
University of Southern Queensland, Australia

ABSTRACT

The idea of feedback in education is accepted as vital in students' learning experience as a key to their success. Moreover, there is a growing recognition that for formative assessment practices to be most effective; data produced should be of a type that can help students improve their learning, and so should be dialogic, and feed forward rather than back. That is, students should have the opportunity to be engaged in critical reflection and dialogue about their performance in relation to such data. This chapter, therefore, presents a framework that positions dialogue at the core of formative assessment practices. It aligns this with Boud and Molloy's "Feedback Mark 2" model and Henderson et al.'s 12 conditions of success within the broader field of formative assessment to present a case for a more fine-grained examination of the concepts involved and the need for a change in mindset. The chapter argues that dialogue is the conduit through which nuanced moments and "feed markers" provide indicators of learning progression, and that how this impacts on the design of formative assessment tasks requires greater scrutiny. It concludes that the nuanced humanistic behaviors of the dialogic experience need further definition and exploration within the feedback space, and that the established narrative around the use of "feedback" needs to change to accommodate the social constructivist view of learning if practices are to be enhanced.

DOI: 10.4018/978-1-7998-0426-0.ch014

INTRODUCTION

This chapter is situated within a *milieu* of feedback literature, as coined by Boud and Molloy (2013), which could be described as a mirroring the milieu of feedback actions themselves. Feedback, by its nature sits within the teaching and learning process, and involves both the teacher and the student, besides some form of clear, consistent communication (Nixon, Brooman, Murphy, Fearon, & Education, 2017). This communication between teachers and students about their learning progress is a vehicle for the provision of formative assessment information to assist students to improve. Traditionally, referred to as 'feedback', as 'assessment data', it is well recognized as that pedagogical feature that has the potential to have the most significant impact on students' learning (Hattie, 2009; Hattie & Timperley, 2007). It is also seen as able to strengthen student evaluative judgement (Henderson, Boud, Molloy, Dawson, Phillips, Ryan, & Mahoney (2018), emerging from the work of Black and Wiliam (1998), Wiliam (2011) and Wiliam and Thompson (2017). Recent studies of the implementation of feedback in the higher education sector, and particularly in initial teacher education, are evident in the work of Carless (2019), Conrad and Dabbagh (2015), and Denton and McIlroy (2017) These researchers moved from investigations of perceptions of the feedback process through to how it can be conceptualization as dialogue (Merry, Price, Carless, & Taras, 2013). Building on the focus on dialogue, Carless (2019) proposes that feedback may involve both short and long feedback loops that are difficult to decipher, thus highlighting the complexity involved. In response, the authors of this chapter argue that a deep analysis of the humanistic operationalization of feedback dialogue and feedback loops is necessary to help address the issues and better define the practices involved. This would require a deep analysis of the humanistic actions and behaviors that emerge from the dialogue and learning loops theorized by Carless et al. (2013). By positioning these humanistic components as documented in the literature in the field of dialogic practice within the feedback practices of higher education programs, the authors seek to answer the questions: "How can higher education institutions best support the student learning experience in their promotion of dialogically rich formative assessment practices?"; "Can we quantify or surface the humanistic element given the complexities and diversification of the individuals involved?" and "Which dialogic feedback behaviors can improve graduate employability, student experience and sustainability in the profession?

Part of a teaching-learning process is the provision of feedback and this chapter proposes a view of these processes in higher education that is characterized by parity between the student and academic perspectives. The following section presents the theoretical perspective of the authors and is followed by a section that presents a theorized framework of five key elements that may influence the resultant enacted curriculum (Remillard, 2018). The authors describe this framework within the higher education space and then align it with the eight characteristics of Feedback Mark 2 (FM2) (Boud & Molloy, 2013) and the twelve conditions of successful feedback reported by Henderson et al. (2018).

THEORETICAL BACKGROUND

This chapter views the milieu of literature and research on feedback through a constructivist, pragmatic and connectivist view of learning activity. This section presents the underlying theoretical position as pragmatism. Social constructivism posits that learning is a cognitive activity that is socially situated (Barak, 2016). Constructivism builds on the premise that learning relies on 'significant relationships with others' that include peers and a skillful teacher (Palincsar, 1998), which are founded in sociocultural theory (Vygotsky, 1978) and Banduras' social cognitive theory (Bandura, 2018). An additional lens adds the theoretical foundations of activity theory (Bligh & Flood, 2017). This is in keeping with the underlying premise that the student and the academic are actively involved, and engaged in the process. Activity theory (Engeström, 2001), along with connectivism (Siemens, 2014), is also relevant to contextualising the dialogic situation. Both the influences of the theory of 'planned behavior' and 'technology acceptance' are also seen as coming into play. However, the chapter scope is unable to expand on these theories other than to consider them within the limited scope of dialogic practices in formative assessment.

The active nature of students and the need to develop parity between the academic/teacher and student brings the foundations of activity theory to the fore. Activity theory provides a theoretical framework to describe the structure, development and context of human activity (Carroll, 2003; Yoon, Ham, & Yoon, 2016). It is within this framework that dialogic theory is ideally placed. Dialogic theory emerged from the work of Buber (1923/37) and was built on by Rogers (1957), then Freire (1970), followed by Gadamer (1991). Each of these theorists contributed actions and key terms to the knowledge base (Kent, 2017) seen as central to unpacking the ongoing relational processes involved in the milieu of the feedback situation. While the humanistic principle of trust (Freire, 1970) is seen as foundational, this chapter proposes that the dialogic theory is more grounded in the practicalities of communication and human activity. However, in any case, in the light of the imposition of technology and the ubiquitous screen culture intervening in the communicative situation, the authors argue that there is a need to closely examine how the dialogic works in technology enhanced formative assessment practices.

Kent and Taylor (2002) developed a set of features of dialogic practice that emerged from their review of the literature. These included mutuality, risk, proquinquity, empathy and commitment, which Kent and Theunissen (2016) present as the pillars of their dialogic model (cited in Kent, 2017). In addition, they present practical interpersonal dispositions, such as being relational, empathetic, and sympathetic. Dialogue is also seen as relating to factual objectivity, cooperativeness, concern for others, open-mindedness, honesty, genuineness, equality and willingness to delay judgement (Johannesen, 1990). The original assertion of Buber (1923/37), that rewarding relationships occur when people treat others as inherently valuable, ties the importance of these communicative features to the concept of relational communication. Interestingly, Rogers (1957) highlighted the dialogic concepts of genuineness, spontaneity and face-to-face interaction, pointing out that true dialogue cannot occur except in a relationship. It becomes evident that humanistic qualities are exposed through dialogic action. Thus, academics working within the complexity of online and blended learning environments are faced with the challenge of meeting these communicative features and building the relationships to support authentic dialogue. It is argued that online and virtual learning spaces change the ecology of the learning environment (Aedo, Peredo, & Schaeffer, 2017; Bamber, 2016; Gergel & Turner, 2017). This changed ecology is seen as being more conceptually refined in terms of being able to capture the elements of dialogic practice (Kent & Taylor, 2002). However, questions arise in terms of its impact on assessment practices and particularly provision of feedback in formative assessment, as well as how the critical features of the relationship might be measured in practice and positioned within feedback theories and frameworks. Gadamer (1991) believes that an openness and willingness of all participants to be actively involved in a conversation or dialogue may not be easy for some. It is pointed out that the need to accept others' viewpoints makes the dialogic practice a more difficult option for all. Carless (2019) examines the actions and behaviors embedded in the dialogic and positions them within models and processes referred to as "feedback loops". He provides a detailed examination of how these apply in practice linking them to spiral forms of learning and iterative cycles of reflecting and engaging with feedback. The next section provides further insights into how these models and processes are being used to surface the human qualities and provide a systematized view of their application to formative assessment.

FEEDBACK MODELS AND LOOPS

There are two approaches to the creation of systematic views of the processes undertaken in feedback. The first is a technological view that is being applied as technologies are becoming more sophisticated and enabling scholars to identify the processes involved. The second approach is a more humanistic approach where the term 'feedback loop' is prominent in the literature. This section presents both views, followed by a section describing three proposed constructs as an additional lens to feedback.

The technological view of feedback implies good intentions to create processes that lead to the enhancement of teaching and learning (Crane-Mitchell & Stafford, 2017). However, these intentions may not necessarily achieve a holistic and positive impact on the teaching-learning process. To address this, recent research has examined how technological advances have impacted on feedback by providing opportunities for tracing the analytics for the purposes of personalization (Pardo, Jovanovic, Dawson, Gašević, & Mirriahi, 2017) and delivery of feedback. Technology-mediated communication of information about a student's work has become a focal point for the use of technology in the feedback process. As Pardo (2018) notes, the application of technology is seen by the authors as retrospective. That is, it has been used to record the evidence and surface after the event to show the interactions and communication where learning outcomes are the focus, rather than illuminating the humanistic characteristics of the interaction itself.

Pardo (2018), proposed a model for data supported feedback through analytics and how an organization might scale up feedback in a personalized manner (Pardo, 2018; Pardo et al., 2017). Pardo (2018) concludes that this model "takes into account other important mediating factors such as the role of the instructor and the need to consider changes in knowledge, beliefs, attitudes, strategies and tactics in the learner as the ultimate goal of the process" (p. 9). The recognition of students as active agents in their own learning is reflective of Henderson et al. (2018), and Boud and Falchikov (2007). The model depersonalizes the relational human factors of the realities faced by the participants in the dialogue, referring to them as 'agent A' (the teacher), and 'agent B', (the student). The Pardo mode infers the influences of online learning spaces (Conrad & Dabbagh, 2015) and interpersonal and personal relationships, and takes the collective view that there is a need to work at several levels (Patrikakou, 2015). For instance, technologies are playing a role in the communication of information to students that traditionally evoked the power, position and intentions of the communicator. When applied to the role of academics this communicator focus places them in the position of power with the student as a receiver of information. This implies a more traditional, monologic practice thus, amplifying the need to understand the human dispositions brought to bear by both agents, and the importance of building the relationship to support dialogue and opportunity for collaboration.

The humanistic view of feedback systems is evidenced in the reference to feedback loops in the literature where Carless (2019) develops the idea of short and long feedback loops in learning. He extends his model to include the concepts of feedback spirals that represent longer term learning, which may include both open and closed feedback loops. Moreover, he positions the spirals within the idea of long term learning, linking it to the concept of sustainable learning (Boud & Dochy, 2010). The introduction of feedback over the long-term implies students will need to "make sense of inputs from a range of sources" (Carless, 2019, p. 8). This has implications for pedagogical approach in Peacock and Cowan's (2017) argument that students and academics need to be working in the context of a community of inquiry, noting also there is a need for a common understanding of feedback. As active agents in the learning process Boud (2015) argues students should also be actively involved with feedback, whereas Wiliam's (2011) position is that feedback should be immediate and directed by the teacher. In the approach taken by Dann and Dann (2018), as shown in Figure 1, they also make a distinction in this regard. The differentiate between the 'professional pathway' of the academic and the 'learning pathway' of the student. Adapted from Dann and Dann's (2018, p. 78) original concept, time is represented as moving from left to right and knowledge levels are depicted between the lines moving from top to bottom.

This helix model also accounts for time when individuals are not directly in communication and acknowledges that students are required to make sense of information, so are cognitively involved. This approach to feedback is the driver behind the 'feedback influencer' framework proposed by the authors and described in the following section.

Feedback Influencer Framework

Following the research by Dann and Dann (2018) further consideration of helix model of feedback (Figure 1) generated a framework with five key elements that are theorized as influencing the resultant enacted curriculum. This, which is often referred to in relation to mathematics teaching, and school based learning (Boesen et al., 2014; Remillard, 2018). These five elements emerged from data about students' and academics' actions and responses to feedback in a 4-year study action research study (Dann & Dann, 2018). The five layers in the framework represent the enacted curriculum, the feedback strategies employed, use of dialogic approaches, academic feedback dispositions, student feedback dispositions

Figure 1. Knowledge gap reducing feedback
(Adapted from Dann & Dann, 2018, p. 78)

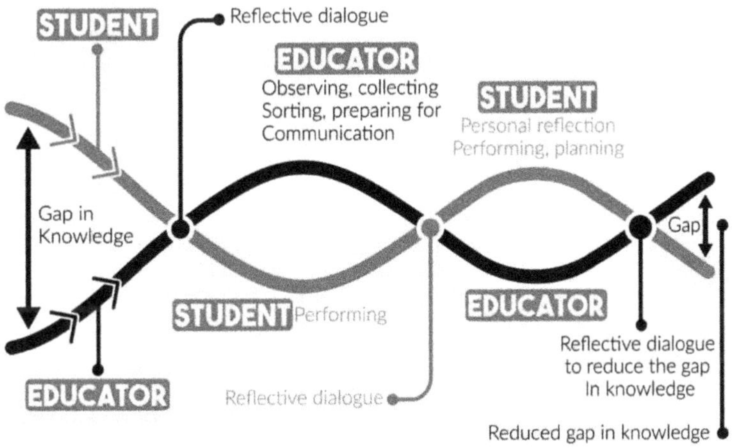

and course design. The authors define the enacted curriculum concept applied to higher education as 'the overall experience of students during their experiences of a course or program of study', although the enacted curriculum is generally associated with school-based curriculum implementation. It is frequently considered in relation to science and mathematics teaching (Remillard, 2018), and is utilized to identify the influencers of feedback, and teachers' and students' behavior. Ultimately, in higher education, it is the behaviors of both students and academics, together with the networks that exist at the formal and informal levels, that influence the enacted curriculum. The feedback influencers' framework was developed to guide a research project into the effects of feedback and how student and academic characteristics impact the enacted higher education. This broadens the view of feedback beyond content to include the five elements of the students' experience outlined in Figure 2 below. This is considered further in the next section.

The further to the right of the 'influencing' continuum the more intense the positive stance towards the element. The left side indicates an increasingly negative stance towards the element. Each element is drawn from Boud and Molloy's (2013) FM2 and relates to dialogic theory and practice. They draw

Figure 2. Enacted curriculum influencers' framework

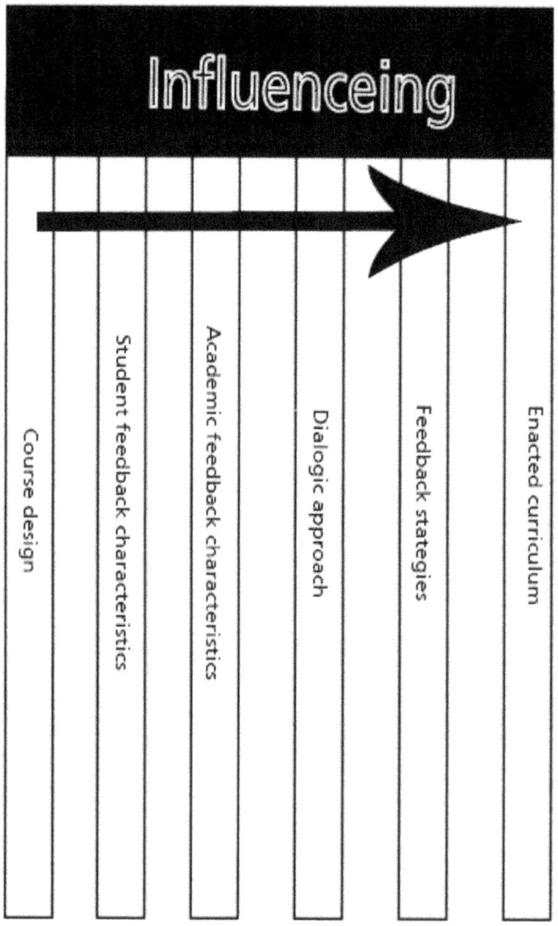

up on dialogism (Benade, 2015), feedback and feedforward practices (Hendry, White, & Herbert, 2016) the student experience and predispositions to feedback (Cole, et al., 2017) as well as course design (Jaggars & Xu, 2016; Jones, Olds, & Lisciandro, 2016; Villarroel, Bloxham, Bruna, Bruna, & Herrera-Seda, 2018). Descriptions are drawn from current evidence in the field (Boud, 2000; Boud & Molloy, 2013; Barnett & Coates, 2005; Boud & Falchikov, 2007; Carless, 2017). Each concept is a continuum on which individuals may place themselves. Each of the five perspectives are briefly described below.

Course Design

Course design refers to the design of learning spaces that are likely to generate new energies amongst student and inspire them, and so inspire their triple engagement – knowing, acting and being, in keeping with the philosophy of Barnett and Coates (2005).

Student Feedback Characteristics

Mulliner and Tucker (2017) report that students in higher education tend not to seek feedback on assessments or take action as a result of it. Carless, Salter, Yang, and Lam (2011, p. 404) emphasize "for feedback to be effective, it needs to place less emphasis on conventional feedback practices and develop further those in which student autonomy and self-monitoring capacities become paramount". Boud and Molley (2013) confirm this by suggesting that to achieve FM2 "students need to be positioned as elicitors of knowledge for improvement not just the recipients of inputs from others" (p. 705). Sadler (1989) identified three conditions for effective feedback. Each of these involve student perceptions. The first requires students to understand and have knowledge of the standards of work to be achieved. The second requires a student to be able to compare those standards of work, and thirdly, students need to be able to take action to close their knowledge gap (see also Dann & Richardson, 2015). Bennett, Dawson, Bearman, Molloy, and Boud (2017) note that Sadler's second and third conditions require the student to actively engage, thus drawing attention to the mind-set of the students regarding feedback (Forsythe & Johnson, 2016). While, feedback that is student-centered has been accepted practice, recent literature advocates for a paradigm shift towards an 'activated' student approach, which is capable of directly impacting on the student disposition and characteristics. This is in keeping with the FM2 framework that directly refers to the learners' orientation to the purpose of feedback, their disposition for seeking feedback, and the learner as a 'seeker and provider'. Further alignment between the student and the FM2 elements is provided at the end of the next section.

Academic Characteristics

Academic perceptions of feedback are varied but are influenced by the environment, career aspirations, personal knowledge, and capability and capacity to enact processes that benefit student learning (Mulliner & Tucker, 2017). The academic's role regarding feedback is deeply embedded in the feedback processes with which they engage. In terms of FM2, other dialogic characteristics may be influential in the orientation of learners and delivery mechanisms used within a course. Pardo (2018) identifies the instructor, the instructors' knowledge, beliefs, attitudes and strategies, and tactics as important for areas for further investigation and clarification. Time is also a consideration as it is influenced by issues such as workload, marking load, type of assignment, links between tasks, student communication and access to resources, all of which form part of the 'milieu' (Boud & Molley, 2013). In O'Neill, Dann and Dann's (2018) description of a technologically enhanced strategy that was piloted in such a milieu, focused on how a technological solution influenced the content of feedback and its design in moving learning forward – feeding forward (Boud & Molloy, 2013; Hendry et al., 2016). They found that collaborative use of video associated with formative assessment criteria that linked to professional standards in initial teacher education engendered critical dialogue about practice.

Dialogic Approach

The dialogic approach in formative assessment in higher education practices remains an emerging construct, although identified as essential for students' learning in the context of social constructivism (Ajjawi & Boud, 2017; Beaumont, Moscrop, & Canning, 2014; Moriarty, Danaher, & Danaher, 2008). The dialogic aspects of a course in technology enhanced learning environments typically include, use of technologies to support communication and interactions, actions or systems that communicate or respond to the needs or requirements of students and educators (Engin & Donanci, 2015). For example, in teacher education Beaumont et al. (2016) explain how a 'learning management system', accessible via the Internet, supports a 'study desk' where the screen displays course information and links to learning resources in-house or external, along with opportunities for communicative interactions. It is seen as a learning space that introduces the tactics used by academics to leverage the systems involved and simultaneously by students for learning. Thus, an understanding and application of dialogic pedagogy is necessary to enhance learning outcomes in this approach (Alexander, 2010; Barnes, 2010).

Feedback Strategies

Feedback strategies in higher education that rely on technological processes are increasingly being used but there is perceived lack of research into their effectiveness regarding the personalization of learning and their adaptability (Goldin, Narciss, Foltz, & Bauer, 2017). Academic's acceptance of technology enhanced feedback strategies, and their ability to use them efficiently and effectively, impact directly on course pedagogy, learning and assessment, and students' experience. The nexus between feedback strategies and formative assessment and the related data contributes the vital information that can help modify course teaching and learning activities (Black & Wiliam, 1998). These strategies provide information for learners' use that is integral to formative processes (Juwah, Macfarlane-Dick, Matthew, Nicol, Ross, & Smith, 2004; Nicol & Macfarlane-Dick, 2004), so essential for students' learning (Brown, 2006). They are seen as one of several features of formative assessment (Black & Wiliam, 1998), that need to co-exist (Brown, 2006; Butt; 2010). As Tara (2003) specifies, they are part of the learning context and are not free standing. Research indicates the need for a greater focus on the design and nature of feedback strategies in relation to the purposes of formative assessment (Furtak et al., 2016; Maier, Wolf, & Randler, 2016). These studies indicate that where formative assessment is being practiced there should be strategies to enhance feedback communication but this is an area that is under-theorized due to its process orientation (Goldin, Narciss, Foltz, & Beauer, 2017; Tomasik, Berger, & Moser, 2018; York, 2003).

Figure 3. Misalignment of elements

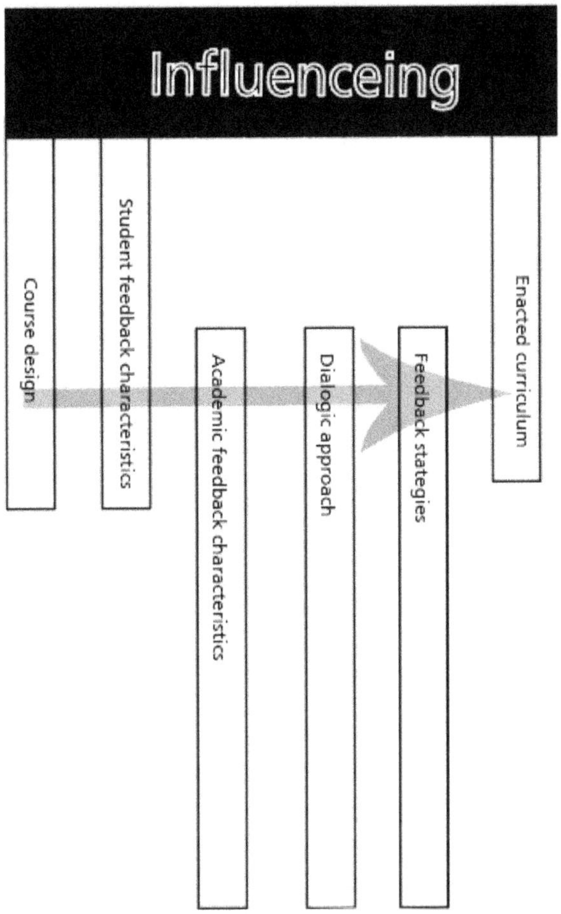

Enacted Curriculum

The enacted curriculum is defined for the purposes of the feedback influencers' framework as the sum total of each element (listed above) as influenced by the variations expressed by students, academics and course/program designers in each of the five areas.

Misalignment

The elements of the feedback influencers' framework (Figure 2) may be misaligned, which would impact the enacted curriculum. In this figure, course design has a nested approach to feedback (Boud & Molloy, 2013), and feedback opportunities are designed into the course structure. For instance, a student with a positive stance towards feedback may have a high level of feedback literacy. But it might be that the academic does not demonstrate an inclination to provide feedback or have the disposition to attempt to communicate feedback on the student's work. This would result in limited dialogic characteristics being presented to students in their learning experiences, and would be expected to reduce the evidence of strategies being employed during the course time line. The negative stance of the three elements when aligned to the student's feedback characteristics and the course design would impact on the enacted curriculum as would other potential scenarios.

NEW PARADIGMS OF FEEDBACK

The socially constructivist view adopted by Boud and Molloy (2013) bodes well with the humanistic view of Carless et al.'s (2015) assertion that higher education should adopt a new approach to feedback. Molloy and Boud (2014) point to eight key elements that comprize the FM2, as noted earlier, displayed in Table 1, which they note is built on the "revolution, sparked by the observations and ideas of Boud, Price, Nicol, and Carless, is starting to hit higher education" (p. 423). This highlights the priority for both educators and students to commit to feedback and the need to move to a more dialogic pedagogical and formative assessment approach.

Curriculum feature characteristics of feedback mark:

1. Learners orientated to the purposes of feedback.
2. Learners participate in activities promoting self-regulation.
3. Learner Disposition for seeking feedback is developed.
4. Opportunities provided for production of work.
5. Collaboration mechanisms.
6. Incremental challenge of tasks.
7. Nested tasks to allow for 'feed forward'.
8. Learner as 'seeker and provider'. (Boud & Molloy, 2013, p. 15-16)

Each of these form part of the enacted curriculum (Barnett & Coates, 2005). This chapter theorizes that the resultant enacted curriculum is impacted by FM2 elements, positioned within dialogue of the course and within the "milieu" (Boud & Molley, 2013, p.708) of higher education assessment, described in part as follows:

Much of the curriculum as enacted is not what is formally documented in learning outcomes and course descriptors, but occurs in the everyday interactions students have with staff, with each other and the context in which they operate. The translation of the curriculum into the everyday learning milieu is crucial in realizing this approach. (Boud & Molloy, 2013, p. 708)

Implementing FM2 requires the academic to have time to complete the collection of data (Boud & Molley, 2017), to conduct marking and moderation, and undertake the release of grades processes (Moore, 2017). There also needs to be time for the one-sided approach of Feedback Mark 1 (Boud & Molloy, 2013), for the academic to receive, view and cognitively assimilate the content of the feedback to conceptualize/formulate a response. The design of formative assessment is a vital part of this such that it needs to be embedded within the structures of the courses and programs. Thus, a more fundamental

Figure 4. Conditions for success categories: capacity, designs and culture
(Henderson et al., 2018, p. 16)

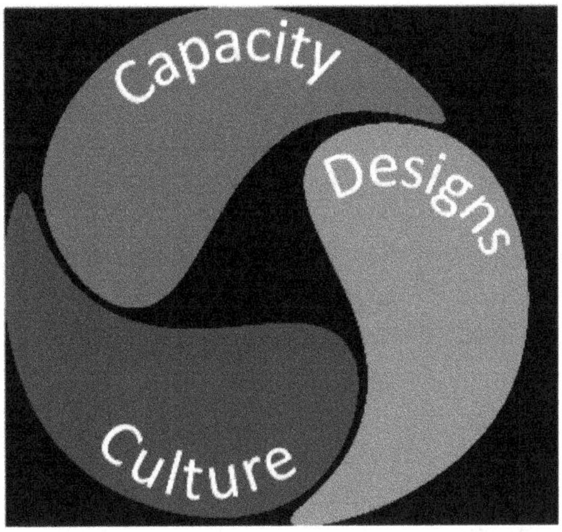

rethinking of formative assessment and feedback is required to move the collective notion of feedback as being corrective to the requirements of FM2 (Boud & Molloy, 2013). This means a change to an acceptance of students as complex humans who are actively engaging with the feedback they receive with a view to enhance their learning. This aligns to the constructivist approach (Rushton, 2005) noted as requirement for deep learning and is supportive of a student-centred approach (Moore, 2017). This represents a move towards change and the transformation of formative assessment practice to enhance learning (Kleinknecht & Schneider, 2013).

CONDITIONS OF SUCCESSFUL FEEDBACK

The central premise, that a student's experience and commitment to their feedback is on parity with the academic's position and experiences, is gaining momentum in the literature. However, while the previous section outlined the eight elements of the FM2 there remains a need to be aware of the conditions that underpin the learning situation for successful feedback to occur. Henderson et al. (2018) are extensive in their work in this area. They present a complex framework that focuses on capacity for feedback, design for feedback and culture for feedback, as shown in Figure 4. While it includes seven challenges that underpin effective feedback, it also identifies twelve conditions for effectiveness, and forty strategies that emerged from case study research. On this basis, this section attempts to align these twelve conditions for effective feedback with the theoretical understanding of the dialogic developed earlier in this chapter.

Henderson et al. (2018, p. 16) define effective feedback "as a process in which learners make sense of the information about their performance and use it to enhance the quality of their work or learning strategies." The conditions for success are organized into three sections, which align to the framework feedback concepts regarding capacity, designs and culture. For the purpose of aligning the conditions to dialogue they are listed below.

Feedback is successful when there is:

1. Capacity for feedback
    a. Learners and educators understand and value feedback.
    b. Learners are active in the feedback process.
    c. Educators seek and use evidence to plan and judge effectiveness.
    d. Learners and educators have access to appropriate space and technology.
2. Design for feedback
    a. Information provided is usable and learners know how to use it.
    b. It is tailored to meet the different needs of learners.
    c. A variety of sources and modes are used as appropriate.
    d. Learning outcomes of multiple tasks are aligned.
3. Culture for feedback
    a. It is a valued and visible enterprise at all levels.
    b. There are processes in place to ensure consistency and quality.
    c. Leaders and educators ensure continuity of vision and commitment.
    d. Educators have flexibility to deploy resources to the best effect.

Henderson et al. (2018, p. 18), however, do not contend that these are the only conditions for successful feedback or that they need to be actioned "simultaneously". Bearing this in mind, the section below sets out to show how the conditions might align with the presence of dialogic dispositions and characteristics drawn from the theoretical perspective described earlier, as this seems necessary if feedback is to be most effective. As the following researchers emphasize to provide students with the best possible chance of success in their higher education learning more effort is required in reforming formative assessment practices and how feedback is provided (Bandiera, Larcinese, & Rasul, 2015; Fry, Ketteridge, & Marshall, 2014; Gallagher & Education, 2017; Sambell, 2016).

Initial reading of the conditions for the presence of interpersonal qualities described in dialogic theory presents a number of interesting conundrums. These are presented in this section which is then followed by a section drawing together the five elements initially identified by the author, the eight elements of Feedback Mark 2 with these 12 conditions for success to argue for an increased focus on the human elements that constitute the phenomenon of feedback.

The first four conditions are grouped within the concept of 'capacity for feedback', which positions them as personally relevant to the individuals involved. This aspect of the framework identifies both the student and the educator as requiring this capacity. The position aligns to the parity required in authentic dialogic behaviors and interactions. These also bode well with the dialogic terms of honesty, equality and relational and a willingness to delay judgement (Johannesen, 1990). The first element positions learners and educators with the same need, while the second and third element positions both learner and educator as actively involved in the effectiveness of feedback. The fourth element shows the relationship between students and academics as democratic. Together they also align with activity theory (Bligh & Flood, 2017).

The second four are positioned with the need for 'design for feedback'. These four elements also call for students to be active learners but goes further to advocate for a coherent community of practice (Wenger, 1998, 2011) to support the implementation of feedback. The emphasis in this grouping is the variety of processes available by design, that need to meet the needs of students and educators at various levels. The authors do not contend that these aspects place the student or the educator in a position of comfort. Rather, they position these elements as tools within the milieu as described by Boud and Molloy (2013) as options for effectively enhancing the learning of the student. The fifth element calls for learners to have access to feedback information and to know how this is to be used. It is at this point that the conditions split from the concept of student satisfaction. Students with limited feedback literacy may not report satisfaction even if their feedback is provided and tailored to meet their needs (Sutton, 2012).

The final four sit within the culture of feedback grouping. This section focuses on the organization's operations around the feedback process. The first element calls for feedback to be valued and visible at all levels of the organization. Further to this there may be competing demands in the Academy. For example, institutions may prioritize reform of certification processes and related concerns before students' learning processes, depending on strategy and governance. The plethora of research on the provision of feedback, together with the need to change mind sets to understand the concept and value of 'feeding forward', in formative assessment practices present a dilemma for future research to address.

DISCUSSION

The milieu of contemporary thought of feedback effectiveness can be distracting and can result in clouded views of the feedback phenomenon. This chapter recognizes this milieu (Boud, 2015) and has been purposefully narrow in its focus in order to ensure a clear view of the role of dialogue in feedback. The chapter contends that the nuanced behaviors of individuals in the formative assessment space, whether they are students, academics or professional staff, requires further investigation. This chapter has highlighted FM2 and the twelve conditions of success, and has attempted to draw them together within a framework of five factors that influence the enacted curriculum of the course of learning. The framework acts as a cohesive tie between the eight elements of FM2 and the twelve conditions of success. In exploring these links and connections, the following discussion argues that humanistic dispositions of academics and students exert an influence on the enacted curriculum.

The first section of this discussion places a lens over the FM2 characteristics, and draws out the humanistic elements. Then it aligns those to the five elements of influence. This is followed by a discussion of the twelve conditions of success in terms of humanistic behavior and how they align to the five elements influencing the enacted curriculum. The discussion concludes with a theoretical response to the guiding questions: "How can higher education institutions best support student learning experience and academic practice of feedback and assessment that is dialogically centered? Which dialogic feedback behaviors can improve graduate employability, student experience and sustainability in the profession?

DIALOGUE IN FEEDBACK MARK 2

The enacted curriculum influencers' framework has two elements that relate directly to the FM2 as presented in the literature. The baseline for the influencers' framework is the course design construct. Course design also provides the context for other influencing factors. The authors contend that course design is directly linked to four of the eight aspects of the FM2. These four are the opportunities for work production, collaborative opportunities, incremental changes in learning and assessment, and nested assessment tasks. The design of assessment to ensure that nested assessments are present for students allows for increased feedback opportunities from one assessment task to the next. Interestingly, this attaches the idea of feedback directly to assessment tasks. The presence of incremental changes that allow feedback to be influential in the learning process also draws on the design of a course. A course with incremental change opportunities will typically have a sequence of smaller learning opportunities within its design, and will afford students a developmental pathway to success, thus reducing their gap in knowledge (Dann & Richardson, 2015). Further to this, the course design will provide opportunities for students to produce work, and the final FM2 element for course design, allows for collaboration between students. This final component is somewhat challenging for academics given the certification focus on grades.

The second layer of the enacted curriculum influences student feedback characteristics. One perception is that this influence may be out of the control of academics, but on closer examination Boud and Molloy (2015) show that academics can influence these. The four remaining include learner orientation, learner participation, learner dispositions and the learner as a seeker of feedback. Learner orientation can be influenced by an academic or course teacher through the processes and expectations set within a course. Learner participation may also be influenced, although it is argued to be centralized around the students' interest in learning activities and personal motivation, which could be intrinsic (love of learning) or extrinsic (access to a job). Students' dialogues highlight their disposition and it is from this that information can begin to be provided, which can be used to make adjustments to improve learning. Attention to such factors and displays of dispositions may provide the academic with information upon which to pivot within the course to meet or enhance feedback to students. Once these motivations are known or surfaced the student can then be assisted to become a seeker of feedback. To seek feedback is only one part of solving the problem. Once a student has feedback there is still a need for the student to understand the content of the feedback, apply the feedback to their learning pathway and then internalize it so that it can be sustained in their lifelong learning journey.

If a student is to internalize the feedback provided, apply it or even at a minimum, understand it, there remains a key aspect to consider. This relates to how the feedback was delivered to the student. The mode of delivery appears central to the discussion on feedback because the mode and tone of feedback impacts on the students' capacity to accept the content in order to link it to future work. The phrasing and tone of feedback may also vary according to the delivery method. Moreover, it is important to remember that verbal feedback tone is distinctly separate from the text-based feedback provided, no matter what the context. While this chapter does not attempt to incorporate the literature on these latter factors, they do highlight the interpersonal skills and sensitivities required to operate in the 'dialogic relationship' (Patrikakou, 2015). Instead, we seek to highlight that if feedback is part of two-way communication then as a dialogue it is the tone and mode of communication about the feedback content that is essentially part of the process reflected in Figure 5.

Figure 5. Influencer relationship to Feedback Mark 2

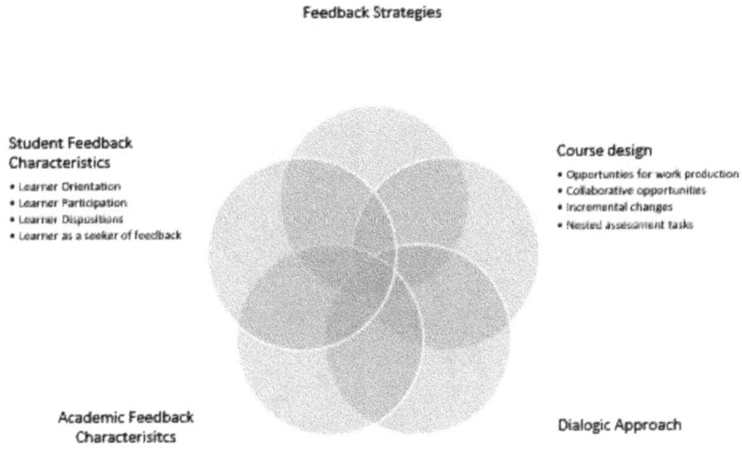

DIALOGUE IN THE 12 CONDITIONS OF SUCCESSFUL FEEDBACK

Twelve conditions of success for feedback were presented above as part of three interrelated concepts of capacity, designs and culture. Four of the twelve conditions were assigned to each of these three concepts, as a high-level graphical organizer, providing a clear framework for ongoing discussion of actions and processes associated with the feedback in higher education. Each of the twelve elements as developed by Henderson et al. (2018) have been aligned to the Feedback Influencer Framework. These five aspects are encompassed by the enacted curriculum and are shown in Figure 6.

Twelve conditions of success for feedback were presented as one part of three inter-related concepts of capacity, designs and culture. Four of the twelve conditions were assigned to each of these three concepts as a high-level graphic organizer providing a clear framework for ongoing discussion of actions and processes associated with feedback in higher education. Each of the twelve elements identified by Henderson et al. (2018) have been aligned to the Feedback Influencer Framework. Academic feedback characteristics are seen as influencing the feedback strategies, and the dialogic approach as a whole, all impact on the student characteristics. This is represented in Figure 7. The entire interplay occurs because of and within the course design, which impacts on the enacted curriculum. The items at each corner are taken from the twelve conditions of success that the authors suggest align to the four central influencers (Henderson et al., 2018). In addition, the authors suggest that course design is impacted by learning outcomes of multiple tasks (condition 8), and that it is a valued and visible enterprise at all levels (condition 9). Then finally, there should be processes in place to ensure consistency and quality (condition 10).

Typically, assessment is the most powerful term in higher education owing to its connection to certification of students' achievement. In addition, feedback as part of formative assessment is the major contributor to improving students' learning. Feedback provided through formative assessment highlights how the design of assessment tasks has the responsibility to provide feed markers that can stimulate critical dialogue to lead students to 'act' on bridging their learning gap. In reiterating that to be effective this practice is envisaged to take place within the context of a community of inquiry where the

Figure 6. Alignment of feedback influencer framework and conditionsof successful feedback

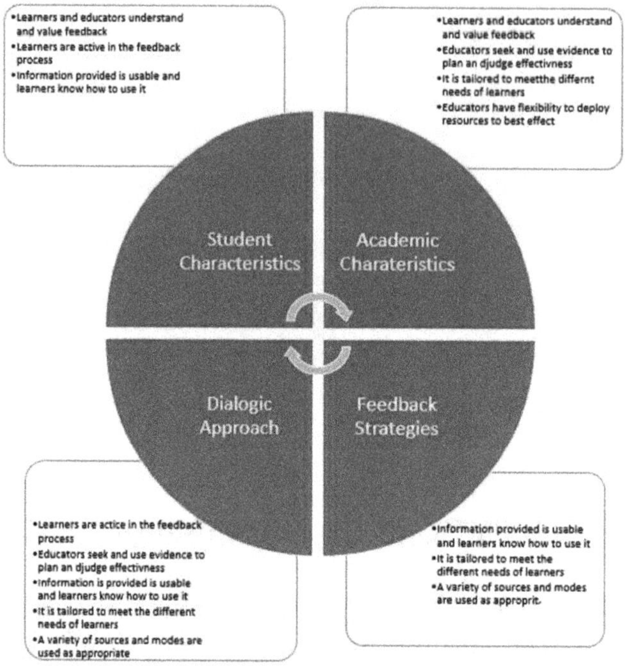

application of the dialogic elements from the three constructs help create a communicative, interactive and supportive learning space. In advocating a move from thinking about feedback on assessment to thoughts of feeding student learning, despite the fact that this may initially sound altruistic the resultant impact on learning needs to be considered. But such a move can only be sought with clear evidence of the humanistic elements influencing the enacted curriculum and the learning experience.

To position attention on feeding student success there is a need to understand the barriers and enhancers relevant to the students' experience of their course. If feeding students information about their learning enhances their learning experience and opportunity for success, then the application of dialogic pedagogy and formative assessment needs to be illuminated rather than cast aside because a monologic approach is perceived as easier and more economic, or the dialogic more time consuming. It is interesting that an analysis of academic nuances and peculiarities to find excellence in dialogic practices with in the three frameworks presented here, may indeed uncover behaviors that are less supportive or even detrimental to student learning and experience. This is raised in light of the concerns that feedback will be uncomfortable for some students, thus, the authors' contest that levels of discomfort may be required to develop effective learning journeys that result in sustained learning.

Our call to examination the following terms is supported by their implied relationship to the conditions of successful feedback and the characteristics of the FM2. Examining the implication that relational, empathetic, and sympathetic characteristics are actually evident in excellent feedback practices, can inform the work of academics and those who may need to be in a position where feedback is to be offered or sought. The authors also seek to better understand the role of factual objectivity, cooperativeness, concern for others, open-mindedness, honesty, genuineness, equality and willingness to delay judgement in formative assessment practices (Johannesen, 1990). Further research is recommended to uncover the presence of these traits and their influence on the relational interactions present in effective communication exchanges between a student and their educational leaders. This examination needs to explore how and when these characteristics emerge and how they exist or manifest in both students and academics. This chapter set out to recognize students and academics as particpants in teaching and learning on equivalent levels in the dialogic space such that this premise needs to be applied in future research into the characteristics of pedagogical dialogue and critical dialogue in formative assessment. The original assertion of Buber (1923/37) that rewarding relationships occur when people treat others as inherently valuable is viewed core to practice. To reiterate, Rogers highlighted the dialogic concepts of genuineness, spontaneity and face-to-face interaction, in the context of true dialogue only occurring in a relationship.

CONCLUSION

This chapter has presented the theorized perspective on the use of feedback in formative assessment within teacher education. The framework presented allows for provider institutions to focus on curriculum characteristics within the milieu described by Boud and Molley (2013) to ensure the promotion of dialogic formative assessment practices can grow and support students' learning and feedback literacy. It has not attempted to align this perspective with the theories that underpin learning and it is acknowledged that this is a limitation of the current research, and so is also recommended for future research. An additional implication is that feedback should not only be examined from the perspective of the student as it needs to take account of the attitudes of educators and others involved. It is also important to consider how these attitudes influence course design and in turn how course design enables 'feedback' strategies that can be built into a dialogic approach where there is 'feedforward'. It is the authors' view that the use of the long-established reference to 'feedback', as the central focus, presents a contradiction in terms. It evokes a traditional conceptualization of learning and practice, that suggests monologic communication, involving the one-way transmission of information from teacher to learner. This narrative runs counter to that associated with a social constructivist view of learning where knowledge is believed to be co-constructed through a dialogic approach that involves two-way communication. In addition, since research shows that effective formative assessment practices depend on identification of 'indicators of learning' that can stimulate discussion to assist learners to take action to better understand, the authors suggest that these 'indicators' represent or provide 'feed'. On this basis 'feed' may may be designed to travel 'forward' or 'backward' according to the purpose of assessment i.e. formative or summative. Moreover, changing the narrative in relation to assessment practices in this way is in keeping with the contemporary shift to dialogic pedagogies where formative assessment is applied in the shorter term, en route to the end of a course of learning. In this way, it supports the expectation that students have an opportunity to improve their understanding during their course. Therefore, as the underpinning element for the successful implementation of any formative assessment approach in higher education, the dialogic approach presents a relatively new catalyst for change. With the availability and application of technology, as a vital element of support, there is an opportunity to significantly improve how learning is experienced, and ultimately improve learning outcomes.

REFERENCES

Aedo, M. P., Peredo, S., & Schaeffer, C. (2017). From an essential being to an actor's becoming: Political ecology transformational learning experiences in adult education. *Environmental Education Research, 25*(1), 1–13.

Ajjawi, R., & Boud, D. (2017). Researching feedback dialogue: An interactional analysis approach. *Assessment & Evaluation in Higher Education, 42*(2), 252–265. doi:10.1080/02602938.2015.1102863

Alexander, R. (2010). *Dialogic teaching essentials*. Singapore: National Institute of Education.

Ally, M., & Wark, N. (2018). *Online student use of mobile devices for learning*. Chicago, IL: Concordia University Chicago. Retrieved from https://www.learntechlib.org/p/184917

Aoun, C., Vatanasakdakul, S., & Ang, K. (2018). Feedback for thought: Examining the influence of feedback constituents on learning experience. *Studies in Higher Education, 43*(1), 72–95. doi:10.1080/03075079.2016.1156665

Bamber, P. M. (2016). *Transformative education through international service-learning: Realising an ethical ecology of learning*. New York: Routledge. doi:10.4324/9781315684970

Bandiera, O., Larcinese, V., & Rasul, I. (2015). Blissful ignorance? A natural experiment on the effect of feedback on students' performance. *Labour Economics, 34*, 13–25. doi:10.1016/j.labeco.2015.02.002

Bandura, A. (2018). Toward a psychology of human agency: Pathways and reflections. *Perspectives on Psychological Science, 13*(2), 130–136. doi:10.1177/1745691617699280 PMID:29592657

Barak, M. (2016). Science teacher education in the twenty-first century: A pedagogical framework for technology-integrated social constructivism. *Research in Science Education, 47*(2), 283–303. doi:10.100711165-015-9501-y

Barnes, D. (2010). Why talk is important. *English Teaching, 9*(2), 7–10.

Barnett, R., & Coates, K. (2005). *Engaging the curriculum in higher education*. Maidenhead, UK: Open University Press.

Beaumont, C., Moscrop, C., & Canning, S. (2016). Easing the transition from school to HE: Scaffolding the development of self-regulated learning through a dialogic approach to feedback. *Journal of Further and Higher Education, 40*(3), 331–350. doi:10.1080/0309877X.2014.953460

Benade, L. (2015). Postgraduate research supervision as a partnership in the context of freire's dialogical pedagogy. *Knowledge Cultures, 3*(5), 102–115.

Bennett, S., Dawson, P., Bearman, M., Molloy, E., & Boud, D. (2017). How technology shapes assessment design: Findings from a study of university teachers. *British Journal of Educational Technology, 48*(2), 672–682. doi:10.1111/bjet.12439

Black, P., & Wiliam, D. (1998). Assessment and classroom learning. *Assessment in Education: Principles, Policy & Practice, 5*(1), 7–74. doi:10.1080/0969595980050102

Bligh, B., & Flood, M. (2017). Activity theory in empirical higher education research: Choices, uses and values. *Tertiary Education and Management*, *23*(2), 125–152. doi:10.1080/13583883.2017.1284258

Boesen, J., Helenius, O., Lithner, J., Bergqvist, E., Bergqvist, T., Palm, T., & Palmberg, B. (2014). Developing mathematical competence: From the intended to the enacted curriculum. *The Journal of Mathematical Behavior*, *33*, 72–87. doi:10.1016/j.jmathb.2013.10.001

Boud, D. (2015). Feedback: Ensuring that it leads to enhanced learning. *The Clinical Teacher*, *12*(1), 3–7. doi:10.1111/tct.12345 PMID:25603699

Boud, D., & Molloy, E. K. (2013). Rethinking models of feedback for learning: The challenge of design. *Assessment & Evaluation in Higher Education*, *38*(6), 698–712. doi:10.1080/02602938.2012.691462

Buber, M. (1958). I and Thou (2nd ed.; R. Gregor Smith, Trans.). New York: Scribner. (Original publication 1923)

Carless, D. (2019). Feedback loops and the longer-term: Towards feedback spirals. *Assessment & Evaluation in Higher Education*, *44*(5), 705–714. doi:10.1080/02602938.2018.1531108

Carless, D., Salter, D., Yang, M., & Lam, J. (2011). Developing sustainable feedback practices. *Studies in Higher Education*, *36*(4), 395–407. doi:10.1080/03075071003642449

Carroll, J. M. (2003). *HCI models, theories, and frameworks: Toward a multidisciplinary science.* Morgan Kaufmann.

Cole, A., Anderson, C., Bunton, T., Cherney, M., Fisher, V. C., Featherston, M., ... Allen, M. (2017). Student predisposition to instructor feedback and perceptions of teaching presence predict motivation toward online courses. *Online Learning Journal*, *21*(4), 245–262.

Conrad, S. S., & Dabbagh, N. (2015). Examining the Factors that influence how instructors provide feedback in online learning environments. *International Journal of Online Pedagogy and Course Design*, *5*(4), 47–66. doi:10.4018/IJOPCD.2015100104

Crane-Mitchell, L., & Stafford, L. (2017). *Enhancing learning outcomes for students with autism through the use of assistive technology: Effect of parent-teacher collaboration.* Paper presented at the Society for Information Technology & Teacher Education International Conference.

Dann, C., & Dann, B. (2018). Exemplifying formative assessment practices in the field through case study: Concepts, issues and challenges. In T. Richardson, B. Dann, C. Dann, & S. O'Neill (Eds.), *Formative assessment practices for pre-srevice teacher practicum feedback: Emerging research and opportunities* (pp. 76–112). IGI Global. doi:10.4018/978-1-5225-2630-8.ch003

Dann, C., & Richardson, T. (2015). Deepening understanding of 'pedagogical outcomes' through video data collection: A catalyst for guided reflective learning conversations. *International Journal of Pedagogies & Learning*, *10*(1), 62–80. doi:10.1080/22040552.2015.1084677

Denton, P., & McIlroy, D. (2017). Response of students to statement bank feedback: The impact of assessment literacy on performance in summative tasks. *Assessment & Evaluation in Higher Education*, *42*(2), 197–206. doi:10.1080/02602938.2017.1324017

Engeström, Y. (2001). Expansive learning: Towards an activity-theoretical reconceptualization. *Journal of Education and Work, 14*(1), 133–156. doi:10.1080/13639080020028747

Engin, M., & Donanci, S. (2015). Dialogic teaching and iPads in the EAP classroom. *Computers & Education, 88*, 268–279. doi:10.1016/j.compedu.2015.06.005

Forsythe, A., & Johnson, S. (2016). Thanks, but no-thanks for the feedback. *Assessment & Evaluation in Higher Education, 42*(6), 850–859. doi:10.1080/02602938.2016.1202190

Freire, P. (1970). *Pedagogy of the oppressed* (M. B. Ramos, Trans.). New York: *Continuum.*

Fry, H., Ketteridge, S., & Marshall, S. (2014). Assessing assessment: new developments in assessment design, feedback practices and marking in higher education. In *A Handbook for Teaching and Learning in Higher Education* (pp. 129–144). New York: Routledge. doi:10.4324/9781315763088

Furtak, E. M., Kiemer, K., Circi, R. K., Swanson, R., de Leon, V., Morrison, D., & Heredia, S. C. (2016). Teachers' formative assessment abilities and their relationship to student learning: Findings from a four-year intervention study. *Instructional Science, 44*(3), 267–291. doi:10.100711251-016-9371-3

Gadamer, H.-G. (1991). *Plato's dialectical ethics: Phenomenological interpretations relating to the Philebus* (R. M. Wallace, Trans.). New Haven, CT: Yale University Press.

Gallagher, G. (2017). Aligning for learning: Including feedback in the Constructive Alignment model. *AISH-J: The All Ireland Journal of Teaching and Learning in Higher Education, 9*(1), 3011–3111.

Gergel, S. E., & Turner, M. G. (2017). *Learning landscape ecology: a practical guide to concepts and techniques.* Springer. doi:10.1007/978-1-4939-6374-4

Giroux, H. A. (2010). Rethinking education as the practice of freedom: Paulo Freire and the promise of critical pedagogy. *Policy Futures in Education, 8*(6), 715–721. doi:10.2304/pfie.2010.8.6.715

Goldin, I., Narciss, S., Foltz, P., & Bauer, M. (2017). New Directions in Formative Feedback in Interactive Learning Environments. *International Journal of Artificial Intelligence in Education, 27*(3), 385–392. doi:10.100740593-016-0135-7

Hattie, J. (2009). *Visible learning: A synthesis of over 800 meta-analyses relating to achievement.* London: Routledge/Taylor & Francis.

Hattie, J., & Timperley, H. (2007). *The power of feedback.* Academic Press.

Henderson, M., Boud, D., Molloy, E., Dawson, P., Phillips, M., Ryan, T., & Mahoney, P. (2018). *Feedback for learning: Closing the assessment loop.* Australian Government. Department of Education and Training.

Hendry, G. D., White, P., & Herbert, C. (2016). Providing exemplar-based 'feedforward' before an assessment: The role of teacher explanation. *Active Learning in Higher Education, 17*(2), 99–109. doi:10.1177/1469787416637479

Jaggars, S. S., & Xu, D. (2016). How do online course design features influence student performance? *Computers & Education, 95*, 270–284. doi:10.1016/j.compedu.2016.01.014

Johannesen, R. L. (1990). Ethics in human communication (3rd ed.). Prospect Heights, IL: Waveland Press.

Jones, A., Olds, A., & Lisciandro, J. (2016). Understanding the Learner: Effective course design in the changing higher education space. *International Studies in Widening Participation, 3*(1), 19–35.

Juwah, C., Macfarlane-Dick, D., Matthew, B., Nicol, D., Ross, D., & Smith, B. (2004). *Enhancing student learning through effective formative feedback.* Higher Education Academy Generic Centre.

Kent, M. L. (2017). Principles of dialogue and the history of dialogic theory in public relations. In X. Chen (Ed.), *Prospect of public relations science* (pp. 105–129). Beijing: Peking University Press. (in Chinese)

Kent, M. L., & Taylor, M. (2002). Toward a dialogic theory of public relations. *Public Relations Review, 28*(1), 21–37. doi:10.1016/S0363-8111(02)00108-X

Kent, M. L., & Theunissen, P. (2016). Elergy for mediated dialogue: Shiva the destroyer and reclaiming our first principles. *International Journal of Communication, 10*, 4040–4054.

Maier, W., Wolf, N., & Randler, C. (2016). Effects of a computer-assisted formative assessment intervention based on multiple-tier diagnostic items and different feedback types. *Computers & Education, 95*, 85–98. doi:10.1016/j.compedu.2015.12.002

Merry, S., Price, M., Carless, D., & Taras, M. (2013). *Reconceptualising feedback in higher education: Developing dialogue with students.* London: Routledge. doi:10.4324/9780203522813

Molloy, E. K., & Boud, D. (2014). Feedback models for learning, teaching and performance. In J. M. Spector, M. D. Merrill, J. Elen & M. J. Bishop (Eds.), Handbook of research on educational communications and technology (4th ed.; pp. 413-424). New York: Springer. doi:10.1007/978-1-4614-3185-5_33

Moriarty, B., Danaher, P., & Danaher, G. (2008). Freire and dialogical pedagogy: A means for interrogating opportunities and challenges in Australian postgraduate supervision. *International Journal of Lifelong Education, 27*(4), 431–442. doi:10.1080/02601370802051447

Mulliner, E., & Tucker, M. (2017). Feedback on feedback practice: Perceptions of students and academics. *Assessment & Evaluation in Higher Education, 42*(2), 266–288. doi:10.1080/02602938.2015.1103365

Nixon, S., Brooman, S., Murphy, B., & Fearon, D. (2017). Clarity, consistency and communication: Using enhanced dialogue to create a course-based feedback strategy. *Journal of Assessment & Evaluation in Higher Education, 42*(5), 812–822. doi:10.1080/02602938.2016.1195333

Palincsar, A. S. (1998). Social constructivist perspectives on teaching and learning. *Annual Review of Psychology, 49*(1), 345–375. doi:10.1146/annurev.psych.49.1.345 PMID:15012472

Pardo, A. (2018). A feedback model for data-rich learning experiences. *Journal of Assessment & Evaluation in Higher Education, 43*(3), 428–433. doi:10.1080/02602938.2017.1356905

Pardo, A., Jovanovic, J., Dawson, S., Gašević, D., & Mirriahi, N. (2017). Using learning analytics to scale the provision of personalised feedback. *British Journal of Educational Technology, 50*(1), 128–138. doi:10.1111/bjet.12592

Patrikakou, E. (2015). Relationships among parents, students, and teachers: The technology wild card. *Procedia: Social and Behavioral Sciences, 174*, 2253–2258. doi:10.1016/j.sbspro.2015.01.883

Peacock, S., & Cowan, J. (2017). Towards online student-directed communities of inquiry. *Journal of Further and Higher Education, 42*(5), 1–16. doi:10.1080/0309877X.2017.1302569

Remillard, J. (2018). *Mapping the relationship between written and enacted curriculum: Examining teachers' decision making*. Paper presented at the Invited Lectures from the 13th International Congress on Mathematical Education. 10.1007/978-3-319-72170-5_27

Rogers, C. R. (1957). The necessary and sufficient conditions of therapeutic personality change. *Journal of Consulting Psychology, 21*(2), 95–103. doi:10.1037/h0045357 PMID:13416422

Sambell, K. (2016). Assessment and feedback in higher education: considerable room for improvement? *Student Engagement in Higher Education, 1*(1).

Siemens, G. (2014). *Connectivism: A learning theory for the digital age*. Elearnspace.

Sutton, P. (2012). Conceptualizing feedback literacy: Knowing, being, and acting. *Innovations in Education and Teaching International, 49*(1), 31–40. doi:10.1080/14703297.2012.647781

Tomasik, M. J., Berger, S., & Moser, U. (2018). On the development of a computer-based tool for formative student assessment: Epistemological, methodological, and practical issues. *Frontiers in Psychology, 9*, 2245. doi:10.3389/fpsyg.2018.02245 PMID:30515119

Villarroel, V., Bloxham, S., Bruna, D., Bruna, C., & Herrera-Seda, C. (2018). Authentic assessment: Creating a blueprint for course design. *Journal of Assessment & Evaluation in Higher Education, 43*(5), 840–854. doi:10.1080/02602938.2017.1412396

Vygotsky, L. S. (1978). *Mind in society: The development of higher mental process*. Cambridge, MA: Harvard University Press.

Wenger, E. (1998). Communities of practice: Learning as a social system. *The Systems Thinker, 9*(5), 2–3.

Wenger, E. (2011). *Communities of practice: A brief introduction*. Watertoen, MA: Harvard Business Press.

Wiliam, D. (2011). *Embedded formative assessment*. Bloomington, IM: Solution Tree Press.

Wiliam, D., & Thompson, M. (2017). Integrating assessment with learning: What will it take to make it work? In *The future of assessment* (pp. 53–82). New York: Routledge. doi:10.4324/9781315086545-3

Yoon, Y. S., Ham, D. H., & Yoon, W. C. (2016). Application of activity theory to analysis of human-related accidents: Method and case studies. *Reliability Engineering & System Safety, 150*, 22–34. doi:10.1016/j.ress.2016.01.013

Compilation of References

Abdous, M., & Yoshimura, M. (2010). Learner outcomes and satisfaction: A comparison of live video-streamed instruction, satellite broadcast instruction, and face-to-face instruction. *Computers & Education, 55*(2), 733–741. doi:10.1016/j.compedu.2010.03.006

Abrahamson, E. (2010). *Assessment through video-feedback on an undergraduate sports rehabilitation programme. Higher Education Academy [HEA] Case Study*. Retrieved from http://www.heacademy.ac.uk/assets/hlst/documents/case_studies/147_abrahamson_video-feedback.pdf

Abramovich, S. (2016). Understanding digital badges in higher education through assessment. *On the Horizon, 24*(1), 126–131. doi:10.1108/OTH-08-2015-0044

Abramovich, S., Schunn, C., & Higashi, R. (2013). Are badges useful in education? It depends upon the type of badge and expertise of learner. *Educational Technology Research and Development, 61*(2), 217–232. doi:10.100711423-013-9289-2

Ackerman, D. S., & Gross, B. L. (2010). Instructor feedback: How much do students really want? *Journal of Marketing Education, 32*(2), 172–181. doi:10.1177/0273475309360159

Adams, R. (2012). *NAPLAN reporting –measures and models for reporting gain over time*. Retrieved from http://www.coagreformcouncil.gov.au/excellence/docs/improvement/CI_NAPLAN_reporting_2012. pdf

Aedo, M. P., Peredo, S., & Schaeffer, C. (2017). From an essential being to an actor's becoming: Political ecology transformational learning experiences in adult education. *Environmental Education Research, 25*(1), 1–13.

Ajjawi, R., & Boud, D. (2017). Researching feedback dialogue: An interactional analysis approach. *Assessment & Evaluation in Higher Education, 42*(2), 252–265. doi:10.1080/02602938.2015.1102863

Akcil, U. (2018). The use of mobile learning for visually impaired learners school in tolerance education contents. *Quality & Quantity, 52*(2), 969–982. doi:10.100711135-017-0552-1

Akerson, V. L., Cullen, T. A., & Hanson, D. L. (2009). Fostering a community of practice through a professional development program to improve elementary teachers' views of nature of science and teaching practice. *Journal of Research in Science Teaching, 46*(10), 1090–1113. doi:10.1002/tea.20303

Akkerman, S. F., & Meijer, P. C. (2011). A dialogical approach to conceptualizing teacher identity. *Teaching and Teacher Education, 27*(2), 308–319. doi:10.1016/j.tate.2010.08.013

Akyol, Z., & Garrison, D. R. (2011). Understanding cognitive presence in an online and blended community of inquiry: Assessing outcomes and processes for deep approaches to learning. *British Journal of Educational Technology, 42*(2), 233–250. doi:10.1111/j.1467-8535.2009.01029.x

Akyol, Z., Garrison, D. R., & Ozden, M. Y. (2009). Online and blended communities of inquiry: Exploring the developmental and perceptional differences. *International Review of Research in Open and Distance Learning*, *10*(6), 65–83. doi:10.19173/irrodl.v10i6.765

Alesandrini, K., & Larson, L. (2002). Teachers bridge to constructivism. *The Clearing House: A Journal of Educational Strategies, Issues and Ideas*, *75*(3), 118–121. doi:10.1080/00098650209599249

Alexander, R. (2010). *Dialogic teaching essentials*. Singapore: National Institute of Education.

Ali, N., Ahmed, L., & Rose, S. (2018). Identifying predictors of students' perception of and engagement with assessment feedback. *Active Learning in Higher Education*, *19*(3), 239–251. doi:10.1177/1469787417735609

Allal, L., & López, L. M. (2015). Formative assessment of learning. A review of publications in French. In OECD (Ed.), *Formative assessment: Improving learning in secondary classrooms* (pp. 241–264). Paris: OECD Publishing.

Allen, I., & Seaman, J. (2014). *Grade change: Tracking online education in the United States*. Retrieved from https://www.onlinelearningsurvey.com/reports/gradechange.pdf

Allen, M. (2014). *Misconceptions in primary science*. McGraw-Hill Education.

Allen, R., & Bentley, R. (2012). *Feedback mechanisms: Efficient and effective use of technology or waste of time and effort*. London, UK: *STEM Annual Conference*.

Alleyne, L. (2013). Breaking sound barriers: Custom integrated audio distribution is taking off in the cloud. *Residential System*, *14*(1), 2–5.

Ally, M., & Wark, N. (2018). *Online Student Use of Mobile Devices for Learning*. Chicago, IL: Concordia University Chicago. https://www.learntechlib.org/p/184917

Ally, M., & Wark, N. (2018). *Online student use of mobile devices for learning*. Chicago, IL: Concordia University Chicago. Retrieved from https://www.learntechlib.org/p/184917

Alpern, M. (1972). Eye movements. In D. Jameson & L. M. Hurvich (Eds.), Handbook of sensory physiology VII/4: Visual psychophysics (pp. 303-330). Berlin: Springer. doi:10.1007/978-3-642-88658-4_12

Alsaleem, B. I. A. (2014). The effect of "WhatsApp" electronic dialogue journaling on improving writing vocabulary word choice and voice of EFL undergraduate Saudi students. *21st Century Academic Forum Conference Proceedings*. Retrieved from http://www.readwritethink.org/lesson_images/lesson782/Rubric.pdf

Anderson, R., Dickey, M., & Perkins, H. (2001). Experiences with tutored video instruction for introductory programming courses. *ACM SIGCSE Bulletin*, *33*(1), 347–351. doi:10.1145/366413.364619

Anderson, R., Feldman, S., & Minstrell, J. (2014). Understanding relationship: Maximizing the effects of science coaching. *Education Policy Analysis Archives*, *22*, 54.

Anderson, T., Rourke, L., Garrison, D. R., & Archer, W. (2001). Assessing teacher presence in a computer conferencing context. *Journal of Asynchronous Learning Networks*, *5*(2), 1–17.

Anderson, T., Rourke, L., Garrison, D. R., & Archer, W. (2001). Assessing teaching presence in a computer conference environment. *Journal of Asynchronous Learning Networks*, *5*(2), 1–1.

Aoun, C., Vatanasakdakul, S., & Ang, K. (2018). Feedback for thought: Examining the influence of feedback constituents on learning experience. *Studies in Higher Education*, *43*(1), 72–95. doi:10.1080/03075079.2016.1156665

Appleton, K. (2006). Science pedagogical content knowledge and elementary school teachers. *Elementary science teacher education: International perspectives on contemporary issues and practice*, 31-54.

Appleton, K. (2002). Science activities that work: Perceptions of primary school teachers. *Research in Science Education, 32*(3), 393–410. doi:10.1023/A:1020878121184

Appleton, K. (2003). How do beginning primary school teachers cope with science? Toward an understanding of science teaching practice. *Research in Science Education, 33*(1), 1–25. doi:10.1023/A:1023666618800

Armstrong, H. (2009). Advanced IT education for the vision impaired via e-learning. *Journal of Information Technology Education, 8*, 243–256. doi:10.28945/691

Asbjørnsen, A. E., Obrzut, J. E., Eikeland, O.-J., & Manger, T. (2010). Can solving of Wordchains be explained by phonological skills alone? *Dyslexia, 16*(1), 24–35. doi:10.1002/dys.394 PMID:19562661

Asperger, H. (1991). Autistic psychopathy' in childhood (U. Frith, Trans.). In U. Frith (Ed.), Autism and Asperger syndrome (pp. 37–92). Cambridge, UK: Cambridge University Press. (Original publication 1944)

Assessment Reform Group. (2002). *Assessment for learning: 10 principles*. Nuffield Foundation. Retrieved from http://www.assessment-reform-group.org/CIE3.pdf

Ata, R. (2016). An exploration of higher education teaching in Second Life in the context of blended learning. *Turkish Online Journal of Educational Technology, 15*(3), 9–26.

Attwood, T. (2000). *Asperger's syndrome: A guide for parents and professionals*. London: Jessica Kingsley.

Attwood, T. (2000). *Asperger's Syndrome: A guide for parents and professionals*. London: Jessica Kingsley.

Australian Institute for Teaching and School Leadership. (2019). *Australian professional standards for teachers*. Retrieved from https://www.aitsl.edu.au/teach/standards

Australian Vice-Chancellors' Committee. (2004). *Guidelines on information access for students with print disabilities*. Retrieved from https://www.adcet.edu.au/download/attachment/5352-1/

Badcock, P. B., Pattison, P. E., & Harris, K. L. (2010). Developing generic skills through university study: A study of arts, science and engineering in Australia. *Higher Education, 60*(4), 441–458. doi:10.100710734-010-9308-8

Baddeley, A. (1992). Working memory. *Science, 255*(5044), 556–559. doi:10.1126cience.1736359 PMID:1736359

Baddeley, A. D., & Hitch, G. (1974). Working memory. In G. H. Bower (Ed.), *The psychology of learning and motivation: Advances in research and theory* (pp. 47–89). New York: Academic Press; doi:10.1016/S0079-7421(08)60452-1

Baecher, L., Kung, S.-C., Jewkes, A. M., & Rosalia, C. (2013). The role of video for self-evaluation in early field experiences. *Teaching and Teacher Education, 36*, 189–197. doi:10.1016/j.tate.2013.08.001

Baladoh, S. M., Elgamal, A. F., & Abas, H. A. (2017). Virtual lab to develop achievement in electronic circuits for hearing-impaired students. *Education and Information Technologies, 22*(5), 2071–2085. doi:10.100710639-016-9532-7

Balamuralithara, B., & Woods, P. C. (2009). Virtual laboratories in engineering education: The simulation lab and remote lab. *Computer Applications in Engineering Education, 17*(1), 108–118. doi:10.1002/cae.20186

Bamber, P. M. (2016). *Transformative education through international service-learning: Realising an ethical ecology of learning*. New York: Routledge. doi:10.4324/9781315684970

Bandiera, O., Larcinese, V., & Rasul, I. (2015). Blissful ignorance? A natural experiment on the effect of feedback on students' performance. *Labour Economics, 34*, 13–25. doi:10.1016/j.labeco.2015.02.002

Bandura, A. (2018). Toward a psychology of human agency: Pathways and reflections. *Perspectives on Psychological Science, 13*(2), 130–136. doi:10.1177/1745691617699280 PMID:29592657

Barak, M. (2016). Science teacher education in the twenty-first century: A pedagogical framework for technology-integrated social constructivism. *Research in Science Education, 47*(2), 283–303. doi:10.100711165-015-9501-y

Baran, E., & Correia, A. P. (2009). Student-led facilitation strategies in online discussions. *Distance Education, 30*(3), 339–361. doi:10.1080/01587910903236510

Barmby, P., Andrà, C., Arzarello, F., Gomez, D. O. A., & Shvarts, A. (2014). The use of eye-tracking technology in mathematics education research. In P. Liljedahl, C. Nicol, S. Oesterle, & D. Allan (Eds.), *Proceedings of the Joint Meeting of PME 38 and PME-NA 36* (Vol. 1, pp. 253-371). Vancouver: Psychology of Mathematics Education.

Barnes, D. (2010). Why talk is important. *English Teaching, 9*(2), 7–10.

Barnett, R., & Coates, K. (2005). *Engaging the curriculum in higher education*. Maidenhead, UK: Open University Press.

Baron-Cohen, S. (2008). *Autism and Asperger's Syndrome*. Oxford, UK: Oxford University Press.

Barrie, S. C. (2006). Understanding what we mean by the generic attributes of graduates. *Higher Education, 51*(2), 215–241. doi:10.100710734-004-6384-7

Barz, M., Daiber, F., & Bulling, A. (2015). *Computational modelling and prediction of gaze estimation error for head-mounted eye trackers*. Retrieved from http://citeseerx.ist.psu.edu/viewdoc/download?doi=10.1.1.717.2872&rep=rep1&type=pdf

Bath, D., Smith, C., Stein, S., & Swann, R. (2004). Beyond mapping and embedding graduate attributes: Bringing together quality assurance and action learning to create a validated and living curriculum. *Higher Education Research & Development, 23*(3), 313–328. doi:10.1080/0729436042000235427

Beaumont, C., Moscrop, C., & Canning, S. (2016). Easing the transition from school to HE: Scaffolding the development of self-regulated learning through a dialogic approach to feedback. *Journal of Further and Higher Education, 40*(3), 331–350. doi:10.1080/0309877X.2014.953460

Bednarik, R., Schulte, C., Budde, L., Heinemann, B., & Vrzakova, H. (2018). Eye-movement modeling examples in source code comprehension: A classroom study. *Proceeding. Koli Calling'18 Proceedingsof the 18th Koli Calling International Conference on Computing Education Research, Article No. 12*. New York: ACM. doi10.1145/3279720.3279722

Belfer, K., & Morgan, T. (2005). Enhancing teaching presence and reducing distance using voice technologies. In *World Conference on E-Learning in Corporate, Government, Healthcare, and Higher Education* (*Vol. 2005*, No. 1, pp. 537-542). Academic Press.

Benade, L. (2015). Postgraduate research supervision as a partnership in the context of freire's dialogical pedagogy. *Knowledge Cultures, 3*(5), 102–115.

Benford, P. (2008). *The use of Internet-based communication by people with autism* (PhD thesis). University of Nottingham. Retrieved from http://etheses.nottingham.ac.uk/661/

Bennett, R. E. (2011). Formative assessment: A critical review. *Assessment in Education: Principles, Policy & Practice, 18*(1), 5–25. doi:10.1080/0969594X.2010.513678

Bennett, S., Dawson, P., Bearman, M., Molloy, E., & Boud, D. (2017). How technology shapes assessment design: Findings from a study of university teachers. *British Journal of Educational Technology, 48*(2), 672–682. doi:10.1111/bjet.12439

Bennett, S., Lockyer, L., & Agostinho, S. (2018). Towards sustainable technology-enhanced innovation in higher education: Advancing learning design by understanding and supporting teacher design practice. *British Journal of Educational Technology, 49*(6), 1014–1026. doi:10.1111/bjet.12683

Bergstrom, J. R., & Schall, A. (2014). *Eye tracking in user experience design.* Amsterdam: Elsevier.

Bernhaupt, R., Eckschlager, M., & Tscheligi, M. (2007). Methods for evaluating games: how to measure usability and user experience in games? *Proceedings of the International Conference on Advances in Computer Entertainment Technology.* 10.1145/1255047.1255142

Bevan, R., Badge, J., Cann, A., Willmott, C., & Scott, J. (2008). Seeing eye-to-eye? Staff and student views on feedback. *Bioscience Education, 12*(1), 1–15. doi:10.3108/beej.12.1

Biggs, J. B. (2003). *Teaching for quality learning at university: What the student does.* Maidenhead, UK: Society for Research into Higher Education & Open University Press.

Biggs, J. B., & Tang, C. (2011). *Teaching for quality learning at university* (4th ed.). New York, NY: McGraw-Hill/ Society for Research in Higher Education/ Open University.

Bindé, J. (Coord.) (2005). *Towards knowledge societies: UNESCO world report. UNESCO reference works series.* Paris: UNESCO. Retrieved from http://unesdoc.unesco.org/images/0014/001418/141843e.pdf

Bindé, J. (Coord.). (2005). *Towards knowledge societies: UNESCO world report. UNESCO reference works series.* Paris: UNESCO. Retrieved from http://unesdoc.unesco.org/images/0014/001418/141843e.pdf

Bishop, D., & Rhind, D. (2011). Barriers and enablers for visually impaired students at a UK Higher Education Institution. *British Journal of Visual Impairment, 29*(3), 177–195. doi:10.1177/0264619611415329

Bista, S., Nepal, S., Colineau, N., & Paris, C. (2012). Using gamification in an online community. *8th International Conference on Collaborative Computing: Networking, Applications and Worksharing.* Retrieved from https://www.infona.pl/resource/bwmeta1.element.ieee-pub-000006415460/tab/bContent

Black, P. E., & Plowright, D. (2010). A multi-dimensional model of reflective learning for professional development. *Reflective Practice, 11*(2), 245–258. doi:10.1080/14623941003665810

Black, P., Harrison, C., Lee, C., Marshall, B., & Wiliam, D. (2004). Working Inside the Black Box: Assessment for Learning in the Classroom. *Phi Delta Kappan, 86*(1), 9–21. doi:10.1177/003172170408600105

Black, P., & Wiliam, D. (1998). Assessment and classroom learning. *Assessment in Education: Principles, Policy & Practice, 5*(1), 7–74. doi:10.1080/0969595980050102

Black, P., & Wiliam, D. (1998). *Inside the black box: Raising standards through classroom assessment.* London: King's College School of Education.

Black, P., & Wiliam, D. (2009). Developing the theory of formative assessment. *Educational Assessment, Evaluation and Accountability, 21*(1), 5–31. doi:10.100711092-008-9068-5

Blair, A. (2017). Understanding first-year students' transition to university: A pilot study with implications for student engagement, assessment, and feedback. *Politics, 37*(2), 215–228. doi:10.1177/0263395716633904

Blake, R. (2013). Brave new digital classroom: Technology and foreign language (2nd. ed.). Georgetown University Press.

Blake, R., Wilson, N. L., Cetto, M., & Pardo-Ballester, C. (2008). Measuring oral proficiency in distance, face-to-face, and blended classrooms. *Language Learning & Technology*, *12*(3), 114–127. Retrieved from http://llt.msu.edu/vol12num3/blakeetal.pdf

Blascheck, T., Burch, M., Raschke, M., & Weiskopf, D. (2015). Challenges and perspectives in big eye-movement data visual analytics. In *Proceedings of the 1st International Symposium on Big Data Visual Analytics*. IEEE. 10.1109/BDVA.2015.7314288

Blayney, P., & Freeman, M. (2004). Automated formative feedback and summative assessment using individualised spreadsheet assignments. *Australasian Journal of Educational Technology*, *20*(2), 209–231. doi:10.14742/ajet.1360

Bligh, B., & Flood, M. (2017). Activity theory in empirical higher education research: Choices, uses and values. *Tertiary Education and Management*, *23*(2), 125–152. doi:10.1080/13583883.2017.1284258

Blomberg, G., Sherin, M. G., Renkl, A., Glogger, I., & Seidel, T. (2013). Understanding video as a tool for teacher education: Investigating instructional strategies to promote reflection. *Instructional Science*, *42*(3), 443–463. doi:10.100711251-013-9281-6

Bloom, B. S. (1969). Some theoretical issues relating to educational evaluation. In R. W. Tyler (Ed.), *Educational evaluation: New roles, new means. The 63rd yearbook of the National Society for the Study of Education, Part 2* (Vol. 69, pp. 26–50). Chicago, IL: University of Chicago Press.

Bloxham, S., & Boyd, P. (2007). *Developing assessment in higher education*. Maidenhead, UK: Open University Press.

Bloxham, S., & Boyd, P. (2007). *Developing effective assessment in higher education: A practical guide*. Maidenhead, UK: Open University Press.

Boesen, J., Helenius, O., Lithner, J., Bergqvist, E., Bergqvist, T., Palm, T., & Palmberg, B. (2014). Developing mathematical competence: From the intended to the enacted curriculum. *The Journal of Mathematical Behavior*, *33*, 72–87. doi:10.1016/j.jmathb.2013.10.001

Bojko, A. A. (2009). Informative or misleading? Heatmaps deconstructed. In J. A. Jacko (Ed.), Lecture Notes in Computer Science: Vol. 5610. *Human-computer interaction: New trends. HCI 2009* (pp. 30–39). Berlin: Springer. doi:10.1007/978-3-642-02574-7_4

Bonk, C. J., Lee, M. M., Kou, X., Xu, S., & Sheu, F.-R. (2015). Understanding the self-directed online learning preferences, goals, achievements, and challenges of MIT OpenCourseWare Subscribers. *Journal of Educational Technology & Society*, *18*(2), 349–368.

Boris, C. (2013). 3 mobile apps for converting voice to text. *Entrepreneur Asia Pacific*. Retrieved from https://www.entrepreneur.com/article/225584

Borup, J., West, R. E., & Thomas, R. (2015). The impact of text versus video communication on instructor feedback in blended courses. *Educational Technology Research and Development*, *63*(2), 161–184. doi:10.100711423-015-9367-8

Boud, D. (2000). Sustainable assessment: Rethinking assessment for the learning society. *Studies in Continuing Education*, *22*(2), 151–167. doi:10.1080/713695728

Boud, D. (2013). *Enhancing learning through self-assessment*. Routledge. doi:10.4324/9781315041520

Boud, D. (2015). Feedback: Ensuring that it leads to enhanced learning. *The Clinical Teacher*, *12*(1), 3–7. doi:10.1111/tct.12345 PMID:25603699

Boud, D., & Falchikov, N. (2005). Redesigning assessment for learning beyond higher education. *Research and Development in Higher Education, 28*, 34–41.

Boud, D., & Falchikov, N. (2006). Aligning assessment with long-term learning. *Assessment & Evaluation in Higher Education, 31*(4), 399–413. doi:10.1080/02602930600679050

Boud, D., Keogh, R., & Walker, D. (2013). Promoting reflection in learning: A model. In D. Boud, R. Keogh, & D. Walker (Eds.), *Reflection: Turning experience into learning* (pp. 18–40). London: Routledge Falmer. doi:10.4324/9781315059051

Boud, D., & Molloy, E. (2013). *Feedback in higher and professional education.* New York: Routledge.

Boud, D., & Molloy, E. (2013). Rethinking models of feedback for learning: The challenge of design. *Assessment & Evaluation in Higher Education, 38*(6), 698–712. doi:10.1080/02602938.2012.691462

Boud, D., & Soler, S. (2016). Sustainable assessment revisited. *Assessment & Evaluation in Higher Education, 41*(3), 400–413. doi:10.1080/02602938.2015.1018133

Bourgault, A., Mundy, C., & Joshua, T. (2013). Comparison of audio vs. written feedback on clinical assignments of nursing students. *Nursing Education Perspectives, 34*(1), 43–46. doi:10.5480/1536-5026-34.1.43 PMID:23586205

Bowen, K., & Thomas, A. (2014). Badges: A common currency for learning. *Change, 46*(1), 21–25. doi:10.1080/00091383.2014.867206

Bower, M., Cavanagh, M., Moloney, R., & Dao, M. M. (2011). Developing communication competence using an online Video Reflection system: Preservice teachers' experiences. *Asia-Pacific Journal of Teacher Education, 39*(4), 311–326. doi:10.1080/1359866X.2011.614685

Boyle, J. (2013). *Great online resources for ESL/EFL teachers. A selection of Internet tools that can be highly useful to teachers of English as a second (or foreign) language.* Retrieved from http://www.emergingedtech.com/2013/07/8-great-online-resources-for-esl-efl-teachers/

Boylestad, R. L. (2015). Introductory circuit analysis (13th ed.). Boston: Pearson.

Bracher, M., Collier, R., Ottewill, R., & Shephard, K. (2005). Accessing and engaging with video streams for educational purposes: Experiences, issues and concerns. *ALT-J. Research in Learning Technology, 13*(2), 139–150. doi:10.3402/rlt.v13i2.10990

British Council. (2017). *IELTS numbers rise to three million a year.* Retrieved from https://www.britishcouncil.org/organisation/press/ielts-numbers-rise-three-million-year

British Dyslexia Association. (2007). *Dyslexia research information - About Dyslexia.* Retrieved from http://www.bdadyslexia.org.uk/about-dyslexia/further-information/dyslexia-research-information-.html

Brouwer, N., Besselink, E., & Oosterheert, I. (2017). The power of video feedback with structured viewing guides. *Teaching and Teacher Education, 66*, 60–73. doi:10.1016/j.tate.2017.03.013

Brown, H. D. (2010). Language assessment, principles and classroom practices (2nd ed.). New York: Pearson Longman.

Brown, A. L. (1992). Design experiments: Theoretical and methodological challenges creating complex interventions in classroom settings. *Journal of the Learning Sciences, 2*(2), 141–178. doi:10.120715327809jls0202_2

Brown, S. (2004-05). The changing nature of assessment. *Learning and Teaching in Higher Education, 1*(1), 81–89.

Brown, S., & Knight, P. (1994). *Assessing learners in higher education.* London: Routledge, Taylor & Francis Group.

Bruneau, D., Sasse, M. A., & McCarthy, J. (2002). The eyes never lie: The use of eyetracking data in HCI Research. In *Proceedings at Conference on Human Factors in Computing Systems.* ACM.

Buber, M. (1958). I and Thou (2nd ed.; R. Gregor Smith, Trans.). New York: Scribner. (Original publication 1923)

Buchem, I., Attwell, G., & Torres, R. (2011), Understanding personal learning environments: Literature review and synthesis through the activity theory lens. *PLE Conference.* Retrieved from http://journal.webscience.org/658/1/PLE_SOU_Paper_ Buchem_Attwell_Torress.doc

Budge, K. (2011). A desire for the personal: Students perceptions of electronic feedback. *International Journal on Teaching and Learning in Higher Education, 23*(3), 342–349.

Budge, K., & Gopal, S. (2009). Feedback: Working from the student perspective. *Proceedings of the Assessment in Different Dimensions, AT Assessment Conference.*

Bulut, O., & Kan, A. (2012). Application of computerized adaptive testing to entrance examination for graduate studies in Turkey. *Eurasian Journal of Educational Research, 49*, 61–80.

Burke, D. (2009). Strategies for using feedback students bring to higher education. *Assessment & Evaluation in Higher Education, 34*(1), 41–50. doi:10.1080/02602930801895711

Buzzi, M., & Leporini. (2012). Investigating mobile learning and accessibility for blind learners. In M. Allegra, M. Arrigo, V. Dal Grande, P. Denaro, D. La Guardia, S. Ottaviano, & G. Todaro (Eds.), Mobile learning for visually impaired people (pp. 26-40). Consiglio Nazionale delle Ricerche – Istituto per le Tecnologie, Didattiche.

Cabraja, A. (2016). *The effects of video games on the receptive vocabulary proficiency of Swedish ESL students.* Retrieved from http://www.diva-portal.org/smash/get/diva2:936329/FULLTEXT01.pdf

Cain, M. (2014). Participants' perceptions of fair and valid assessment practices in higher music education. In D. Lebler, S. Harrison, & G. Carey (Eds.), *Assessment in music: Principles and Practices* (pp. 87–106). UK: Springer.

Cain, M., & Fanshawe, M. (2019). Opening eyes to vision impairment: Inclusion is just another way of seeing. In S. Carter (Ed.), *Opening eyes onto inclusion and diversity.* Pressbooks.

Çakir, R., Korkmaz, Ö., Bacanak, A., & Arslan, Ö. (2016). An exploration of the relationship between students' preferences for formative feedback and self-regulated learning skills. *Malaysian Online Journal of Educational Sciences, 4*(4), 14-30. Retrieved from http://ezproxy.usq.edu.au/login?url=http://search.ebscohost.com/login.aspx?direct=true&db=eric&AN=EJ1116318&site=ehost-live

Cambridge Centre for Sixth-form Studies. (2017). *Teaching methods: Traditional vs modern.* Retrieved from https://www.ccss.co.uk/news/traditional-vs-modern-teaching/

Campion, G. C. (2016). The learning never ends: Exploring teachers' views on the transition from General English to EAP. *Journal of English for Academic Purposes, 3*, 59–70. doi:10.1016/j.jeap.2016.06.003

Cann, A. (2015). Engaging students with audio feedback. *Bioscience Education, 22*(1), 31-41. doi-org.ezproxy.usq.edu.au/10.11120/beej.2014.00027

Cann, A. (2007). Podcasting is dead. Long live video! *Bioscience Education, 10*(1), 1–4. doi:10.3108/beej.10.c1

Cano, F., Martin, A., Ginns, P., & Berbén, A. (2018). Students' self-worth protection and approaches to learning in higher education: Predictors and consequences. *Higher Education, 76*(1), 163–181. doi:10.100710734-017-0215-0

Carey, K. (1997). Comparative disadvantage and special opportunities and the information technology revolution: General considerations and the examination of visual impairment as a case study. *Aslib Proceedings*, *4*(49), 77–81. doi:10.1108/eb051452

Carey, K., & Stefaniak, J. (2018). An exploration of the utility of digital badging in higher education settings. *Educational Technology Research and Development*, *66*(5), 1211–1229. doi:10.100711423-018-9602-1

Carless, D. (2016). Feedback as dialogue. Encyclopedia of Educational Philosophy and Theory, 1-6. doi:10.1007/978-981-287-532-7_389-1

Carless, D. R. (2016). Feedback as Dialogue. Encyclopedia of Educational Philosophy and Theory. doi:10.1007/978-981-287-532-7_389

Carless, D. (2013). Trust and its role in facilitating dialogic feedback. In D. Boud & E. Molloy (Eds.), *Feedback in higher and professional education: Understanding it and doing it well* (pp. 90–103). London: Routledge.

Carless, D. (2019). Feedback loops and the longer-term: Towards feedback spirals. *Assessment & Evaluation in Higher Education*, *44*(5), 705–714. doi:10.1080/02602938.2018.1531108

Carless, D. R. (1998). A case study of curriculum implementation in Hong Kong. *System*, *26*(3), 353–368. doi:10.1016/S0346-251X(98)00023-2

Carless, D., & Boud, D. (2018). The development of student feedback literacy: Enabling uptake of feedback. *Assessment & Evaluation in Higher Education*, *43*(8), 1315–1325. doi:10.1080/02602938.2018.1463354

Carless, D., Salter, D., Yang, M., & Lam, J. (2011). Developing sustainable feedback practices. *Studies in Higher Education*, *36*(4), 395–407. doi:10.1080/03075071003642449

Carrio-Pastor, M., & Skorczynska, H. (2015). Collaborative learning and communication technologies in teaching business English. *Procedia: Social and Behavioral Sciences*, *178*, 32–37. doi:10.1016/j.sbspro.2015.03.142

Carroll, J. M. (2003). *HCI models, theories, and frameworks: Toward a multidisciplinary science.* Morgan Kaufmann.

Carver, M. (2017). Feedback, feedforward, or dialogue?: Defining a model for self-regulated learning. In E. Cano & G. Ion (Eds.), *Innovative practices for higher education assessment and measurement* (pp. 1–18). Hershey, PA: IGI Global. doi:10.4018/978-1-5225-0531-0.ch001

Casilli, C., & Hickey, D. (2016). Transcending conventional credentialing and assessment paradigms with information-rich digital badges. *The Information Society*, *32*(2), 117–129. doi:10.1080/01972243.2016.1130500

Cassidy, S. (2011). Self-regulating learning in higher education: Identifying key component process. *Studies in Higher Education*, *36*(8), 989–1000. doi:10.1080/03075079.2010.503269

Chan, C., & Fok, W. (2009). Evaluating learning experiences in virtual laboratory training through student perceptions: A case study in Electrical and Electronic Engineering at the University of Hong Kong. *Engineering Education*, *4*(2), 70–75. doi:10.11120/ened.2009.04020070

Chandler, P., & Sweller, J. (1991). Cognitive load theory and the format of instruction. *Cognition and Instruction*, *8*(4), 293–332. doi:10.12071532690xci0804_2

Chang, N., Watson, A. B., Bakerson, M., Williams, E. E., McGoron, F. X., & Spitzer, B. (2012). Electronic feedback or handwritten feedback: What do undergraduate students prefer and why? *Journal of Teaching and Learning with Technology*, *1*(1), 1–13.

Chang, V., Gütl, C., Kopeinik, S., & Williams, R. (2009). Evaluation of collaborative learning settings in 3D virtual worlds. *International Journal of Emerging Technologies in Learning*, *4*(3), 6–17. doi:10.3991.ijet.v453.1112

Chapelle, C. A., & Voss, E. (2016). 20 years of technology and language assessment in *Language Learning & Technology, 20*(2), 116–128. Retrieved from http://llt.msu.edu/issues/june2016/chapellevoss.pdf

Charlop, M. H., & Milstein, J. P. (1989). Teaching autistic children conversational speech using video modeling. *Journal of Applied Behavior Analysis*, *22*(3), 275–285. doi:10.1901/jaba.1989.22-275 PMID:2793634

Chen, J., & Zhang, J. (2016). Design collaborative formative assessment for sustained knowledge building using idea thread mapper. In *Proceedings of the International Conference of the Learning Sciences* (ICLS 2016). Singapore: International Society of the Learning Sciences.

Chen, C.-F. E., & Cheng, W.-Y. E. (2008). Beyond the design of automated writing evaluation: Pedagogical practices and perceived learning effectiveness in EFL writing classes. *Language Learning & Technology*, *12*(2), 94–112. Retrieved from http://llt.msu.edu/vol12num2/chencheng.pdf

Cheng, Y. (2018, October 17). *China Daily*. Retrieved from http://global.chinadaily.com.cn/a/201810/17/WS5bc6a025a310eff303282da3.html

Cheng, G., & Chau, J. (2009). Digital video for fostering self-reflection in an ePortfolio environment. *Learning, Media and Technology*, *34*(4), 337–350. doi:10.1080/17439880903338614

Cheng, Z., Watson, S., & Newby, T. (2018). Goal Setting and Open Digital Badges in Higher Education. *TechTrends*, *62*(2), 190–196. doi:10.100711528-018-0249-x

Chiasson, K., Terras, K., & Smart, K. (2015). Faculty perceptions of moving a face-to-face course to online instruction. *Journal of College Teaching and Learning*, *12*(3), 231–240. doi:10.19030/tlc.v12i3.9315

Chou, P. N., Chang, C. C., & Lu, P. F. (2015). Prezi versus PowerPoint: The effects of varied digital presentation tools on students' learning performance. *Computers & Education*, *91*, 73–82. doi:10.1016/j.compedu.2015.10.020

Christ, T., Arya, P., & Chiu, M. M. (2017). Video use in teacher education: An international survey of practices. *Teaching and Teacher Education*, *63*, 22–35. doi:10.1016/j.tate.2016.12.005

Cimer, A. (2007). Effective teaching in science. A review of literature. *Journal of Turkish Science Education*, *4*(1), 20–44.

Clark, J. M., & Paivio, A. (1991). Dual coding theory and education. *Educational Psychology Review*, *3*(3), 149–210. doi:10.1007/BF01320076

Cobb, P., diSessa, A., Lehrer, R., & Schauble, L. (2003). Design experiments in educational research. *Educational Researcher*, *32*(1), 9–13. doi:10.3102/0013189X032001009

Cochran-Smith, M., & Lytle, S. L. (2009). *Inquiry as stance: Practitioner research for the next generation*. New York: Teachers College Press.

Cochran-Smith, M., & Villegas, A. M. (2014). Framing Teacher Preparation Research. *Journal of Teacher Education*, *66*(1), 7–20. doi:10.1177/0022487114549072

Cole, A., Anderson, C., Bunton, T., Cherney, M., Fisher, V. C., Featherston, M., ... Allen, M. (2017). Student predisposition to instructor feedback and perceptions of teaching presence predict motivation toward online courses. *Online Learning Journal*, *21*(4), 245–262.

Collins, A. (1992). Towards a design science of education. In E. Scanlon & T. O'Shea (Eds.), *New directions in educational technology* (pp. 15–22). Berlin: Springer. doi:10.1007/978-3-642-77750-9_2

Collins, J. L., Cook-Cottone, C. P., Robinson, J. S., & Sullivan, R. R. (2005). Technology and new directions in professional development: Applications of digital video, peer review, and self-reflection. *Journal of Educational Technology Systems, 33*(2), 131–146. doi:10.2190/G9E9-LCQ1-H50X-4N23

Common European Framework of Reference (CEFR). (n.d.). *Common European framework of reference for languages: Learning, teaching and assessment.* Strasbourg: Cambridge University Press. Retrieve from https://rm.coe.int/1680459f97

Commonwealth of Australia. (1992). *Disability discrimination act*. Canberra: Author.

Compton, J. (2016). Flipped learning with Turkish students: Issues of dissonance to possible harmonization. *The Global eLearning Journal, 5*(22), 1-7.

Conrad, S. S., & Dabbagh, N. (2015). Examining the Factors that influence how instructors provide feedback in online learning environments. *International Journal of Online Pedagogy and Course Design, 5*(4), 47–66. doi:10.4018/IJOPCD.2015100104

Cooperman, N., & Berenato, C. L. (2014). Students' perceptions of electronic feedback as an alternative to handwritten feedback: One University's inquiry. *Journal of Education and Learning, 3*(1), 79–85.

Corter, J. E., Esche, S. K., Chassapis, C., Ma, J., & Nickerson, J. V. (2011). Process and learning outcomes from remotely-operated, simulated, and hands-on student laboratories. *Computers & Education, 57*(3), 2054–2067. doi:10.1016/j.compedu.2011.04.009

Council of Chief State School Officers. (2012). *Distinguishing formative assessment from other educational assessment labels*. Washington, DC: Author. Retrieved from http://www.ccsso.org/Documents/FASTLabels.pdf

Cowie, B., Harrison, C., & Willis, J. (2018). Supporting teacher responsiveness in assessment for learning through disciplined noticing. *Curriculum Journal, 29*(4), 1–15. doi:10.1080/09585176.2018.1481442

Craig, S. D., Sullins, J., Witherspoon, A., & Gholson, B. (2006). The deep-level-reasoning-question effect: The role of dialogue and deep-level-reasoning questions during vicarious learning. *Cognition and Instruction, 24*(4), 565–591. doi:10.12071532690xci2404_4

Cramp, A. (2011). Developing first year engagement with written feedback. *Active Learning in Higher Education, 12*(2), 112–124. doi:10.1177/1469787411402484

Crane-Mitchell, L., & Stafford, L. (2017). *Enhancing learning outcomes for students with autism through the use of assistive technology: Effect of parent-teacher collaboration.* Paper presented at the Society for Information Technology & Teacher Education International Conference.

Crisp, B. R. (2007). Is it worth the effort? How feedback influences students' subsequent submission of assessable work. *Assessment & Evaluation in Higher Education, 32*(5), 571–581. doi:10.1080/02602930601116912

Croft, N., Dalton, A., & Grant, M. (2010). Overcoming isolation in distance learning: Building a learning community through time and space. *UK Journal for Education in the Built Environment, 5*(1), 27–64. doi:10.11120/jebe.2010.05010027

Crompton, H., & Burke, D. (2018). The use of mobile learning in higher education: A systematic review. *Computers & Education, 123*, 53–64. doi:10.1016/j.compedu.2018.04.007

Crook, A., Mauchline, A., Maw, S., Lawson, C., Drinkwater, R., Lundqvist, K., ... Park, J. (2012). The use of video technology for providing feedback to students: Can it enhance the feedback experience for staff and students? *Computers & Education*, *58*(1), 386–396. doi:10.1016/j.compedu.2011.08.025

Crossland, M., Silva, S., & Macedo, A. (2014). Smartphone, tablet computer and e-reader use by people with vision impairment (report). *Ophthalmic & Physiological Optics*, *34*(5), 552–557. doi:10.1111/opo.12136 PMID:25070703

Crossouard, B., Pryor, J., & Torrance, H. (2004). Creating an alternative assessment regime with online formative assessment: Developing a researcher identity. In *European Conference on Educational Research* (pp. 22-25). Academic Press.

Dann, B. (2016). *Implementation of the new Australian Curriculum: Science: A study of the experiences of a range of Queensland schools and teachers* (Doctoral thesis). University of the Sunshine Coast, Queensland, Australia.

Dann, B., Dann, C., & O'Neill, S. (2017). Formative assessment via video feedback on practicum: Implications for higher education and professional teacher accreditation bodies. In T. Richardson, B. Dann, C. Dann, & S. O'Neill (Eds.), *Formative assessment practices for pre-service teacher practicum feedback: Emerging research and opportunities* (pp. 158–183). Hershey, PA: IGI Global.

Dann, C. (2014). Mobile video feedback: When 'seeing' informs reflection and learning. *International Journal for Cross-Disciplinary Subjects in Education*, *5*(3), 1727–1734. doi:10.20533/ijcdse.2042.6364.2014.0241

Dann, C., & Allen, B. (2015). Mobile video collection in preservice teacher practicum placements. *Journal of Technology and Teacher Education*, *23*(1), 5–27.

Dann, C., & Dann, B. (2018). Exemplifying formative assessment practices in the field through case study: Concepts, issues and challenges. In T. Richardson, B. Dann, C. Dann, & S. O'Neill (Eds.), *Formative assessment practices for pre-srevice teacher practicum feedback: Emerging research and opportunities* (pp. 76–112). IGI Global. doi:10.4018/978-1-5225-2630-8.ch003

Dann, C., & O'Neill, S. (2018). Positioning preservice teacher formative assessment in the literature. In T. Richardson, B. Dann, C. Dann, & S. O'Neill (Eds.), *Formative assessment practices for pre-service teacher practicum feedback: Emerging research and opportunities* (pp. 1–48). Hershey, PA: IGI Global. doi:10.4018/978-1-5225-2630-8.ch001

Dann, C., & Richardson, T. (2015). Deepening understanding of 'pedagogical outcomes' through video data collection: A catalyst for guided reflective learning conversations. *International Journal of Pedagogies & Learning*, *10*(1), 62–80. doi:10.1080/22040552.2015.1084677

Darling-Hammond, L. (2010). *Evaluating teacher effectiveness: How teacher performance assessments can measure and improve teaching*. Center for American Progress.

Datta, P., & Talukdar, J. (2016). The impact of support services on students' test anxiety and/or their ability to submit assignments: A focus on vision impairment and intellectual disability. *International Journal of Inclusive Education*, *21*(2), 160–171. doi:10.1080/13603116.2016.1193561

Debnath, T., Paul, M., Chakraborty, S., Podder, P. K., Gururajan, R., & Hafeez-Baig, A. (2016). *An experimental analysis of assessor specific bias in a programming assessment in multi-assessor scenarios utilizing an eye tracker.* Paper presented at the London International Conference on Education: Proceedings (LICE-2016).

DeCarlo, D., McGwin, G., Bixler, M., Wallander, J., & Owsley, C. (2012). Impact of paediatric vision impairment on daily life: Results of focus groups. *Optometry and Vision Science: Official Publication of the American Academy of Optometry*, *89*(9), 1409–1416. doi:10.1097/OPX.0b013e318264f1dc PMID:22863790

Deci, E., Ryan, R., Hunsley, J., Pelletier, L., Vallerand, R., & Koestner, R. (2008). Self-Determination theory: A macrotheory of human motivation, development, and health. *Canadian Psychology*, *49*(3), 182–185. doi:10.1037/a0012801

Deen, M. (2015). *G.A.M.E., Games autonomy motivation and education: How autonomy-supportive game design may improve motivation to learn.* Eindhoven: Technische Universiteit Eindhoven.

Deglau, D., Ward, P., O'Sullivan, M., & Bush, K. (2006). Chapter 5: Professional dialogue as professional development. *Journal of Teaching in Physical Education*, *25*(4), 413–427. doi:10.1123/jtpe.25.4.413

DeHaan, J. (2005). Learning language through video games: A theoretical framework, an evaluation of game genres and questions for future research. In S. P. Schaffer & M. L. Price (Eds.), *Interactive convergence: Critical issues in multimedia* (pp. 229-239). London: Inter-disciplinary Press. Retrieved from http://interdisciplinary.net/publishing/id-press/ebooks/interactive-convergence-criticalissues-inmultimedia

DeHaan, J., Reed, W. M., & Kuwada, K. (2010). The effect of interactivity with a music video game on second language vocabulary recall. *Language Learning & Technology*, *14*(2), 74–94.

Deniz, D. Z., Bulancak, A., & Ozcan, G. (2003, November). A novel approach to remote laboratories. In 33rd Annual Frontiers in Education, 2003. FIE 2003 (Vol. 1, pp. T3E-T3E). IEEE. doi:10.1109/FIE.2003.1315161

Denton, P., & McIlroy, D. (2017). Response of students to statement bank feedback: The impact of assessment literacy on performance in summative tasks. *Assessment & Evaluation in Higher Education*, *42*(2), 197–206. doi:10.1080/02602938.2017.1324017

Denton, P., & McIlroy, D. (2017). Response of students to statement bank feedback: The impact of assessment literacy on performances in summative tasks. *Assessment & Evaluation in Higher Education*, 1–10.

Denzin, N. K. (2001). The reflexive interview and a performative social science. *Qualitative Research*, *1*(1), 23–46. doi:10.1177/146879410100100102

Department of Education and Training (DET). (2003). *Quality teaching in NSW public schools: A classroom practice guide.* Department of Education and Training, Professional Support and Curriculum Directorate.

Design Based Research Collective. (2003). Design based research: An emerging paradigm in educational inquiry. *Educational Researcher*, *32*(1), 5–8. doi:10.3102/0013189X032001005

Dessus, P., Cosnefroy, O., & Luengo, V. (2016). "Keep your eyes on'em all!": A mobile eye-tracking analysis of teachers' sensitivity to students. In K. Verbert M. Sharples & T. Klobučar (Eds.), *Adaptive and adaptable learning. Proc. 11th European Conf. on Technology Enhanced Learning (EC-TEL 2016)* (pp. 72–84). New York: Springer. 10.1007/978-3-319-45153-4_6

DeVaney, T. A. (2009). Impact of Video Tutorials in an Online Educational Statistics Course. *MERLOT Journal of Online Learning and Teaching*, *5*(4), 600–608.

Dewey, J. (1929). My pedagogic creed. *Journal of the National Educational Association*, *18*(9), 291–295.

Dewey, J. (1944). *Democracy and education.* New York: Free Press.

Dewey, J. (1944). *Democracy and Education.* New York: Free Press.

Dillenbourg, P. (1999). What do you mean by collaborative learning? In P. Dillenbourg (Ed.), *Collaborative-learning: Cognitive and computational approaches* (pp. 1–19). Oxford, UK: Elselvier Publishing.

Dillenbourg, P., Järvelä, S., & Fischer, F. (2009). The evolution of research on computer-supported collaborative learning: from design to orchestration. In N. Balacheff, S. Ludvigsen, T. de John, T. Lazonder, & S. Barnes (Eds.), *Technology-enhanced learning: Principles and products* (pp. 3–19). Heidelberg, Germany: Springer. doi:10.1007/978-1-4020-9827-7_1

Dittmar, R., & Kahlcke, T. (2016). A lab for undergraduate control engineering education equipped with industrial distributed control systems. *Computer Applications in Engineering Education*, *24*(2), 288–296. doi:10.1002/cae.21708

Dixon, S. (2015). The pastoral potential of audio feedback: A review of the literature. *Pastoral Care in Education*, *33*(2), 96–104. doi:10.1080/02643944.2015.1035317

Doherty-Sneddon, G. (2008). Eyes: A new window on mental disorders. *Scientific American*. Retrieved from https://www.scientificamerican.com/article/eyes-mental-disorders/

Doolan, M. A., & Simpson, M. (2010). Engaging tutors and learners through audio supported pedagogy. In *Proceedings of the 5th Annual Blended Learning Conference*. Hatfield, UK: University of Hertfordshire.

Doolittle, P. E. (1995). *Understanding cooperative learning through Vygotsky's Zone of Proximal Development.* Paper presented at Lilly Conference on Excellence in College Teaching, Columbia, SC.

Dossel, S. (2016). Maths anxiety. *Australian Mathematics Teacher*, *72*(3). Retrieved from https://search.informit.com.au/documentSummary;dn=351118154615118;res=IELHSS

Dowden, T., Pittaway, S., Yost, H., & McCarthy, R. (2013). Students' perceptions of written feedback in teacher education: Ideally feedback is a continuing two-way communication that encourages progress. *Assessment & Evaluation in Higher Education*, *38*(3), 349–362. doi:10.1080/02602938.2011.632676

Dowling-Hetherington, L., & Glowatz, M. (2017). The usefulness of digital badges in higher education: Exploring the students' perspectives. *Irish Journal of Academic Practice*, *6*(1). Retrieved from https://arrow.dit.ie/cgi/viewcontent.cgi?article=1062&context=ijap

Drummond, I., Nixon, I., & Wiltshire, J. (1998). Personal transferable skills in higher education: The problems of implementing good practice. *Quality Assurance in Education*, *6*(1), 19–27. doi:10.1108/09684889810200359

Duchowski, A. T. (2007). Eye tracking methodology: Theory and Practice (2nd ed.). Dordrecht: Springer.

Duncan, N. (2007). "Feed forward": Improving use of tutors' comments. *Assessment & Evaluation in Higher Education*, *32*(3), 271–283. doi:10.1080/02602930600896498

Dyjur, P., & Lindstrom, G. (2017). Perceptions and uses of digital badges for professional learning development in higher education. *TechTrends*, *61*(4), 386–392. doi:10.100711528-017-0168-2

Edwards-Groves, C., Anstey, M., & Bull, G. (2014). Classroom talk: Understanding dialogue, pedagogy and practice. Newtown, Australia: Primary English Teaching Association Australia (PETAA).

Elawady, Y., & Tolba, A. S. (2009). *Educational objectives of different laboratory types: A comparative study.* arXiv preprint arXiv:0912.0932

Elmahdi, I., Abdulghani, A.-H., & Fawzi, H. (2018). Using technology for formative assessment to improve students' learning. *TOJET: Turkish Online Journal of Educational Technology*, *17*(2), 182–188.

Emery, R., & Atkinson, A. (2009*). Group assessment feedback: 'The good, the bad and the ugly'*. Paper presented at the 1st Annual Meeting of a Word in Your Ear, Sheffield, UK.

Emery, R., & Atkinson, A. (2009). Group assessment feedback: The good, the bad and the ugly. In *A word in your ear conference proceedings* (pp. 1–7). Retrieved from http://research.shu.ac.uk/lti/awordinyourear2009/docs/emery-atkinson-Solent_Audio_Feedback_paper.pdf

Engeström, Y. (2001). Expansive learning: Towards an activity-theoretical reconceptualization. *Journal of Education and Work, 14*(1), 133–156. doi:10.1080/13639080020028747

Engin, M., & Donanci, S. (2015). Dialogic teaching and iPads in the EAP classroom. *Computers & Education, 88*, 268–279. doi:10.1016/j.compedu.2015.06.005

Entwistle, N. J., Hounsell, C. J., Macaulay, C., Situnayake, G., & Tait, H. (1989). *The Performance of Electrical Engineers in Scottish Education. Report to the Scottish Education Department, Centre for Research on Learning and Instruction Department of Education.* University of Edinburgh.

Ericsson, K. A., & Simon, H. A. J. P. (1980). Verbal reports as data. *Psychological Review, 87*(3), 215–251. doi:10.1037/0033-295X.87.3.215

Ernst, E. W. (1883). A new role for the undergraduate engineering laboratory. *IEEE Transactions on Education, 26*(2), 49-51.

Estriegana, R., Medina-Merodio, J.-A., & Barchino, R. (2019). Student acceptance of virtual laboratory and practical work: An extension of the technology acceptance model. *Computers & Education, 135*, 1–14. doi:10.1016/j.compedu.2019.02.010

Evans, C. (2013). Making sense of assessment feedback in Higher Education. *Review of Educational Research, 83*(1), 70–120. doi:10.3102/0034654312474350

Evans, K. M., Jacobs, R. A., Tarduno, J. A., & Pelz, J. B. (2012). Collecting and analyzing eye tracking data in outdoor environments. *Journal of Eye Movement Research, 5*(2), 6.

Facey-Shaw, L., Specht, M., Bartley-Bryan, J., & van Rosmalen, P. (2018). Technological and implementation issues in moodle-based digital badge system. *Proceedings of the European Conference on Games Based Learning*, 82–90. Retrieved from http://ezproxy.usq.edu.au/login?url=http://search.ebscohost.com/login.aspx?direct=true&db=ehh&AN=132746038&site=ehost-live

Fahimirad, M., & Kotamjani, S. S. (2018). A review of application of artificial intelligence in teaching and learning in educational contexts. *International Journal of Learning and Development, 8*(4), 106–118. doi:10.5296/ijld.v8i4.14057

Falloon, G. (2010). Using avatars and virtual environments in learning: What do they have to offer? *British Journal of Educational Technology, 41*(1), 108–122. doi:10.1111/j.1467-8535.2009.00991.x

Fanfarelli, J. R., & McDaniel, R. (2017). Exploring digital badges in university courses: Relationships between quantity, engagement, and performance. *Online Learning, 21*(2). Retrieved from http://ezproxy.usq.edu.au/login?url=http://search.ebscohost.com/login.aspx?direct=true&db=eric&AN=EJ1149350&site=ehost-live

Fanshawe, M. (2015). The implications of monocular vision on orientation and mobility. *Journal of the South Pacific Educators in Vision Impairment, 8*(1), 77–86.

Farnsworth, B. (2018a). *Eye tracking: The complete pocket guide*. Available at https://imotions.com/blog/eye-tracking/

Farnsworth, B. (2018b). *7 most used eye tracking metrics and terms*. Retrieved from https://imotions.com/blog/7-terms-metrics-eye-tracking/

FarnsworthB. (2017). Retrieved from https://imotions.com/blog/top-eye-tracking-hardware-companies/

Farooq, M. U., & Soomro, A. F. (2018). Teachers and technology: Trends in English language teaching in Saudi Arabia. *International Journal of English Linguistics*, *8*(5), 10–19. doi:10.5539/ijel.v8n5p10

Fawcett, H., & Oldfield, J. (2018). Investigating expectations and experiences of audio and written assignment feedback in first-year undergraduate students. *Teaching in Higher Education*, *28*, 1–15.

Feilzer, Y. (2010). Doing mixed methods research pragmatically: Implications for the rediscovery of pragmatism as a research paradigm. *Journal of Mixed Methods Research*, *4*(1), 6–16. doi:10.1177/1558689809349691

Feisel, L. D., & Rosa, A. J. (2005). The role of the laboratory in undergraduate engineering education. *Journal of Engineering Education*, *94*(1), 121–130. doi:10.1002/j.2168-9830.2005.tb00833.x

Ferguson, P. (2011). Student's perceptions of quality feedback in teacher education. *Assessment & Evaluation in Higher Education*, *36*(1), 51–62. doi:10.1080/02602930903197883

Finkelstein, J., Knight, E., & Manning, S. (2013). The potential and value of using digital badges for adult learners final report. American Institutes for Research, 16.

Fitzgerald, A., Dawson, V., & Hackling, M. (2013). Examining the beliefs and practices of four effective Australian primary Science teachers. *Research in Science Education*, *43*(3), 981–1003. doi:10.100711165-012-9297-y

Fleming, N. D., & Mills, C. (1992). Not another inventory, rather a catalyst for reflection. *To Improve the Academy*, *11*, 137-55.

Foldnes, N. (2016). The flipped classroom and cooperative learning: Evidence from a random experiment. *Active Learning in Higher Education*, *17*(1), 39–49. doi:10.1177/1469787415616726

Ford, E. (2017). To badge or not to badge? *College & Research Libraries News*, *78*(1), 20–21. doi:10.5860/crln.78.1.9602

Ford, E., Izumi, B., Lottes, J., & Richardson, D. (2015). Badge it!: A collaborative learning outcomes based approach to integrating information literacy badges within disciplinary curriculum. *RSR. Reference Services Review*, *43*(1), 31–44. doi:10.1108/RSR-07-2014-0026

Foreman, P., & Arthur-Kelly, M. (2017). *Inclusion in action* (5th ed.). Cengage.

Forsythe, A., & Johnson, S. (2016). Thanks, but no-thanks for the feedback. *Assessment & Evaluation in Higher Education*, *42*(6), 850–859. doi:10.1080/02602938.2016.1202190

Francis, R., Shannon, S., & Murison, S. (2008). *Delivery of online digital feedback and assessment for design and engineering (architectural) students.* Paper presented at the 46th Annual Conference of the Architectural Science Association, Griffith University. Retrieved from http://anzasca.net/wp-content/uploads/2014/08/p201.pdf

Freire, P. (1970). *Pedagogy of the oppressed* (M. B. Ramos, Trans.). New York: *Continuum.*

Fry, H., Ketteridge, S., & Marshall, S. (2014). Assessing assessment: new developments in assessment design, feedback practices and marking in higher education. In *A Handbook for Teaching and Learning in Higher Education* (pp. 129–144). New York: Routledge. doi:10.4324/9781315763088

Fu, I., Chuang, Y., Fu, S., & Beasley, R. (2012). Implementing Cloud technology of NETPAW at I-Shou University. *The APAMALL Higher Education Journal of Language Learning Technologies*, *2*(1), 91–119.

Furtak, E. M., Kiemer, K., Circi, R. K., Swanson, R., de Leon, V., Morrison, D., & Heredia, S. C. (2016). Teachers' formative assessment abilities and their relationship to student learning: Findings from a four-year intervention study. *Instructional Science*, *44*(3), 267–291. doi:10.100711251-016-9371-3

Gadamer, H.-G. (1991). *Plato's dialectical ethics: Phenomenological interpretations relating to the Philebus* (R. M. Wallace, Trans.). New Haven, CT: Yale University Press.

Gadzhunov, S., & Nafalski, A. (2010). Pedagogical effectiveness of remote laboratories for measurement and control. *World Transactions on Engineering and Technology Education, 8*(2), 162–167.

Gage, N. L. (1989). The paradigm wars and their aftermath: A "historical" sketch of research on teaching since 1989. *Educational Researcher, 18*(7), 4–10.

Gagné, N. (2009). *Cooperative learning and oral interaction activities in an intensive Grade 6 ESL program in Quebec* (Unpublished master's thesis). Université Laval, Quebec City, Canada.

Gagnon, G. W., & Collay, M. (2006). *Constructivist learning design.* Corwin Press.

Galdon, P., Madrid, R., Rubia, E., Cuestas, E., Diaz, A., Estrella, A., & Gonzalez, L. (2013). Enhancing mobile phones for people with visual impairments through haptic icons, the effect of learning processes. *Assistive Technology, 25*(2), 80–87. doi:10.1080/10400435.2012.715112 PMID:23923690

Gallagher, G. (2017). Aligning for learning: Including feedback in the Constructive Alignment model. *AISH-J: The All Ireland Journal of Teaching and Learning in Higher Education, 9*(1), 3011–3111.

Galusha, J. (1998). *Barriers to learning in distance education.* Hattiesburg, MS: The University of Southern Mississippi (ERIC Document Reproduction No. ED 416 377). Retrieved from http://files.eric.ed.gov/fulltext/ED416377.pdf

Gamrat, C., Zimmerman, H., Dudek, J., & Peck, K. (2014). Personalised workplace learning: An exploratory study on digital badging within a teacher professional development program. *British Journal of Educational Technology, 45*(6), 1136–1148. doi:10.1111/bjet.12200

Garrison, D., & Akyol, Z. (2013). *The community of inquiry theoretical framework.* Retrieved from https://www.researchgate.net/publication/284306348_The_Community_of_Inquiry_Theoretical_Framework

Garrison, D. R., Anderson, T., & Archer, W. (1999). Critical inquiry in a text-based environment: Computer conferencing in higher education. *The Internet and Higher Education, 2*(2), 87–105. doi:10.1016/S1096-7516(00)00016-6

Garrison, D. R., Anderson, T., & Archer, W. (2001). Critical thinking, cognitive presence, and computer conferencing in distance education. *American Journal of Distance Education, 15*(1), 7–23. doi:10.1080/08923640109527071

Garrison, D. R., Cleveland-Innes, M., & Fung, T. (2010). Exploring causal relationships among teaching, cognitive and social presence: Student perceptions of the community of inquiry framework. *The Internet and Higher Education, 13*(1), 31–36. doi:10.1016/j.iheduc.2009.10.002

Garrison, D. R., & Kanuka, H. (2004). Blended learning: Uncovering its transformative potential in higher education. *The Internet and Higher Education, 7*(2), 95–105. doi:10.1016/j.iheduc.2004.02.001

Gaudin, C., & Chaliès, S. (2015). Video viewing in teacher education and professional development: A literature review. *Educational Research Review, 16*, 41–67. doi:10.1016/j.edurev.2015.06.001

Gedye, S. (2010). Formative assessment and feedback: A review. *Planet, 23*(1), 40–45. doi:10.11120/plan.2010.00230040

Gee, J. P. (2003). *What video games have to teach us about learning and literacy.* Basingstoke, UK: Palgrave Macmillan. doi:10.1145/950566.950595

Gee, J. P. (2007). *Good video games and good learning: Collected essays on video games, learning and literacy.* New York: Peter Lang Publishers. doi:10.3726/978-1-4539-1162-4

Geoghegan, D., O'Neill, S., & Petersen, S. (2013). Metalanguage: The 'teacher talk' of explicit literacy teaching in practice. *Improving Schools*, *16*(2), 119–129. doi:10.1177/1365480213493707

Gergel, S. E., & Turner, M. G. (2017). *Learning landscape ecology: a practical guide to concepts and techniques*. Springer. doi:10.1007/978-1-4939-6374-4

Gibbings, P. (2014). Qualitatively different ways students experience remote access laboratories. *International Journal of Engineering Education*, *30*(5), 1120–1129.

Gibbs, G., & Simpson, C. (2004). Conditions under which assessment supports students' learning. *Learning and Teaching in Higher Education, 1*(1), 3-31.

Gibbs, G., & Simpson, C. (2004). Conditions under which assessment supports students' learning. *Learning and Teaching in Higher Education, 1*, 1–31. Retrieved from http://eprints.glos.ac.uk/3609/1/LATHE%201.%20Conditions%20Under%20Which%20Assessment%20Supports%20Students'%20Learning%20Gibbs_Simpson.pdf

Gibbs, G., & Simpson, C. (2004). Conditions under which assessment supports learning. *Learning and Teaching in Higher Education*, *1*(1), 3–31.

Gibbs, G., & Simpson, C. (2004). Conditions under which assessment supports students' learning. *Learning and Teaching in Higher Education, 1*(1), 3–31.

Gibson, D., Ostashewski, N., Flintoff, K., Grant, S., & Knight, E. (2013). Digital badges in education. *Education and Information Technologies*, *20*(2), 403–410. doi:10.100710639-013-9291-7

Gibson, J. J. (1977). The theory of affordances. In R. Shaw & J. Bransford (Eds.), *Perceiving, acting, and knowing: Toward an ecological psychology* (pp. 67–82). Hillsdale, NJ: Lawrence Erlbaum Associates.

Gibson, K. M. (2013). Fostering collaboration and learning in asynchronous online environments. *Journal of Teaching and Learning with Technology*, *2*(2), 60–78.

Gikandi, J., Morrow, D., & Davis, N. (2011). Online Formative Assessment in higher education: A review of the literature. *Computers & Education*, *57*(4), 2333–2351. doi:10.1016/j.compedu.2011.06.004

Gilakjani, A. P. (2017). A review of the literature on the integration of technology into the learning and teaching of English language skills. *International Journal of English Linguistics*, *7*(5), 95–106. doi:10.5539/ijel.v7n5p95

Gilbert, L., Whitelock, D., & Gale, V. (2011). *Synthesis report on assessment and feedback with technology enhancement*. Southampton, UK: University of Southampton.

Giroux, H. A. (2010). Rethinking education as the practice of freedom: Paulo Freire and the promise of critical pedagogy. *Policy Futures in Education*, *8*(6), 715–721. doi:10.2304/pfie.2010.8.6.715

Glaser, B. G., & Strauss, A. L. (1967). *The discovery of grounded theory: Strategies for qualitative research*. New York: Aldine de Gruyter.

Glover, C., & Brown, E. (2006). Written feedback for students: Too much, too detailed or too incomprehensible to be effective? *Bioscience Education*, *7*(3), 1–15. doi:10.3108/beej.2006.07000004

Gokhale, A. (1995). Collaborative learning enhances critical thinking. *Journal of Technology Education*, *7*(1), 22–30. doi:10.21061/jte.v7i1.a.2

Goldin, I., Narciss, S., Foltz, P., & Bauer, M. (2017). New Directions in Formative Feedback in Interactive Learning Environments. *International Journal of Artificial Intelligence in Education*, *27*(3), 385–392. doi:10.100740593-016-0135-7

Gonzalez, J. (2018). *Moving from feedback to feedforward: Cult of pedagogy*. Retrieved from https://www.cultofpedagogy.com/feedforward/

Goodyear, M., & Nathan-Roberts, D. (2017). Gamification and The Design of Badges in Relation to Educational Achievement. *Proceedings of the Human Factors and Ergonomics Society Annual Meeting*, *61*(1), 1229–1233. doi:10.1177/1541931213601789

Gould, J., & Day, P. (2013). Hearing you loud and clear: Student perspectives of audio feedback in higher education. *Assessment & Evaluation in Higher Education*, *38*(5), 554–566. doi:10.1080/02602938.2012.660131

Grainger, P. R., Heck, D., & Carey, M. D. (2018). Are assessment exemplars perceived to support self-regulated learning in teacher education? *Frontiers in Education*, *3*, 60. doi:10.3389/feduc.2018.00060

Green, A. (2006). Washback to the learner: Learner and teacher perspectives on IELTS preparation course expectations and outcomes. *Assessing Writing*, *11*(2), 113–134. doi:10.1016/j.asw.2006.07.002

Griffiths, M. E., & Graham, C. R. (2009). Using asynchronous video in online classes: Results from a pilot study. *International Journal of Instructional Technology & Distance Learning*, *6*(3), 65–76.

Groff, J. (2017). *Personalized learning: The state of the field & future directions*. Center for Curriculum Redesign.

Gürışık, A., & Demirkan, Ö. (2019). Opinion of high school students about plickers. *International Journal of Scientific Research and Innovative Technology*, *1*(6), 12–25.

Haefner, L., & Zembal-Saul, C. (2004). Learning by doing? Prospective elementary teachers' developing understandings of scientific inquiry and science teaching and learning. *International Journal of Science Education*, *26*(13), 1653–1674. doi:10.1080/0950069042000230709

Hall, M., Nix, I., & Baker, K. (2012). "Why should I?" Engaging learners in digital literacy skills development. *Proceedings of the 11th European Conference on e-Learning*, 220-229. Retrieved from http://oro.open.ac.uk/35574/1/ProceedingsECEL2012-MHall.pdf

Hamari, J. (2017). Do badges increase user activity? A field experiment on the effects of gamification. *Computers in Human Behavior*, *71*, 469–478. doi:10.1016/j.chb.2015.03.036

Hamari, J., Shernoff, D. J., Rowe, E., Coller, B., Asbell-Clarke, J., & Edwards, T. (2016). Challenging games help students learn: An empirical study on engagement, flow and immersion in game-based learning. *Computers in Behavior*, *54*, 170–179. doi:10.1016/j.chb.2015.07.045

Hammerness, K., Darling-Hammond, L., Bransford, J., Berliner, D., Cochran-Smith, M., McDonald, M., & Zeichner, K. (2005). How teachers learn and develop. In *Teachers for a Changing World: What Teachers Should Learn and Be Able to Do* (p. 1). New York: John Wiley & Sons.

Hamp-Lyons, L. (2007). The impact of testing practices on teaching. In E. Hinkel (Ed.), *The handbook of research in second language teaching and learning* (pp. 487–504). London: Lawrence Erlbaum Associates. doi:10.1007/978-0-387-46301-8_35

Hardin, J. R., & Ellington, A. J. (2005). Using multimedia to facilitate software instruction in an introductory modeling course. *INFORMS Transactions on Education, 5*(2).

Hari, R., Renvall, H., & Tanskanen, T. (2001). Left minineglect in dyslexic adults. *Brain*, *124*(7), 1373–1380. doi:10.1093/brain/124.7.1373 PMID:11408332

Harks, B., Rakoczy, K., Hattie, J., Besser, M., & Klieme, E. (2014). The effects of feedback on achievement, interest and self-evaluation: The role of feedback's perceived usefulness. *Educational Psychology*, *34*(3), 269–290. doi:10.1080/01443410.2013.785384

Harlen, W., & James, M. (1997). Assessment and learning: Differences and relationships between formative and summative assessment. *Assessment in Education: Principles, Policy & Practice*, *4*(3), 365–379. doi:10.1080/0969594970040304

Harpur, P. (2010). Accessible or not? Academics' hand written comments on assessment items written by students with print disabilities. *Disability & Society*, *25*(1), 121–122. doi:10.1080/09687590903363472

Harpur, P., & Loudoun, R. (2011). The barrier of the written word: Analysing universities' policies to students with print disabilities. *Journal of Higher Education Policy and Management*, *33*(2), 153–167. doi:10.1080/1360080X.2011.550088

Harpur, P., & Suzor, N. (2013). Copyright protections and disability rights: Turning the page to a new international paradigms. *The University of New South Wales Law Journal*, *36*(3), 745–778.

Harris, L., Brown, G., & Harnett, J. A. (2014). Understanding classroom feedback practices: A study of New Zealand student experiences, perceptions, and emotional responses. *Educational Assessment, Evaluation and Accountability*, *26*(2), 107–133. doi:10.100711092-013-9187-5

Harvey, L., Geall, V., & Moon, S. (1997). Graduates' work: Implications of organizational change for the development of student attributes. *Industry and Higher Education*, *11*(5), 287–296. doi:10.1177/095042229701100504

Hashemi, M., Azizinezhad, M., Najafi, V., & Nesari, A. J. (2011). What is mobile learning? Challenges and capabilities. *Procedia: Social and Behavioral Sciences*, *30*, 2477–2481. doi:10.1016/j.sbspro.2011.10.483

Haston, W. (2007). Teacher modeling as an effective teaching strategy. *Music Educators Journal*, *93*(4), 26–30. doi:10.1177/002743210709300414

Hatch, T., Shuttleworth, J., Jaffee, A. T., & Marri, A. (2016). Videos, pairs, and peers: What connects theory and practice in teacher education? *Teaching and Teacher Education*, *59*, 274–284. doi:10.1016/j.tate.2016.04.011

Hattie, J., & Timperley, H. (2007). *The power of feedback*. Academic Press.

Hattie, J. (2009). *Visible learning: A synthesis of over 800 meta-analyses relating to achievement*. London: Routledge/Taylor & Francis.

Hattie, J., & Timperley, H. (2007). The power of feedback. *Review of Educational Research*, *77*(1), 81–112. doi:10.3102/003465430298487

Hegarty, M., & Just, M.-A. (1993). Constructing mental models of machines from text and diagrams. *Journal of Memory and Language*, *32*(6), 717–742. doi:10.1006/jmla.1993.1036

Hemmi, A., Bayne, S., & Land, R. (2009). The appropriation and repurposing of social technologies in higher education. *Journal of Computer Assisted Learning*, *25*(1), 19–30. doi:10.1111/j.1365-2729.2008.00306.x

Henderson, M., Boud, D., Molloy, E., Dawson, P., Phillips, M., Ryan, T., & Mahoney, P. (2018). *Feedback for learning: Closing the assessment loop*. Australian Government. Department of Education and Training.

Henderson, M., Boud, D., Molloy, E., Dawson, P., Phillips, M., Ryan, T., & Mahoney, P. (2018). *Feedback for learning: Closing the assessment loop: Final report*. Victoria: Monash University.

Henderson, M., & Phillips, M. (2015). Video-based feedback on student assessment: Scarily personal. *Australasian Journal of Educational Technology*, *31*(1), 51–56. doi:10.14742/ajet.1878

Hendry, G. D., White, P., & Herbert, C. (2016). Providing exemplar-based 'feedforward' before an assessment: The role of teacher explanation. *Active Learning in Higher Education, 17*(2), 99–109. doi:10.1177/1469787416637479

Hensiek, S., DeKorver, B., Harwood, C., Fish, J. O., Shea, K., & Towns, M. (2017). Digital badges in science: A novel approach to the assessment of student learning. *Journal of College Science Teaching, 46*(3), 28–33. doi:10.2505/4/jcst17_046_03_28

Her Majesty's Stationary Office (HMSO). (*2001*). *The Special Educational Needs and Disabilities Act.* Retrieved from http://www.legislation.gov.uk/ukpga/2001/10/pdfs/ukpga_20010010_en.pdf

Her Majesty's Stationary Office (HMSO). (2010). *The Equality Act.* Retrieved from http://www.legislation.gov.uk/ukpga/2010/15/pdfs/ukpga_20100015_en.pdf

Her Majesty's Stationery Office (HMSO). (*2001*). *The Special Educational Needs and Disabilities Act.* Retrieved from http://www.legislation.gov.uk/ukpga/2001/10/pdfs/ukpga_20010010_en.pdf

Her Majesty's Stationery Office (HMSO). (2005). *The Disabilities Discrimination Act.* Retrieved from http://www.legislation.gov.uk/ukpga/2005/13/pdfs/ukpga_20050013_en.pdf

Her Majesty's Stationery Office (HMSO). (2010). *The Equality Act.* Retrieved from http://www.legislation.gov.uk/ukpga/2010/15/pdfs/ukpga_20100015_en.pdf

Herrington, J., Reeves, T. C., & Oliver, R. (2006). Authentic tasks online: A synergy among learner, task, and technology. *Distance Education, 27*(2), 233–247. doi:10.1080/01587910600789639

Herrington, J., Reeves, T. C., & Oliver, R. (2010). *A guide to authentic e-learning.* New York: Routledge.

Hesketh, A. J. (2000). Recruiting an elite? Employers' perceptions of graduate education and training. *Journal of Education and Work, 13*(3), 245–271. doi:10.1080/713676992

Hessels, R. S., Kemner, C., van den Boomen, C., & Hooge, I. T. (2016). The area-of-interest problem in eyetracking research: A noise-robust solution for face and sparse stimuli. *Behavior Research Methods, 48*(4), 1694–1712. doi:10.375813428-015-0676-y PMID:26563395

Hewett, R., Douglas, G., McLinden, M., & Keil, S. (2017). Developing an inclusive learning environment for students with visual impairment in higher education: Progressive mutual accommodation and learner experiences in the United Kingdom. *Journal European Journal of Special Needs Education, 32*(1), 89–109. doi:10.1080/08856257.2016.1254971

Higgins, R., Hartley, P., & Skelton, A. (2001). Getting the message across: The problem of communicating assessment feedback. *Teaching in Higher Education, 6*(2), 269–274. doi:10.1080/13562510120045230

Higgins, R., Hartley, P., & Skelton, A. (2002). The conscientious consumer: Reconsidering the role of assessment feedback in student learning. *Studies in Higher Education, 27*(1), 53–64. doi:10.1080/03075070120099368

Higher Education Funding Council for England (HEFCE). (2011). *National student survey: Findings and trends 2006 to 2010.* Retrieved from http://www.hefce.ac.uk/media/hefce/content/pubs/2011/201111/11_11.pdf

Higher Education Statistics Agency (HESA). (2013). *Higher education statistics for the United Kingdom, 2012-13.* Retrieved from: http://www.hesa.ac.uk/pubs/heuk

Hill, P., & Barber, M. (2014). *Preparing for a Renaissance in Assessment.* Retrieved from https://www.pearson.com/content/dam/one-dot-com/one-dot-com/uk/documents/educator/primary/preparing_for_a_renaissance_in_assessment_and_summary_text_december_2014.pdf

Hills, L., & Hughes, J. (2016). Assessment worlds colliding? Negotiating between discourses of assessment on an online open course. *Open Learning: The Journal of Open, Distance and e-Learning, 31*(2), 108-115. doi:10.1080/02680513.2016.1194747

Hismanoglu, M. (2012). Prospective EFL teachers' perceptions of ICT integration: A study of distance higher education in Turkey. *Journal of Educational Technology & Society, 15*(1), 185–196.

Hoffman, J. E., & Subramaniam, B. (1995). The role of visual attention in saccadic eye movements. *Perception & Psychophysics, 57*(6), 787–795. doi:10.3758/BF03206794 PMID:7651803

Hofmann, J. (2014). *Solutions to the top 10 challenges of blended learning. Top 10 challenges of blended learning.* Available on cedma-europe.org

Holmes, L. (2001). Reconsidering graduate employability: The 'graduate identity' approach. *Quality in Higher Education, 7*(2), 111–119. doi:10.1080/13538320120060006

Hooijdonk, R. v. (2016). *The magic of biometric eye-tracking in education.* Retrieved from https://www.emergingedtech.com/2016/06/the-magic-of-biometric-eye-tracking-in-education/

Hope, S. (2010). *Giving audio-visual feedback using Jing and GradeMark.* Retrieved from http://www.bioscience.heacademy.ac.uk/ftp/casestudies/hope.pdf

Hounsell, D. (1987). Essay writing and the quality of feedback. In d. J.T.E. Richardson, M.W. Eysenck, & D. Warren-Piper (Eds.), Student learning: Research in education and cognitive psychology (pp. 109–19). Milton Keynes, UK: Open University Press and Society for Research into Higher Education.

Hounsell, D. (2003). Student feedback, learning and development. In M. Slowey & D. Watson (Eds.), *Higher education and the life course* (pp. 67–78). Maidenhead, UK: Open University Press.

Hounsell, D. (2007). Towards a more sustainable feedback to students. In D. Boud & N. Falchikov (Eds.), *Rethinking assessment in higher education* (pp. 101–113). London: Routledge.

Hounsell, D., McCune, V., Hounsell, J., & Litjens, J. (2008). The quality of guidance and feedback to students. *Higher Education Research & Development, 27*(1), 55–67. doi:10.1080/07294360701658765

Howanitz, G. (2013). 'Metro 2033' – More than a cinegame? *Digital Icons: Studies in Russian, Eurasian and Central European New Media, 8*, 97-116. Retrieved from http://www.digitalicons.org/issue08/gernot-howanitz/

Hrastinski, S., Keller, C., & Carlsson, S. A. (2010). Design exemplars for synchronous e-learning: A design theory approach. *Computers & Education, 55*(2), 652–662. doi:10.1016/j.compedu.2010.02.025

Huang, Y.-M. (2015). Exploring the factors that affect the intention to use collaborative technologies: The differing perspectives of sequential/global learners. *Australasian Journal of Educational Technology, 31*(3), 278–292. doi:10.14742/ajet.1868

Hudson, P. B. (2010). Mentors report on their own mentoring practices. *Australian Journal of Teacher Education, 35*(7), 30–42. doi:10.14221/ajte.2010v35n7.3

Hullen, J. (2012). The relevance of e-learning-accessibility for people with vision impairment. In M. Allegra, M. Arrigo, V. Dal Grande, P. Denaro, D. La Guardia, S. Ottaviano, & G. Todaro (Eds.), *Mobile learning for visually impaired people* (pp. 72–79). Consiglio Nazionale delle Ricerche – Istituto per le Tecnologie, Didattiche.

Humfrey, P. (2010). Digital 'deviants' and the spirit of '68. *Times Higher Education.* Retrieved from http://www.timeshighereducation.co.uk/story.asp?sectioncode=26&storycode=414268&c=1

Hussain, I. (2012). Use of constructivist approach in higher education: An instructor's observation. *Creative Education, 3*(02), 179–184. doi:10.4236/ce.2012.32028

Ibabe, I., & Jauregizar, J. (2010). Online self-assessment with feedback and metacognitive knowledge. *Higher Education: The International Journal of Higher Education and Educational Planning, 59*(2), 243–258. doi:10.100710734-009-9245-6

Ibbotson, M., & Krekelberg, B. (2011). Visual perception and saccadic eye movements. *Current Opinion in Neurobiology, 21*(4), 553–558. doi:10.1016/j.conb.2011.05.012 PMID:21646014

Ice, P., Curtis, R., Phillips, P., & Wells, J. (2007). Using asynchronous audio feedback to enhance teaching presence and students' sense of community. *Journal of Asynchronous Learning Networks, 11*(2), 3–25.

Ihrig, C. (2013). Clinical pearls. *Optometry and Vision Science, 90*(3), 302–303. doi:10.1097/OPX.0b013e31828b8189 PMID:23400025

Imotions. (2017). Retrieved from https://imotions.com/blog/eye-tracking/

Ingram, D. E., & Wylie, E. (1979/1999). *The International Second Language Proficiency Ratings (ISLPR).* Brisbane: Centre for Applied Linguistics and Languages, Griffith University.

Inman, C., Wright, V. H., & Hartman, J. A. (2010). Use of Second Life in K-12 and higher education: A review of research. *Journal of Interactive Online Learning, 9*, 44–63.

International Language Testing System (ELTS). (2017). *IELTS.* London: British Council. Available at https://www.ielts.org/

Internet World Stats. (2017). *Usage and population studies.* Retrieved from www.internetworldstats.com/stats.htm

Iordache, D., Pop, D., Samoila, C., & Ursutia, D. (2011). Multifunctional laboratories based on Agilent USB technology LabView generated web interface. In J. G. Zubía & G. R. Alves (Eds.), *Using remote labs in education: Two little ducks in remote experimentation* (pp. 135–156). Bilbao, Spain: University of Deusto.

Ippoliti, C., & Baeza, V. (2017). Using digital badges to organize student learning opportunities. *Journal of Electronic Resources Librarianship, 29*(4), 221–235. doi:10.1080/1941126X.2017.1378541

Irons, A. (2008). *Enhancing learning through formative assessment and feedback. Key guides for effective teaching in higher education.* Abingdon, UK: Routledge.

Iwata, J., Telloyan, J., Murphy, L., Wang, S., & Clayton, J. (2017). The use of a digital badge as an indicator and a motivator. *International Association for Development of the Information Society*, 5. Retrieved from http://ezproxy.usq.edu.au/login?url=http://search.ebscohost.com/login.aspx?direct=true&db=eric&AN=ED579293&site=ehost-live

Jabbar, A., Gasser, R., & Lodge, J. (2016). Can new digital technologies support parasitology teaching and learning? *Trends in Parasitology, 32*(7), 522–530. doi:10.1016/j.pt.2016.04.004 PMID:27131629

Jacobs, V. R., Lamb, L. L., & Philipp, R. A. (2010). Professional noticing of children's mathematical thinking. *Journal for Research in Mathematics Education, 41*(2), 169–202.

Jaggars, S. S., & Xu, D. (2016). How do online course design features influence student performance? *Computers & Education, 95*, 270–284. doi:10.1016/j.compedu.2016.01.014

Jarodzka, H., Holmqvist, K., & Gruber, H. (2017). Eye tracking in educational science: Theoretical frameworks and research agendas. *Journal of Eye Movement Research, 10*(1), 1–18.

Järvelä, S., Näykki, P., Laru, J., & Luokkanen, T. (2007). Structuring and regulating collaborative learning in higher education with wireless networks and mobile tools. *Journal of Educational Technology & Society, 10*(4), 71–79.

Jiang, H., Gonnot, T., Yi, W., & Saniie, J. (2017). Computer vision and text recognition for assisting visually impaired people using android smartphone. *2017 IEEE International Conference on Electro Information Technology*. 10.1109/EIT.2017.8053384

Jirgensons, M., & Kapenieks, J. (2018). Blockchain and the future of digital learning credential assessment and management. *Journal of Teacher Education for Sustainability*, *20*(1), 145–156. doi:10.2478/jtes-2018-0009

Johannesen, R. L. (1990). Ethics in human communication (3rd ed.). Prospect Heights, IL: Waveland Press.

Johnson, R. T., & Johnson, D. W. (2008). Active learning: Cooperation in the classroom. **教育心理学年報**, *47*, 29-30. doi:10.5926/arepj1962.47.0_29

Jones, A., Olds, A., & Lisciandro, J. (2016). Understanding the Learner: Effective course design in the changing higher education space. *International Studies in Widening Participation*, *3*(1), 19–35.

Jones, M. G., & Ennes, M. (2018). High-stakes testing. In A. Hynds (Ed.), *Oxford Bibliographies in Education* (pp. 1–12). New York: Oxford University Press. doi:10.1093/OBO_dataset_home

Jones, W., Hope, S., & Adams, B. (2018). Teachers' perceptions of digital badges as recognition of professional development. *British Journal of Educational Technology*, *49*(3), 427–438. doi:10.1111/bjet.12557

Jönsson, E. (2005). *If looks could kill–an evaluation of eye tracking in computer games* (Unpublished Master's Thesis). Royal Institute of Technology (KTH), Stockholm, Sweden.

Joosten, T. (2012). *Social media for educators' strategies and best practices*. San Francisco, CA: John Wiley & Sons.

Jordan, L. (2012). Video for peer feedback and reflection: Embedding mainstream engagement into learning and teaching practice. *Research in Learning Technology*, *20*(sup1), 16–25. doi:10.3402/rlt.v20i0.19192

Joseph, N. (2009). Metacognition needed: Teaching middle and high school students to develop strategic learning skills. *Preventing School Failure*, *54*(2), 99–103. doi:10.1080/10459880903217770

Joseph, N. (2010). Practical intelligence: Do students have it? Helping them develop it through metacognitive learning. *Language Arts Journal of Michigan*, *26*(1), 38–42. doi:10.9707/2168-149X.1016

Juwah, C., Macfarlane-Dick, D., Matthew, B., Nicol, D., Ross, D., & Smith, B. (2004). *Enhancing student learning through effective formative feedback*. The Higher Education Academy.

Kafai, Y. B., & Burke, Q. (2015). Constructionist gaming: Understanding the benefits of making games for learning. *Educational Psychologist*, *50*(4), 313–334. doi:10.1080/00461520.2015.1124022 PMID:27019536

Kahl, S. (2005). Where in the world are formative tests? Right under your nose. *Education Week*, *25*(9), 38.

Kanwal, F., & Rehman, M. (2017). Factors affecting e-Learning adoption in developing countries: Empirical evidence from Pakistan's higher education sector. *IEEE Access: Practical Innovations, Open Solutions*, *5*, 10968–10978. doi:10.1109/ACCESS.2017.2714379

Kar, A., & Corcoran, P. (2017). A review and analysis of eye-gaze estimation systems, algorithms and performance evaluation methods in consumer platforms. *IEEE Access: Practical Innovations, Open Solutions*, *5*, 16495–16519. doi:10.1109/ACCESS.2017.2735633

Kawalkar, A., & Vijapurkar, J. (2013). Scaffolding science talk: The role of teachers' questions in the inquiry classroom. *International Journal of Science Education*, *35*(12), 2004–2027. doi:10.1080/09500693.2011.604684

Keller, J. M. (1983). Motivational design of instruction. In C. M. Reigeluth (Ed.), *Instructional design theories and models: An overview of their current status* (pp. 383–434). Hillsdale, NJ: Erlbaum.

Kemmis, S., McTaggart, R., & Nixon, R. (2014). *The action research planner: Doing critical participatory action research.* Singapore: Springer; doi:10.1007/978-981-4560-67-2

Kennepohl, D. K., & Shaw, L. (2010). *Accessible elements: teaching science online and at a distance.* Athabasca University Press.

Kent, M. L. (2017). Principles of dialogue and the history of dialogic theory in public relations. In X. Chen (Ed.), *Prospect of public relations science* (pp. 105–129). Beijing: Peking University Press. (in Chinese)

Kent, M. L., & Taylor, M. (2002). Toward a dialogic theory of public relations. *Public Relations Review, 28*(1), 21–37. doi:10.1016/S0363-8111(02)00108-X

Kent, M. L., & Theunissen, P. (2016). Elergy for mediated dialogue: Shiva the destroyer and reclaiming our first principles. *International Journal of Communication, 10*, 4040–4054.

Kermit, P., & Holiman, S. (2018). Inclusion in Norwegian higher education: Deaf students' experiences with lecturers. *Social Inclusion, 6*(4), 158–167. doi:10.17645i.v6i4.1656

Kerr, W., & McLaughlin, P. (2008). The benefit of screen recorded summaries in feedback for work submitted electronically. In F. Khandia (Ed.), *Proceedings of the CAA International Computer Assisted Assessment Conference* (pp. 153-168). Loughborough University.

King, A. (1993). From sage on the stage to guide on the side. *College Teaching, 41*(1), 30–36. doi:10.1080/87567555.1993.9926781

King, D., McGugan, S., & Bunyan, N. (2008). Does it make a difference? Replacing text with audio feedback. *Practice and Evidence of Scholarship of Teaching and Learning in Higher Education, 3*(2), 145–163.

Kleinknecht, M., & Schneider, J. (2013). What do teachers think and feel when analyzing videos of themselves and other teachers' teaching? *Teaching and Teacher Education, 33*, 13–23. doi:10.1016/j.tate.2013.02.002

Klenowski, V. (2013). Investigating the complexity of judgement practice. *Assessment in Education: Principles, Policy & Practice, 20*(1), 1–4. doi:10.1080/0969594X.2012.745702

Klin, A., & Volkmar, F. R. (2000). Treatment and intervention guidelines for individuals with Asperger's Syndrome. In A. Klin, F. R. Volkmar, & S. Sparrow (Eds.), *Asperger Syndrome* (pp. 340–366). New York: The Guildford Press.

Kluger, A. N., & DeNisi, A. (1996). The effects of feedback interventions on performance: A historical review, a meta-analysis, and a preliminary feedback intervention theory. *Psychological Bulletin, 119*(2), 254–284. doi:10.1037/0033-2909.119.2.254

Kneale, P. (1997). The rise of the 'strategic student': How can we adapt to cope. In S. Armstrong, G. Thompson, & S. Brown (Eds.), *Facing up to radical changes in universities and colleges* (pp. 119–130). London: Kogan Page.

Knight, P. (1995). *Assessment for learning in higher education.* London: Kogan Page.

Knight, P. (2002). Summative assessment in higher education: Practices in disarray. *Studies in Higher Education, 27*(3), 275–286. doi:10.1080/03075070220000662

Knight, P., & Saunders, M. (1999). Understanding teachers' professional cultures through interview: A constructivist approach. *Evaluation and Research in Education, 13*(3), 144–156. doi:10.1080/09500799908666954

Koller, M., Salzberger, T., Brenner, G., & Walla, P. (2012). Broadening the range of applications of eye-tracking in business research. *Proto Alegre*, *23*(1), 71–77.

Kollöffel, B., & Jong, T. (2013). Conceptual understanding of electrical circuits in secondary vocational engineering education: Combining traditional instruction with inquiry learning in a virtual lab. *Journal of Engineering Education*, *102*(3), 375–393. doi:10.1002/jee.20022

Kounin, J. S. (1970). *Discipline and group management in classrooms*. New York: Holt, Rinehard & Winston.

Krashen, S. (2003). *Explorations in language acquisition and use: The Taipei lectures*. Portsmouth, NH: Heinemann.

Krassanakis, V., Filippakopoulou, V., & Nakos, B. (2014). EyeMMV toolbox: An eye movement post-analysis tool based on a two-step spatial dispersion threshold for fixation identification. *Journal of Eye Movement Research*, *7*(1), 1–10.

Kravcik, M., & Klamma, R. (2012), Supporting self-regulation by personal learning environments. In I. Aedo, R.-M. Bottino, N.-S. Chen, C. Giovannella, D. G. Sampson, & Kinshuk (Eds.), *Proceedings of 12th IEEE International Conference on Advanced Learning Technologies* (pp. 710-711). Rome: IEEE Computer Society. 10.1109/ICALT.2012.192

Krueger, R., Koch, S., & Ertl, T. (2016). SaccadeLenses: Interactive exploratory fltering of eye tracking trajectories. Presentation at the *2016 Conference at IEEE Second Workshop on Eye Tracking and Visualisation*, Baltimore, MD. 10.1109/ETVIS.2016.7851162

Kukulska-Hulme, A. (2012). Prospects for inclusive mobile learning. In M. Allegra, M. Arrigo, V. Dal Grande, P. Denaro, D. La Guardia, S. Ottaviano, & G. Todaro (Eds.), *Mobile learning for visually impaired people* (pp. 13–25). Consiglio Nazionale delle Ricerche-Istituto per le Tecnologie, Didattiche.

Kun, A., Paek, T., Medenica, Z., Memarovic, N., & Palinko, O. (2009). Glancing at personal navigation devices can affect driving: Experimental results and design implications. In *Proceedings of the 1st International Conference on Automotive User Interfaces and Interactive Vehicular Applications* (pp. 129-136). Academic Press. 10.1145/1620509.1620534

Kunze, K., Iwamura, M., Kise, K., Uchida, S., & Omachi, S. (2013). Activity recognition for the mind: Toward a cognitive "quantified self". *Computer*, *46*(10), 105–108. doi:10.1109/MC.2013.339

Kusschau, R. (2003). *Understanding psychology*. Columbus, OH: Glencoe and McGraw Hill Co. Inc.

LaBanca, F., Worwood, M., Schauss, S., LaSala, J., & Donn, J. (2013). *Blended instruction: Exploring student centered pedagogical strategies to promote a technology-enhanced learning environment*. Litchfield, CT: Education Connection.

Lafuente, M., Remesal, A., & Álvarez Valdivia, I. (2013). Assisting learning in e-assessment: A closer look at educational supports. *Assessment & Evaluation in Higher Education*, *39*(4), 443–460. doi:10.1080/02602938.2013.848835

Lai, M.-L., Tsai, M.-J., Yang, F.-Y., Hsu, C.-Y., Liu, T.-C., Lee, S. W.-Y., ... Tsai, C.-C. J. E. r. (2013). A review of using eye-tracking technology in exploring learning from 2000 to 2012. *Educational Research*, *10*, 90–115.

LaMagna, M. (2017). Placing digital badges and micro-credentials in context. *Journal of Electronic Resources Librarianship*, *29*(4), 206–210. doi:10.1080/1941126X.2017.1378538

Lambert, M. (2000). Using cognitive and metacognitive learning strategies in the classroom. *Preventing School Failure*, *44*, 81–82. doi:10.1080/10459880009599788

Lamb, T. E., Hatoss, A., & O'Neill, S. (in press). Challenging social justice in superdiverse contexts through 'activist' languages education. In R. Papa (Ed.), *Handbook on Promoting Social Justice in Education*. Springer.

Lang, R., Shogren, K. A., Machalicek, W., Rispoli, M., O'Reilly, M., Baker, S., & Regester, A. (2009). Video self-modeling to teach classroom rules to two students with Asperger's. *Research in Autism Spectrum Disorders*, *3*(2), 483–488. doi:10.1016/j.rasd.2008.10.001

Larsen-Freeman, D., & Anderson, M. (2011). *Techniques and principles in language teaching*. Oxford, UK: Oxford University Press.

Laurillard, D. (2002). Rethinking university teaching: A conversational framework for the effective use of learning technologies (2nd ed.). London: Routledge Falmer.

Laurillard, D. (1993). *Rethinking university teaching: A framework for the effective use of educational technology*. London: Routledge.

Lawson, G., Salanitri, D., & Waterfield, B. (2016). Future directions for the development of virtual reality within an automotive manufacturer. *Applied Ergonomics*, *53*, 323–330. doi:10.1016/j.apergo.2015.06.024 PMID:26164106

Leach, J., & Scott, P. (2002). Designing and evaluating science teaching sequences: An approach drawing upon the concept of learning demand and social constructivist perspective on learning. *Studies in Science Education*, *38*(1), 115–142. doi:10.1080/03057260208560189

Lea, M., & Street, B. (1998). Student writing in higher education: An academic literacies approach. *Studies in Higher Education*, *23*(2), 157–172. doi:10.1080/03075079812331380364

Lebler, D. (2008). Perspectives on assessment in the learning of music. In D. Bennett & M. Hannan (Eds.), *Inside, outside, downside up: Conservatoire training and musicians' work* (pp. 181–193). Perth, Australia: Black Swan Press.

Lebler, D., Harrison, S., Carey, G., & Cain, M. (2014). *Assessment in music: An approach to aligning assessment with learning outcomes in the higher music education*. Queensland: Griffith University.

Lee, M. J., Pradhan, S., & Dalgarno, B. (2008). The effectiveness of screencasts and cognitive tools as scaffolding for novice object-oriented programmers. *Journal of Information Technology Education*, *7*, 61–80. doi:10.28945/179

Lehman, R., & Conceicao, S. (2014). *Motivating and retaining online students: Research based strategies that work*. San Francisco, CA: Jossey-Bass Press.

Leibold, N., & Schwarz, L. M. (2015). The art of giving online feedback. *The Journal of Effective Teaching*, *15*(1), 34–46.

Leininger, J., & Robinson, J. (1995, November). Combining groupware and total quality management principles to improve engineering students' teamwork skills. In *Proceedings Frontiers in Education 1995 25th Annual Conference. Engineering Education for the 21st Century* (Vol. 2, pp. 4c3-10). IEEE.

Liddicoat, A. J., Papademetre, L., Scarino, A., & Kohler, M. (2003). *Report on intercultural language learning*. Canberra: Department of Education, Science and Training.

Lin, W., & Yang, S. (2011). Exploring students' perceptions of integrating Wiki technology and peer feedback into English writing courses. *English Teaching*, *10*(2), 88–103.

Litzier, E., & Young, J. (2012). Understanding the risk of attrition in undergraduate Engineering: Results from the project to assess climate in engineering. *Journal of Engineering Education*, *101*(2), 319–345. doi:10.1002/j.2168-9830.2012.tb00052.x

Litzler, M. F., & García Laborda, J. (2016). *Students' opinions about ubiquitous delivery of standardized English exams*. Retrieved from https://files.eric.ed.gov/fulltext/ED577152.pdf

Lizzio, A., & Wilson, K. (2008). Feedback as assessment: Students' perceptions of quality and effectiveness. *Assessment & Evaluation in Higher Education*, *33*(3), 263–275. doi:10.1080/02602930701292548

London, J. E. (2018). Use of synchronous audio feedback in discussion boards with online RN-BSN Students. *Nurse Educator*, *1-4*. doi:10.1097/NNE.0000000000000633 PMID:30585883

Lopez-Fernandez, O., Kuss, D. J., Romo, L., Morvan, Y., Kern, L., Graziani, P., ... Rousseau, A. (2017). Self-reported dependence on mobile phones in young adults: A European cross-cultural empirical survey. *Journal of Behavioral Addictions*, *6*(2), 168–177. doi:10.1556/2006.6.2017.020 PMID:28425777

Louca, L. T., Zacharia, Z. C., & Constantinou, C. P. (2011). In quest of productive modeling-based learning discourse in elementary school science. *Journal of Research in Science Teaching*, *48*(8), 919–951. doi:10.1002/tea.20435

Loughran, J. J. (2002). Effective reflective practice: In search of meaning in learning about teaching. *Journal of Teacher Education*, *53*(1), 33–43. doi:10.1177/0022487102053001004

Lowe, D. B., Berry, C., Murray, S., & Lindsay, E. (2009). Adapting a remote laboratory architecture to support collaboration and supervision. *International Journal of Engineering*, *5*(S1), 51–56.

Lunt, T., & Curran, J. (2010). "Are you listening please?" The advantages of electronic audio feedback compared to written feedback. *Assessment & Evaluation in Higher Education*, *35*(7), 759–769. doi:10.1080/02602930902977772

Lupu, R. G., & Ungureanu, F. (2013). A survey of eye tracking methods and applications. *Buletinul Institutului Politehnic din Iasi. Automatic Control and Computer Science Section*, *3*, 72–86.

Magnan, S. S., & Back, M. (2007). Social interaction and linguistic gain during study abroad. *Foreign Language Annals*, *40*(1), 43–61. doi:10.1111/j.1944-9720.2007.tb02853.x

Maier, W., Wolf, N., & Randler, C. (2016). Effects of a computer-assisted formative assessment intervention based on multiple-tier diagnostic items and different feedback types. *Computers & Education*, *95*, 85–98. doi:10.1016/j.compedu.2015.12.002

Male, S. A., & Bennett, D. (2014). Don't know what we'll be doing yet: Enhancing career preview and engagement among undergraduate engineering students. In *25th Annual Conference of the Australasian Association for Engineering Education: Engineering the Knowledge Economy: Collaboration, Engagement & Employability* (pp.1387-1394). Academic Press.

Manary, M. A., Flannagan, C. A. C., Reed, M. P., & Schneider, L. W. (1998). Development of an improved driver eye position model. Technical Paper 980856. *SAE Transactions: Journal of Passenger Cars*, 107.

Martinez, M. (2006). What is Metacognition? *Phi Delta Kappan*, *87*(9), 696–699. doi:10.1177/003172170608700916

Masats, D., & Dooly, M. (2011). Rethinking the use of video in teacher education: A holistic approach. *Teaching and Teacher Education*, *27*(7), 1151–1162. doi:10.1016/j.tate.2011.04.004

Massis, B. E. (2012). Librarians and faculty collaboration – partners in student success. *New Library World*, *113*(1/2), 90–93. doi:10.1108/03074801211199077

Masui, C., & De Corte, E. (2005). Learning to reflect and to attribute constructively as basic components of self-regulated learning. *The British Journal of Educational Psychology*, *75*(3), 351–372. doi:10.1348/000709905X25030 PMID:16238871

Mathieson, K. (2012). Exploring student perceptions of audiovisual feedback via screencasting in online courses. *American Journal of Distance Education*, *26*(3), 143–156. doi:10.1080/08923647.2012.689166

Mathisen, P. (2012). Video feedback in higher education. A contribution to improving the quality of written feedback. *Nordic Journal of Digital Literacy*, *7*(2), 97–116.

Mavodza, J. (2011). The academic librarian and the academe. *New Library World*, *112*(9/10), 446–451. doi:10.1108/03074801111182030

Mayer, J. H., & Land, R. (2005). Threshold concepts and troublesome knowledge (2): Epistemological considerations and a conceptual framework for teaching and learning. *Higher Education*, *49*(3), 373–388. doi:10.100710734-004-6779-5

Mayer, R. E. (2001). *Multimedia learning*. Cambridge University Press. doi:10.1017/CBO9781139164603

Mayer, R. E. (2002). Cognitive theory and the design of multimedia instruction: An example of the two-way street between cognition and instruction. *New Directions for Teaching and Learning*, *89*(89), 55–71. doi:10.1002/tl.47

Mayer, R. E. (2002). Rote versus meaningful learning. *Theory into Practice*, *41*(4), 226–232. doi:10.120715430421tip4104_4

Mayer, R. E., & Moreno, R. (2003). Nine ways to reduce cognitive load in multimedia learning. *Educational Psychologist*, *38*(1), 43–52. doi:10.1207/S15326985EP3801_6

Mayer, S. J. (2012). *Classroom discourse and democracy: Making meanings together*. New York, NY: Peter Lang.

McCarthy, J. (2015). Evaluating written, audio and video feedback in higher education summative assessment tasks. *Issues in Educational Research*, *25*(2), 153–169.

McConnell, D. (2006). *E-learning groups and communities*. Maidenhead, UK: Open University Press.

McConnell, T. J., Parker, J. M., & Eberhardt, J. (2013). Assessing teachers' science content knowledge: A strategy for assessing depth of understanding. *Journal of Science Teacher Education*, *24*(4), 717–743. doi:10.100710972-013-9342-3

McCune, V. (2004). Development of first-year students' conceptions of essay writing. *Higher Education*, *47*(3), 257–282. doi:10.1023/B:HIGH.0000016419.61481.f9

McDaniel, R., & Fanfarelli, J. (2016). Building Better Digital Badges: Pairing Completion Logic With Psychological Factors. *Simulation & Gaming*, *47*(1), 73–102. doi:10.1177/1046878115627138

McDowell, J. (2010). *Anarchy in the universities: Beyond the student-teacher hierarchy*. Paper presented at the Learning Futures Festival 2010. University of Leicester. Retrieved from http://eprints.hud.ac.uk/id/eprint/10802/1/McDowell_J_2010_AnarchyInTheUniversities-BeyondTheStudentTeacherHierarchy.pdf

McDowell, J. (2011). Using asynchronous video to promote learner engagement through the enhancement of assessment and feedback. In *Proceedings of the 6th International Blended Learning Conference* (pp. 69-78). University of Hertfordshire.

McDowell, J. (2012). Cultivating a viral community of practice to drive institutional enhancement through the promotion of video-enhanced learning, feedback and assessment. In *Proceedings of the 7th International Blended Learning Conference*. University of Hertfordshire.

McDowell, J. (2015). A black swan in a sea of white noise: Using technology-enhanced learning to afford educational inclusivity for learners with Asperger's Syndrome. *Social Inclusion*, *3*(6), 7–15. doi:10.17645i.v3i6.428

McGarvey, D., & Haxton, K. (2011). Using audio for feedback on assessments: Tutor and student experiences. *New Directions*, *7*(7), 5–9. doi:10.29311/ndtps.v0i7.459

McInnes Bowers, C. V., Chew, B., Bowers, M. R., Ford, C. E., Smith, C., & Herrington, C. (2009). Interdisciplinary synergy: A partnership between business and library faculty and its effects on students' information literacy. *Journal of Business & Finance Librarianship*, *14*(2), 110–127. doi:10.1080/08963560802362179

Meadows, M. (2006). *Can we predict who will be a reliable marker.* AQA Centre for Education Research and Policy.

Mele, M. L., Federici, S. J. D., & Technology, R. A. (2012). A psychotechnological review on eye-tracking systems: Towards user experience. *Disability and Rehabilitation. Assistive Technology*, *7*(4), 261–281. doi:10.3109/17483107.2011.635326 PMID:22117604

Mellati, M., Khademi, M., & Shirzad, A. (2015). The relationships among sources of teacher pedagogical beliefs, teaching experiences, and student outcomes. *International Journal of Applied Linguistics and English Literature*, *4*(2), 177–184. doi:10.7575/aiac.ijalel.v.4n.2p.177

Mena, J., García, M., Clarke, A., & Barkatsas, A. (2015). An analysis of three different approaches to student teacher mentoring and their impact on knowledge generation in practicum settings. *European Journal of Teacher Education*, *39*(1), 53–76. doi:10.1080/02619768.2015.1011269

Merry, S., Price, M., Carless, D., & Taras, M. (2013). *Reconceptualising feedback in higher education: Developing dialogue with students.* London: Routledge. doi:10.4324/9780203522813

Middleton, A., Nortcliffe, A., & Owens, R. (2009). *iGather: Learners as responsible audio collectors of tutor, peer and self-reflection.* Paper presented at the A Word in User Ear – Audio Feedback Conference. Retrieved from http://research.shu.ac.uk/lti/awordinyourear2009/docs/Middleton-Nortcliffe-Owens-iGather_final.pdf

Ministerial Council on Education, Employment, Training and Youth Affairs (MCEETYA). (2008). *Melbourne declaration on educational goals for your Australians.* Retrieved from http://www.curriculum.edu.au/verve/_resources/National_Declaration_on_the_Educational_Goals_for_Young_Australians.pdf

Mndzebele, N. (2013). Teachers' readiness in using ICT in the classroom: The case of a developing country. *International Journal of Information and Education Technology (IJIET)*, *3*(4), 409–412. doi:10.7763/IJIET.2013.V3.309

Mohamed, A. O., Da Silva, M. P., & Courboulay, V. (2007). A history of eye gaze tracking. *Laboratoire Informatique, Image et Interaction.* Rochelle: Universite de La Rochelle.

Moje, E. B., Ciechanowski, K. M., Kramer, K., Ellis, L., Carrillo, R., & Collazo, T. (2004). Working toward third space in content area literacy: An examination of everyday funds of knowledge and discourse. *Reading Research Quarterly*, *39*(1), 38–70. doi:10.1598/RRQ.39.1.4

Molloy, E. K., & Boud, D. (2014). Feedback models for learning, teaching and performance. In J. M. Spector, M. D. Merrill, J. Elen, & M. J. Bishop (Eds.), *Handbook of research on educational communications and technology* (pp. 413–424). New York: Springer. doi:10.1007/978-1-4614-3185-5_33

Moore, D. J., & Voltmer, D. R. (2003). Curriculum for an engineering renaissance. *IEEE Transactions on Education*, *46*(4), 452–455. doi:10.1109/TE.2003.818754

Moreno, R., & Mayer, R. E. (2000). A learner-centered approach to multimedia explanations: Deriving instructional design principles from cognitive theory. *Interactive Multimedia Electronic Journal of Computer-Enhanced Learning*, *2*(2), 12–20.

Moriarty, B., Danaher, P., & Danaher, G. (2008). Freire and dialogical pedagogy: A means for interrogating opportunities and challenges in Australian postgraduate supervision. *International Journal of Lifelong Education*, *27*(4), 431–442. doi:10.1080/02601370802051447

Mory, E. H. (2003). Feedback research revisited. In D. H. Jonassen (Ed.), *Handbook of research for educational communications and technology* (pp. 745–783). New York: Macmillan.

Mulholland, J., & Wallace, J. (2005). Growing the tree of teacher knowledge: Ten years of learning to teach elementary science. *Journal of Research in Science Teaching, 42*(7), 767–790. doi:10.1002/tea.20073

Mulliner, E., & Tucker, M. (2017). Feedback on feedback practice: Perceptions of students and academics. *Assessment & Evaluation in Higher Education, 42*(2), 266–288. doi:10.1080/02602938.2015.1103365

Murphy, C., Neil, P., & Beggs, J. (2007). Primary science teacher confidence revisited: Ten years on. *Educational Research, 49*(4), 415–430. doi:10.1080/00131880701717289

Murray, I. (2008). Instructional eLearning technologies for the vision impaired. *Curtin University espace.* http://hdl.handle.net/20.500.11937/645

Murtagh, L., & Baker, N. (2009). Feedback to "feed forward": Student response to tutors' written comments on assignments. *Practitioner Research in Higher Education, 3*(1), 20–28.

Mutch, A. (2003). Exploring the practice of feedback to students. *Active Learning in Higher Education, 4*(1), 24–38. doi:10.1177/1469787403004001003

Myllyäki, S. (2013). Incorporation of continuous assessment into lectures in engineering education. *European Journal of Engineering Education, 1*(1), 1–9.

NAEYC & Fred Rogers Center for Early Learning and Children's Media. (2012). *Technology and interactive media as tools in early childhood programs serving children from birth through age 8.* Joint position statement. Washington, DC: NAEYC; Latrobe, PA: Fred Rogers Center at St. Vincent College. Retrieved from www.naeyc.org/content/technology-and-young-children

Narciss, S. (2008). Feedback strategies for interactive learning tasks. In J. M. Spector, M. D. Merrill, J. Van Merrienboer, & M. P. Driscoll (Eds.), *Handbook of research on educational communications and technology* (3rd ed.; pp. 125–143). New York: Erlbaum.

National Council of Assessment and Feedback. (2015). *Focus on learning; formative feedback.* Retrieved from https://www.ncca.ie/media/1925/assessment-booklet-3_en.pdf

National Instruments. (2019). *Learn LabView.* Retrieved from http://www.ni.com/academic/students/learn-labview/

National Student Forum. (2009). *National Student Forum Annual Report.* Author. Retrieved from https://webarchive.nationalarchives.gov.uk/20121106154314/http://www.bis.gov.uk/assets/biscore/higher-education/docs/n/09-p83-national-student-forum-annual-report-09

National Student Forum. (2009). *National Student Forum Annual Report.* Retrieved from http://www.bis.gov.uk/assets/biscore/higher-education/docs/n/09-p83-national-student-forum-annual-report-09

Nelson, M., & Schunn, C. (2009). The nature of feedback: How different types of peer feedback affect writing performance. *Instructional Science, 37*(4), 375–401. doi:10.100711251-008-9053-x

Neupokoeva, E. E., Chapaev, N. K., Akimova, O. B., Shcherbin, M. D., & Borovikov, E. A. (2016). Open access using educational technology in applications as element of teaching for special discipines. *International Journal of Environmental and Science Education, 11*(16), 8858–8872.

Nicol, D., & Milligan, C. (2006). Rethinking technology-supported assessment practices in relation to the seven principles of good feedback practice. *Innovative Assessment in Higher Education*, 64-77.

Nicol, D. (2010). From monologue to dialogue: Improving written feedback processes in mass higher education. *Assessment & Evaluation in Higher Education, 35*(5), 501–517. doi:10.1080/02602931003786559

Nicol, D., & Macfarlane-Dick, D. (2006). Formative assessment and self-regulated learning: A model and seven principles of good feedback practice. *Studies in Higher Education, 31*(2), 199–218. doi:10.1080/03075070600572090

Nixon, S., Brooman, S., Murphy, B., & Fearon, D. (2017). Clarity, consistency and communication: Using enhanced dialogue to create a course-based feedback strategy. *Journal of Assessment & Evaluation in Higher Education, 42*(5), 812–822. doi:10.1080/02602938.2016.1195333

Noland, R. B., Weiner, M. D., Gao, D., Cook, M. P., & Nelessen, A. (2017). Eye-tracking technology, visual preference surveys, and urban design: Preliminary evidence of an effective methodology. *Journal of Urbanism, 10*(1), 98–110. doi:10.1080/17549175.2016.1187197

Norling, E. (2009). *Modelling Human Behaviour with BDI Agents* (PhD thesis). Engineering - Computer Science and Software Engineering, The University of Melbourne.

Nortcliffe, A., & Middleton, A. (2007). *Audio Feedback for the iPod Generation.* Paper presented at the International Conference on Engineering Education – ICEE 2007, Coimbra, Portugal. Retrieved from http://ineerweb.osanet.cz/Events/ICEE2007/papers/489.pdf

Nortcliffe, A., & Middleton, A. (2008). A three year case study of using audio to blend the engineer's learning environment. *Engineering Education, 3*(2), 45–57. doi:10.11120/ened.2008.03020045

Nortcliffe, A., & Middleton, A. (2011). Smartphone feedback: Using an iPhone to improve the distribution of audio feedback. *International Journal of Electrical Engineering Education, 48*(3), 280–293. doi:10.7227/IJEEE.48.3.6

Nyström, M., & Holmqvist, K. (2010). An adaptive algorithm for fixation, saccade, and glissade detection in eyetracking data. *Behavior Research Methods, 42*(1), 188–204. doi:10.3758/BRM.42.1.188 PMID:20160299

O'Donoghue, M., & Cochrane, T. (2010). The role of live video capture production in the development of student communication skills. *Learning, Media and Technology, 35*(3), 323–335. doi:10.1080/17439884.2010.509350

O'Neill, S. (2012a). Implementing NETPAW's test of English proficiency in Australia: A case study. [Winter.]. *Language Learning Technologies: An International Online Journal, 2*(1), 121–144.

O'Neill, S. (2012b). Teaching and assessment of persuasive writing: Juggling the language features and grasping the metalanguage. *International Journal of Pedagogies and Learning, 7*(1), 84–98. doi:10.5172/ijpl.2012.7.1.84

O'Neill, S. (2015). EFL teachers' professional learning needs: Working with multimedia and the CLOUD. In L. T. Wong & A. Dubey-Jhaveri (Eds.), *English language education in a global world: Practices, issues and challenges* (pp. 47–60). New York: Nova Science Publishers.

O'Neill, S. (2016). Educational growth and the semiotics of the tree: Exploring the OAK as a framework for deep learning. In F. V. Tochon (Ed.), *Deep education in the disciplines and beyond* (pp. 43–69). Blue Mounds, WI: Deep University Press.

O'Neill, S. (2018). Building students' capacity to write English for academic purposes: Pedagogy and the demands of writing persuasively. In L. T. Wong & W. L. Wong (Eds.), *Teaching and learning English for Academic Purposes: Current research and practices* (pp. 69–96). New York: Nova Science Publishers.

O'Neill, S. (2019). New opportunities for languages learning through 21st Century knowledge building communities. In R. Arber, M. Weinmann, & J. Blackmore (Eds.), *Rethinking languages education: Directions, challenges and innovations*. New York: Routledge.

O'Neill, S., & Dann, C. (2018). Video feedback on practicum: Demonstrating effective pedagogy and and making children's progress explicit. In T. Richardson, B. Dann, C. Dann, & S. O'Neill (Eds.), *Formative assessment practices for pre-service teacher practicum feedback: Emerging research and opportunities* (pp. 113–139). IGI Global. doi:10.4018/978-1-5225-2630-8.ch004

Odeh, S., Mckenna, S., & Abu-Mulaweh, H. (2017). A unified first-year engineering design-based learning course. *International Journal of Mechanical Engineering Education, 45*(1), 47–58. doi:10.1177/0306419016674508

O-K, S., & Hong-Fa, H. (2014). Students' perceptions towards the use of dialogic feedback in mobile applications for students' writing: A qualitative case study. *Journal of e-Learning and Knowledge Society, 10*(3).

Orsmond, P., & Merry, S. (2008). Students' attitudes to and usage of academic feedback provided by audio files. *Bioscience Education, 11*(3). doi:10.3108/beej.11.3

Orsmond, P., & Merry, S. (2011). Feedback alignment: Effective and ineffective links between tutors' and students' understanding of coursework feedback. *Assessment & Evaluation in Higher Education, 36*(2), 125–136. doi:10.1080/02602930903201651

Orsmond, P., Merry, S., & Callaghan, A. C. (2004). Implementation of a formative assessment model incorporating peer and self assessment. *Innovations in Education & Training International, 41*(3), 273–290. doi:10.1080/14703290410001733294

Orsmond, P., Merry, S., & Reiling, K. (2002). The use of exemplars and formative feedback when using student derived marking criteria in peer and self assessment. *Assessment & Evaluation in Higher Education, 27*(4), 309–323. doi:10.1080/0260293022000001337

Orsmond, P., Merry, S., & Reiling, K. (2005). Biology students' utilisation of tutors' formative feedback: A qualitative interview study. *Assessment & Evaluation in Higher Education, 30*(4), 369–386. doi:10.1080/02602930500099177

Osberg, K. M. (1997). *Constructivism in practice: The case for meaning-making in the virtual world* (Unpublished Doctoral Dissertation). University of Washington, Washington, DC.

Osborne, J., Simon, S., & Collins, S. (2003). Attitudes towards science: A review of the literature and its implications. *International Journal of Science Education, 25*(9), 1049–1079. doi:10.1080/0950069032000032199

Osen, O. L., & Bye, R. T. (2017). Reflections on teaching electrical and computer engineering courses at the bachelor level. In *Proceedings of the 9th International Conference on Computer Supported Education (CSEDU 2017)* (vol. 2, pp. 57-68). Academic Press. 10.5220/0006359000570068

Osipova, A., Prichard, B., Boardman, A. G., Kiely, M. T., & Carroll, P. E. (2011). Refocusing the lens: Enhancing elementary special education reading instruction through video self-reflection. *Learning Disabilities Research & Practice, 26*(3), 158–171. doi:10.1111/j.1540-5826.2011.00335.x

Osman, K., & Bakar, N. A. (2013). Teachers and students as game designers: Designing games for new pedagogical approaches in game enhanced learning. *Curriculum Integration*, 102-113.

Pacheeco, E., Lips, M., & Yoong, P. (2018). Transition 2.0. Digital technologies, higher education and vision impairment. *The Internet and Higher Education, 37*, 1–10. doi:10.1016/j.iheduc.2017.11.001

Pais Marden, M., & Herrington, J. (2011). Supporting interaction and collaboration in the language classroom through computer mediated communication. In T. Theo Bastiaens & M. Ebner (Eds.), *ED-MEDIA, World Conference on Educational Multimedia, Hypermedia, and Telecommunications* (pp. 1161-1168). AACE.

Paivio, A. (1990). *Mental representations: A dual coding approach.* Oxford University Press. doi:10.1093/acprof:oso/9780195066661.001.0001

Palincsar, A. S. (1998). Social constructivist perspectives on teaching and learning. *Annual Review of Psychology*, *49*(1), 345–375. doi:10.1146/annurev.psych.49.1.345 PMID:15012472

Palinko, O., Kun, A. L., Shyrokov, A., & Heeman, P. (2010). Estimating cognitive load using remote eye tracking in a driving simulator. *Proceedings of the 2010 symposium on eye-tracking research & applications*. 10.1145/1743666.1743701

Palloff, R. M., & Pratt, K. (2005). *Collaborating online: Learning together in community*. San Francisco, CA: Jossey-Bass.

Palmquist, S. (2007). *Active project-based learning in structural analysis: Field inspection of steel truss bridge.* Paper presented at the American Society for Engineering Education, Honolulu, HI.

Pamplin, M., & Poirier, M. (2012). A Learner-centred induction to moodle. *Proceedings of the European Conference on E-Learning*, 641-644. Retrieved from http://ezproxy.usq.edu.au/login?url=http://search.ebscohost.com/login.aspx?direct=true&db=ehh&AN=101746542&site=ehost-live

Panadero, P., Andrade, H., & Brookhart, S. (2018). Fusing self-regulated learning and formative assessment: A roadmap of where we are, how we got here, and where we are going. *Australian Educational Researcher*, *45*(1), 13–31. doi:10.100713384-018-0258-y

Panse, A., Alcock, L., & Inglis, M. (2018). Reading proofs for validation and comprehension: An expert-novice eye-movement study. *International Journal of Research in Undergraduate Mathematics Education*, *4*(3), 357–375. doi:10.100740753-018-0077-6

Pardo, A. (2018). A feedback model for data-rich learning experiences. *Journal of Assessment & Evaluation in Higher Education*, *43*(3), 428–433. doi:10.1080/02602938.2017.1356905

Pardo, A., Jovanovic, J., Dawson, S., Gašević, D., & Mirriahi, N. (2017). Using learning analytics to scale the provision of personalised feedback. *British Journal of Educational Technology*, *50*(1), 128–138. doi:10.1111/bjet.12592

Paris, S., & Winograd, P. (1990). How metacognition can promote academic learning and instruction. In B. Jones & L. Idol (Eds.), *Dimensions of thinking and cognitive instruction* (pp. 15–52). Lawrence Erlbaum Associates.

Parker, M., Patton, K., & O'Sullivan, M. (2016). Signature pedagogies in support of teachers' professional learning. *Irish Educational Studies*, *35*(2), 137–153. doi:10.1080/03323315.2016.1141700

Parsayi, F., & Soyoof, A. (2018). Video games: The interface between language learning and storytelling. *International Journal of Pedagogies and Learning*, *13*(2), 103–118.

Pask, G. (1976). *Conversation theory: Applications in education and epistemology*. Amsterdam: Elsevier.

Patchan, M., & Schunn, C. (2016). Understanding the effects of receiving peer feedback for text revision: Relations between author and reviewer ability. *Journal of Writing Research*, *8*(2), 227–265. doi:10.17239/jowr-2016.08.02.03

Patrikakou, E. (2015). Relationships among parents, students, and teachers: The technology wild card. *Procedia: Social and Behavioral Sciences*, *174*, 2253–2258. doi:10.1016/j.sbspro.2015.01.883

Payne, T., Kirkpatrick, D., Goodacre, C., & McClean, P. (2006). *Creating accessible teaching and support for students with disabilities. Carrick Institute for Learning and Teaching in Higher Education* [Report]. University of Tasmania.

Peacock, S., & Cowan, J. (2017). Towards online student-directed communities of inquiry. *Journal of Further and Higher Education*, *42*(5), 1–16. doi:10.1080/0309877X.2017.1302569

Pea, R. D. (2004). The social and technological dimensions of scaffolding and related theoretical concepts for learning, education, and human activity. *Journal of the Learning Sciences*, *13*(3), 423–451. doi:10.120715327809jls1303_6

Peregoy, S., & Boyle, O. (2012). *Reading, writing and learning in ESL: A resource book for teachers*. New York: Allyn & Bacon.

Permvattana, R., Armstrong, H., & Murray, I. (2013). eLearning for the vision impaired: A holistic perspective. *International Journal of Cyber Society and Education*, *6*(1), 15–30. doi:10.7903/ijcse.1029

Peterson, P. M., & Helms, R. M. (2013). *Challenges and opportunities for the global engagement of higher education*. Occasional Paper produced for CIGE: American Council on Education. Washington, DC: American Council of Education.

Petty, G. (2009). *Evidence-based teaching: A practical approach*. Cheltenham, UK: Nelson Thornes.

Pierce, D. (2015). Say what? *Wired*, *23*(10), 86.

Pieri, M., & Diamantini, D. (2012). Approaches to mobile learning accessibility. In M. Allegra, M. Arrigo, V. Dal Grande, P. Denaro, D. La Guardia, S. Ottaviano, & G. Todaro (Eds.), *Mobile learning for visually impaired people* (pp. 52–63). Consiglio Nazionale delle Ricerche – Istituto per le Tecnologie, Didattiche.

Plano Clark, V. L., & Creswell, J. W. (2008). *The mixed methods reader*. Thousand Oaks, CA: Sage.

Poth, C.-A. (2018). The contributions of mixed insights to advancing technology-enhanced formative assessments within higher education learning environments: An illustrative example. *International Journal of Educational Technology in Higher Education*, *15*(9). doi:10.118641239-018-0090-5

Poulos, A., & Mahony, M. (2008). Effectiveness of feedback: The students' perspective. *Assessment & Evaluation in Higher Education*, *33*(2), 143–154. doi:10.1080/02602930601127869

Povah, C., & Vaukins, S. (2018). *Generation Z is starting university – but is higher education ready?* Retrieved from https://www.theguardian.com/higher-educationnetwork/2017/jul/10/generation-z-starting-university-highereducation-ready

Pretorius, M., & Calitz, A. (2010). *The Development of a Usability Methodology Incorporating Eye Tracking for Developing Countries*. Fourth International Business Conference, Victoria Waterfalls, Zambia.

Price, M. (2007). Should we be giving less written feedback? Centre for Bioscience Bulletin, 22(9).

Price, M., Handley, K., Millar, J., & O'Donovan, B. (2010). Feedback: All that effort, but what is the effect? *Assessment & Evaluation in Higher Education*, *35*(3), 277–289. doi:10.1080/02602930903541007

Prosser, J., & Loxley, A. (2007). Enhancing the contribution of visual methods to inclusive education. *Journal of Research in Special Education*, *7*(1), 55–68.

Pryor, J., & Crossouard, B. (2008). A socio-cultural theorisation of formative assessment. *Oxford Review of Education*, *34*(1), 1–20. doi:10.1080/03054980701476386

Quaye, S. J., & Harper, S. R. (Eds.). (2015). *Student engagement in higher education: Theoretical perspectives and practical approaches for diverse populations* (2nd ed.). New York, NY: Routledge.

Quinton, S., & Smallbone, T. (2010). Feeding forward: Using feedback to promote student reflection and learning-a teaching model. *Innovations in Education and Teaching International*, *47*(1), 125–135. doi:10.1080/14703290903525911

Race, A., & Williams, M. (2018). Digital audio feedback technology: Nursing students' perspectives. *Computers, Informatics, Nursing*, *36*(2), 63–67. doi:10.1097/CIN.0000000000000418 PMID:29401070

Ramsden, P. (1992). *Learning to teach in higher education*. New York, NY: Routledge Press.

Rashbass, C., & Westheimer, G. (1961). Disjunctive eye movements. *The Journal of Physiology*, *159*(2), 339–360. doi:10.1113/jphysiol.1961.sp006812 PMID:14490420

Rasi, P., & Vuojarvi, H. (2017). Toward personal and emotional connectivity in mobile higher education through asynchronous formative audio feedback. *British Journal of Educational Technology*, *49*(2), 292–304. doi:10.1111/bjet.12587

Rayner, C., Denholm, C., & Sigafoos, J. (2009). Video-based intervention for individuals with autism: Key questions that remain unanswered. *Research in Autism Spectrum Disorders*, *3*(2), 291–303. doi:10.1016/j.rasd.2008.09.001

Rayner, K., Chace, K. H., Slattery, T. J., & Ashby, J. (2006). Eye movements as reflections of comprehension processes in reading. *Scientific Studies of Reading*, *10*(3), 241–255. doi:10.12071532799xssr1003_3

Rayner, K., Inhoff, A. W., Morrison, R. E., Slowiaczek, M. L., & Bertera, J. H. (1981). Masking of foveal and parafoveal vision during eye fixations in reading. *Journal of Experimental Psychology. Human Perception and Performance*, *7*(1), 167–179. doi:10.1037/0096-1523.7.1.167 PMID:6452494

Reeves, T. C. (2000). *Enhancing the worth of instructional technology research through "design experiments" and other development research strategies*. Paper presented at the Annual Meeting of the American Educational Research Association, New Orleans, LA. Retrieved from http://treeves.coe.uga.edu/AERA2000Reeves.pdf

Remillard, J. (2018). *Mapping the relationship between written and enacted curriculum: Examining teachers' decision making*. Paper presented at the Invited Lectures from the 13th International Congress on Mathematical Education. 10.1007/978-3-319-72170-5_27

Ribchester, C., France, D., & Wheeler, A. (2008). Podcasting: A tool for enhancing assessment feedback? In E. O'Doherty (Eds.), *The Fourth Education in a Changing Environment Conference Book 2007* (pp. 119-136). Santa Rosa, CA: Informing Science Press.

Richardson, A., Hazzard, M., Challman, S. D., Morgenstein, A. M., & Brueckner, J. K. (2011). A "second life" for gross anatomy: Applications for multiuser virtual environments in teaching the anatomical sciences. *Anatomical Sciences Education*, *4*(1), 39–43. doi:10.1002/ase.195 PMID:21265036

Richardson, J. T., & Wydell, T. N. (2003). The representation and attainment of students with dyslexia in UK higher education. *Reading and Writing*, *16*(5), 475–503. doi:10.1023/A:1024261927214

Richardson, T., Dann, B., Dann, C., & O'Neill, S. (2018). *Formative assessment practices for pre-service teacher practicum feedback: Emerging research and opportunities*. IGI Global. doi:10.4018/978-1-5225-2630-8

Rogers, C. R. (1957). The necessary and sufficient conditions of therapeutic personality change. *Journal of Consulting Psychology*, *21*(2), 95–103. doi:10.1037/h0045357 PMID:13416422

Rogers, C., & Scott, K. (2008). The development of the personal self and professional identity in learning to teach. In M. Cochran-Smith, S. Feiman-Nemser, D. McIntyre, & K. Demers (Eds.), *Handbook of research on teacher education* (pp. 732–755). New York: Routledge.

Rosaen, C. L., Lundeberg, M., Cooper, M., Fritzen, A., & Terpstra, M. (2008). Noticing noticing: How does investigation of video records change how teachers reflect on their experiences? *Journal of Teacher Education*, *59*(4), 347–360. doi:10.1177/0022487108322128

Roschelle, J., & Teasley, S. D. (1995). The construction of shared knowledge in collaborative problem solving. In C. E. O'Malley (Ed.), *Computer-supported collaborative learning* (pp. 69–197). Berlin: Springer-Verlag. doi:10.1007/978-3-642-85098-1_5

Roth, A., Ogrin, S., & Schmitz, B. (2016). Assessing self-regulated learning in higher education: A systematic literature review of self-report instruments. *Educational Assessment, Evaluation and Accountability, 28*(3), 225–250. doi:10.100711092-015-9229-2

Rotheram, B. (2009). *Sounds good.* JISC Final Report. Retrieved from. http://www.jisc.ac.uk/publications/reports/2009/soundsgoodfinalreport.aspx

Rotherham, R. (2009). *Sounds good: Quicker, better assessment using audio feedback.* Retrieved from http://sites.google.com/site/soundsgooduk/downloads/

Rowe, A. D., & Wood, L. N. (2008). Student perceptions and preferences for feedback. *Asian Social Science, 4*(3), 78–88.

Rowe, A., Fitness, J., & Wood, L. (2014). The role and functionality of emotions in feedback at university: A qualitative study. *Australian Educational Researcher, 41*(3), 283–309. doi:10.100713384-013-0135-7

Rowe, K. (2005). *Teaching reading: Report of the national inquiry into the teaching of literacy.* Canberra: Department of Education, Science and Training. Retrieved from http://www.dest.gov.au/nitl/documents/report_recommendations.pdf

Rule, A., Stefanich, G., Boody, R., & Peiffer, B. (2011). Impact of adaptive materials on teachers and their students with visual impairments in secondary science and mathematics classes. *International Journal of Science Education, 33*(6), 865–887. doi:10.1080/09500693.2010.506619

Rumble, G. (2001). Re-inventing distance education, 1971-2001. *International Journal of Lifelong Education, 20*(1-2), 31–43.

Rushton, A. (2005). Formative assessment: A key to deep learning? *Medical Teacher, 27*(6), 509–513. doi:10.1080/01421590500129159 PMID:16199357

Rushton, D., & Lahlafi, A. (2013). The value and impact of cross professional collaborations in developing student information and academic literacy skills at Sheffield Hallam University, UK. *Nordic Journal of Information Literacy in Higher Education, 5*(1), 38–43. doi:10.15845/noril.v5i1.178

Rust, C. (2002). The impact of assessment on student learning: How can the research literature practically help to inform the development of departmental assessment strategies and learner-centred assessment practices? *Active Learning in Higher Education, 3*(2), 145–158. doi:10.1177/1469787402003002004

Rust, C., O'Donovan, B., & Price, M. (2005). A social constructivist assessment process model: How the research literature shows us this could be best practice. *Assessment & Evaluation in Higher Education, 30*(3), 231–240. doi:10.1080/02602930500063819

Sadler, D. R. (2015). Backwards assessment explanations: Implications for teaching and assessment practice. In Assessment in music: From policy to practice (pp. 9-19). Cham, Switzerland: Springer.

Sadler, D. R. (1998). Formative assessment: Revisiting the territory. *Assessment in Education: Principles, Policy & Practice, 5*(1), 77–85. doi:10.1080/0969595980050104

Sadler, D. R. (2010). Beyond feedback: Developing student capability in complex appraisal. *Assessment & Evaluation in Higher Education, 35*(5), 535–550. doi:10.1080/02602930903541015

Sadler, D. R. (2013). Opening up feedback: Teaching learners to see. In S. Merry, M. Price, D. Carless, & M. Taras (Eds.), *Reconceptualising feedback in higher education: Developing dialogue with students* (pp. 54–63). London: Routledge.

Sadler, R. (1989). Formative assessment and the design of instructional systems. *Instructional Science, 18*(2), 119–144. doi:10.1007/BF00117714

Salmon, G. (2000). *E-Moderating: The key to teaching and learning online.* Kogan Page Limited.

Salvucci, D. D., & Goldberg, J. H. (2000). Identifying fixations and saccades in eye-tracking protocols. *Proceedings of the 2000 symposium on Eye tracking research & applications.* 10.1145/355017.355028

Salvucci, D. D., & Anderson, J. R. J. H.-C. I. (2001). Automated eye-movement protocol analysis. *Human-Computer Interaction, 16*(1), 39–86. doi:10.1207/S15327051HCI1601_2

Sambell, K. (2016). Assessment and feedback in higher education: considerable room for improvement? *Student Engagement in Higher Education, 1*(1).

Sánchez-Vera, M. M., & Prendes-Espinosa, M. P. (2015). Beyond objective testing and peer assessment: Alternative ways of assessment in MOOCs. RUSC. *Universities and Knowledge Society Journal, 12*(1), 119–130. doi:10.7238/rusc.v12i1.2262

Sankalpani, P., Wijesinghe, I., Jeewani, I., Anooj, R., Mahadikaara, M., & Jayakody, J. (2018). Smart assistant: A solution to facilitate vision impaired individuals. *2018 National Information Technology Conference (NITC).* 10.1109/NITC.2018.8550059

Saunders, W. L. (1992). The constructivist perspective: Implications and teaching strategies for science. *School Science and Mathematics, 92*(3), 136–141. doi:10.1111/j.1949-8594.1992.tb12159.x

Saville, B., Lawrence, N., & Jakobsen, K. (2012). Creating learning communities in the classroom. *New Directions for Teaching and Learning, 2012*(132), 57–69. doi:10.1002/tl.20036

Schaffer, D., Doube, W., & Tuovinen, J. (2003). Applying cognitive load theory to computer science education. In M. Petre & D. Budgen (Eds.), *Proceedings of the Joint Conference for Evaluation and Assessment in Software Engineering and the Psychology of Programming Interest Group 2003* (pp. 333-346). Academic Press.

Schaffhauser, D. (2012). As high-stakes online testing approaches, Will your iPads work? *Assessment, Digital Edition.* Retrieved from http://thejournal.com/Articles/2012/10/11/You-May-Now-Open-Your-Test-Tablets.aspx

Scherr, R. E., Laugero, K. D., Graham, D. J., Cunningham, B. T., Jahns, L., Lora, K. R., ... Mobley, A. R. (2017). Innovative techniques for evaluating behavioral nutrition interventions. *Advances in Nutrition, 8*(1), 113–125. doi:10.3945/an.116.013862 PMID:28096132

Schkoda, R. F., Schweisinger, T. A., & Wagner, J. R. (2012). An improved undergraduate mechanical engineering laboratory structure and curriculum: Design and assessment. *International Journal of Mechanical Engineering, 40*(3), 182–195.

Scriven, M. (1967). The methodology of evaluation. In R. Tyler, R. Gagne, & M. Scriven (Eds.), Perspectives on curriculum evaluation (pp. 39-83). Chicago: Rand McNally.

Sekovanić, V., Vukovac, D. P., & Podbojec, Z. (2012). Usability case study of an adapted e-learning course for dyslexic students. *Proceedings of the 5th International Conference of Education, Research and Innovation* 1998-2007.

Séror, J. (2013). Show me! Enhanced feedback through screencasting technology. *TESL Canada Journal, 30*(1), 104–116. doi:10.18806/tesl.v30i1.1128

Shephard, K. L. (2002). Integrating national initiatives and local developments: streaming video to support nurse education in the UK. In *Winds of change in the sea of learning. Proceedings of the 19th Annual* Auckland, New Zealand.

Sherin, M. G. (2017). Exploring the boundaries of teacher noticing: Commentary. In E. Schack, M. Fisher, & J. Wilhelm (Eds.), *Teacher Noticing: Bridging and Broadening Perspectives, Contexts, and Frameworks* (pp. 401–408). Cham: Springer International Publishing. doi:10.1007/978-3-319-46753-5_23

Shipley-Benamou, R., Lutzker, J. R., & Taubman, M. (2002). Teaching daily living skills to children with autism through instructional video modeling. *Journal of Positive Behavior Interventions*, *4*(3), 166–177. doi:10.1177/109830070200 40030501

Shokri, H., & Abdolmanafi-Rokni, S. J. (2014). The impact of computer games on EFL learners' spelling: A qualitative study. *Studies in English Language Teaching*, *2*(3), 266–274. doi:10.22158elt.v2n3p266

Shore, J. R., Wolf, M.-K., & Heritage, M. (2016). A case study of formative assessment to support teaching of reading comprehension for English learners. *Journal of Educational Research and Innovation*, *5*(2), 4. Retrieved from http://digscholarship.unco.edu/jeri/vol5/iss2/4

Shumaker, D. (2010). A wide range of approaches. *Information Outlook*, *14*(1), 10–11.

Shute, V. J. (2008). Focus on formative feedback. *Review of Educational Research*, *78*(1), 153–189. doi:10.3102/0034654307313795

Shweta, P., & Shri Ram, P. (2009). *Developing a more effective and flexible learning management system (LMS) for the academic institutions using moodle*. ICAL.

Siemens, G. (2014). *Connectivism: A learning theory for the digital age*. Elearnspace.

Simon, S. E. (2013). Chaos of textures or 'tapisserie'? A model for creative teacher education curriculum design. *Australian Journal of Teacher Education*, *38*(11), 6. doi:10.14221/ajte.2013v38n11.2

Sitzmann, T., Ely, K., Brown, K. G., & Bauer, K. N. (2010). Self-assessment of knowledge: A cognitive learning or affective measure? *Academy of Management Learning & Education*, *9*(2), 169–191. doi:10.5465/amle.9.2.zqr169

Skamp, K., & Peers, S. (2012). *Implementation of science based on the 5E learning model: Insights from teacher feedback on trial Primary Connections units*. Paper presented at the The 43th Annual Conference of the Australasian Science Education Research Association (ASERA), University of the Sunshine Coast, Australia.

Skamp, K., & Mueller, A. (2001). Student teachers' conceptions about effective primary science teaching: A longitudinal study. *International Journal of Science Education*, *23*(4), 331–351. doi:10.1080/09500690119248

Sloan Consortium. (2019). *Online learning consortium*. Retrieved from http://sloanconsortium.org/

Smith, T. (2008). Integrating undergraduate peer mentors into liberal arts courses: A pilot study. *Innovative Higher Education*, *33*(1), 49–63. doi:10.100710755-007-9064-6

Smith, T. M., Desimone, L. M., Zeidner, T. L., Dunn, A. C., Bhatt, M., & Rumyantseva, N. L. (2007). Inquiry-oriented instruction in science: Who teaches that way? *Educational Evaluation and Policy Analysis*, *29*(3), 169–199. doi:10.3102/0162373707306025

Smyth, G. (2005). Wireless technologies bridging the digital divide in education. In *ECEL2006-5th, European Conference on e-learning: ECEL2006* (pp. 407-415). Reading, UK: Academic Conferences Ltd.

Soliman, N. (2014). Using e-Learning to develop EFL students' language skills and activate their independent learning. *Creative Education.*, *5*(10), 752–757. doi:10.4236/ce.2014.510088

Soyoof, A. (2018). Video games: A springboard for enhancing students' L2C. *International Journal of Pedagogies and Learning*, *13*(2), 137–151.

Stake, R. (1995). *The art of case research*. Newbury Park, CA: Sage Publications.

Stannard, R. (2007). Using screen capture software in student feedback. *The Higher Education Academy English Centre Commissioned Case Studies.* Retrieved from http://www.english.heacademy.ac.uk/explore/publications/casestudies/technology/camtasia.php

Star, J. R., & Strickland, S. K. (2008). Learning to observe: Using video to improve preservice mathematics teachers' ability to notice. *Journal of Mathematics Teacher Education, 11*(2), 107–125. doi:10.100710857-007-9063-7

Starkey, L. (2010). Teachers' pedagogical reasoning and action in the digital Age. *Teachers and Teaching, 16*(2), 233–244. doi:10.1080/13540600903478433

Stellmack, M. A., Konheim-Kalkstein, Y. L., Manor, J. E., Massey, A. R., & Schmitz, J. A. P. (2009). An assessment of reliability and validity of a rubric for grading APA-style introductions. *Teaching of Psychology, 36*(2), 102–107. doi:10.1080/00986280902739776

Stewart, W. (2009). *Audio supported enhanced learning. JISC Final Report.* Retrieved from http://reports.jiscemerge.org.uk/Download-document/16-ASEL-Audio-Supported-Enhanced-Learning.html

Stockwell, G., & Hubbard, P. (2013). *Some emerging principles for mobile-assisted language learning.* Monterey, CA: The International Research Foundation for English Language Education; Retrieved from http://www.tirfonline.org/english-in-the-workforce/mobile-assisted-language-learning

Stošić, L. (2015). The importance of educational technology in teaching. *International Journal of Cognitive Research in Science Engineering and Education, 3*(1), 111–114.

Suppasetseree, S., & Dennis, N. K. (2010). *The use of Moodle for teaching and learning English at tertiary level in Thailand. The International Journal of Humanities.*

Sutton, P. (2012). Conceptualizing feedback literacy: Knowing, being, and acting. *Innovations in Education and Teaching International, 49*(1), 31–40. doi:10.1080/14703297.2012.647781

Tabatabaei, M., & Gui, Y. (2011). The impact of technology on teaching and learning languages. In A. Méndez-Vilas (Ed.), *Education in a technological world: Communicating current and emerging research and technological efforts* (pp. 513–517). Retrieved from http://www.formatex.info/ict/book/isbn-contents.pdf

Tang, J. (2013). The research on blended learning of ESL based on a Moodle platform. *Studies in Literature and Language, 6*(2), 30–34. doi:10.3698/j.sll.1923156320130602.3612

Tang, J. (2013). The research on blended learning of ESL based on Moodle platform. *Studies in Literature and Language, 6*(2), 30–34. doi:10.3968/j.sll.1923156320130602.3612

Taras, M. (2005). Assessment – summative and formative – some theoretical reflections. *British Journal of Educational Studies, 55*(4), 466–478. doi:10.1111/j.1467-8527.2005.00307.x

Taylor, D. 1994). Inauthentic authenticity or authentic inauthenticity? *TESL-EJ, 1*(2). Retrieved from http://www.tesl-ej.org/wordpress/issues/volume1/ej02/ej02a1/

Teo, A. (2012). Promoting EFL students' inferential reading skills through computerized dynamic assessment. *Language Learning & Technology, 16*(3), 10–20. Retrieved from http://llt.msu.edu/issues/october2012/action.pdf

Thienthong, A., & Lian, A. (2014). The use of Internet resources and applications for Language instruction. *Voices in Asia., 1*, 107–116.

Thomas, L. (2012). *Building student engagement and belonging in Higher Education at a time of change; Final report from What Works? Student retension and success programme.* Retrieved from https://www.heacademy.ac.uk/system/files/what_works_final_report.pdf

Thompson, N., & Pascal, J. (2012). Developing critically reflective practice. *Reflective Practice, 13*(2), 311–325. doi:10.1080/14623943.2012.657795

Thompson, P. (2013). The digital natives as learners: Technology use patterns and approaches to learning. *Computers & Education, 65*, 12–33. doi:10.1016/j.compedu.2012.12.022

Thompson, P., Vogler, J., & Xiu, Y. (2017). Strategic tooling: Technology for constructing a community of inquiry. *Journal of Educators Online, 14*(2), 117–124. doi:10.9743/jeo.2017.14.2.10

Thorne, S. L., Black, R. W., & Sykes, J. M. (2009). Second language use, socialization, and learning in internet interest communities and online gaming. *Modern Language Journal, 93*, 802–821. doi:10.1111/j.1540-4781.2009.00974.x

Thornhill, S., Asensio, M., & Young, C. (2002). *Video streaming: A guide for educational development.* Manchester, UK: JISC Click and Go Project, UMIST.

Thurston, A. (2014). Disability, power and equality in the school. *International Journal of Disability Development and Education, 61*(2), 105–107. doi:10.1080/1034912X.2014.905048

Tillema, H. H., Smith, K., & Leshem, S. (2011). Dual roles-conflicting purposes: A comparative study on perceptions of assessment in mentoring relations during practicum. *European Journal of Teacher Education, 34*(2), 139–159. doi:10.1080/02619768.2010.543672

Tochon, T., Karaman Cendel, A., & Ökten, C. (2014). Online instructional personal environment for deep language learning. *International Online Journal of Education and Teaching, 1*(2), 147–173.

Tomasik, M. J., Berger, S., & Moser, U. (2018). On the development of a computer-based tool for formative student assessment: Epistemological, methodological, and practical issues. *Frontiers in Psychology, 9*, 2245. doi:10.3389/fpsyg.2018.02245 PMID:30515119

Tomitsch, M., Curwood, J., Thomson, K., & Hendry, G. (2016). *Professional development for academics: Evaluating a personalised mobile website for learning from student feedback. Final report.* Sydney, NSW: University of Sydney.

Tomlinson, B. (2009). *Materials development in language teaching.* Cambridge, UK: Cambridge University Press.

Torrance, H. (2007). Assessment as learning? How the use of explicit learning objectives, assessment criteria and feedback in post-secondary education and training can come to dominate learning. *Assessment in Education: Principles, Policy & Practice, 14*(3), 281–294. doi:10.1080/09695940701591867

Torrance, H., & Pryor, J. (1998). *Investigating formative assessment. Teaching, learning and assessment in the classroom.* Buckingham, UK: Open University Press.

Torres, J. O. (2019). Positive impact of utilizing more formative assessment over summative assessment in the EFL/ESL classroom. *Open Journal of Modern Linguistics, 9*(01), 1–11. doi:10.4236/ojml.2019.91001

Towns, S. G., & Loo, D. B. (2012). Current issues in the development ofMultimedia language learning software for mobile devices: Platforms to reach Thai university students. *AsiaCALL 2011 Proceedings and Presentations*, 1-8.

Tripp, T., & Rich, P. (2012). Using video to analyze one's own teaching. *British Journal of Educational Technology, 43*(4), 678–704. doi:10.1111/j.1467-8535.2011.01234.x

Trochim, W. M. K. (2006). *Construct validity*. Web Centre for Social Research Methods. Retrieved from http://www.socialresearchmethods.net/kb/constval.php

Tsang, H. Y., Tory, M., & Swindells, C. (2010). eSeeTrack: Visualizing sequential fixation patterns. *IEEE Transactions on Visualization and Computer Graphics*, *16*(6), 953–962. doi:10.1109/TVCG.2010.149 PMID:20975132

Tschofen, C., & Mackness, J. (2012). Connectivism and dimensions on individual experiences. *International Review of Research in Open and Distance Learning*, *13*(1), 124–143. doi:10.19173/irrodl.v13i1.1143

Tseng, C., Wu, B., Morrison, A. M., Zhang, J., & Chen, Y. (2015). Travel blogs on China as a destination image formation agent: A qualitative analysis using Leximancer. *Tourism Management*, *46*, 347–358. doi:10.1016/j.tourman.2014.07.012

Tucci, V. K. (2011). Faculty/librarian collaboration: Catalyst for student learning and librarian growth. *Science & Technology Libraries*, *30*(3), 292–305. doi:10.1080/0194262X.2011.596792

United Nations Educational, Scientific and Cultural Organization (UNESCO). (1994). *The Salamanca statement and framework for action on special needs education*. Adopted by the World Conference on Special Needs Education: Access and Quality, Salamanca, Spain.

United Nations Educational, Scientific and Cultural Organization (UNESCO). (2012). *Mobile learning for teachers in Europe: exploring the potential of mobile technologies to support teachers and improve practice*. Author.

United Nations General Assembly (2007). *Convention on the Rights of Persons with Disabilities: Resolution adopted by the General Assembly*. A/RES/61/106. Author.

Vail Lowery, N. (2002). Construction of teacher knowledge in context: Preparing elementary teachers to teach mathematics and science. *School Science and Mathematics*, *102*(2), 68–83. doi:10.1111/j.1949-8594.2002.tb17896.x

van Gog, T., & Jarodzka, H. (2013). Eye tracking as a tool to study and enhance cognitive and metacognitive processes in computer-based learning environments. In *International Handbook of Metacognition and Learning Technologies* (pp. 143–156). Sydney: Springer. doi:10.1007/978-1-4419-5546-3_10

Van Tol, C. (2016). The global English language job market is changing: what this means for ESL teachers. *University of Toronto TEFL Online*. Retrieved from https://medium.com/accelerated/the-global-english-language-job-market-is-changing-what-this-means-for-esl-teachers-92c8d87cf8eb

Veenman, M., Van Hout-Wolters, B., & Afflerbach, P. (2006). Metacognition and learning: Conceptial and methodological considerations. *Metacognition and Learning*, *1*(1), 3–14. doi:10.100711409-006-6893-0

Veronikas, S. W., & Maushak, N. (2005). Effectiveness of Audio on Screen Captures in Software Application Instruction. *Journal of Educational Multimedia and Hypermedia*, *14*(2), 199–205.

Vidal, M., Turner, J., Bulling, A., & Gellersen, H. (2012). Wearable eye tracking for mental health monitoring. *Computer Communications*, *35*(11), 1306–1311. doi:10.1016/j.comcom.2011.11.002

Villarroel, V., Bloxham, S., Bruna, D., Bruna, C., & Herrera-Seda, C. (2018). Authentic assessment: Creating a blueprint for course design. *Journal of Assessment & Evaluation in Higher Education*, *43*(5), 840–854. doi:10.1080/02602938.2017.1412396

Volpentesta, A. P., & Frega, N. (2007). Developing a blended-learning community in a university setting. *International Journal of Web Based Communities*, *3*(2), 134–150. doi:10.1504/IJWBC.2007.014076

Vygotsky, L. S. (1978). *Mind in society: The development of higher mental process*. Cambridge, MA: Harvard University Press.

Walker, M. (2009). An investigation into written comments on assignments: Do students find them usable? *Assessment & Evaluation in Higher Education, 34*(1), 67–78. doi:10.1080/02602930801895752

Walsh, M. (2006). The "textual shift": Examining the reading process with visual and multimodal texts. *Australian Journal of Language and Literacy, 29*(1), 24–37.

Walsh, S. (2006). Talking the talk of the TESOL classroom. *ELT Journal, 60*(2), 133–141. doi:10.1093/elt/cci100

Wang, K. (2013). Showing or telling a story: A comparative study of public education texts in multimodality and mono-modality. *English Language Teaching, 6*(2), 136. doi:10.5539/elt.v6n2p136

Wang, Q., Yang, S., Liu, M., Cao, Z., & Ma, Q. (2014). An eye-tracking study of website complexity from cognitive load perspective. *Decision Support Systems, 62*, 1–10. doi:10.1016/j.dss.2014.02.007

Wang, S., Wu, C., & Wang, Y. (2009). Investigating the determinants and age and gender differences in the acceptance of mobile learning. *British Journal of Educational Technology, 40*(1), 92–118. doi:10.1111/j.1467-8535.2007.00809.x

Ware, P. (2011). Computer generated feedback in student writing. *TESOL Quarterly, 45*, 769-774.

Weasmer, J., & Woods, A. (2003). The role of the host teacher in the student teaching experience. *The Clearing House: A Journal of Educational Strategies, Issues and Ideas, 76*(4), 174–177. doi:10.1080/00098650309601998

Weaver, M. (2006). Do students value feedback? Student perceptions of tutors' written responses. *Assessment & Evaluation in Higher Education, 31*(3), 379–394. doi:10.1080/02602930500353061

Wedel, M., & Pieters, R. (2008). A review of eye-tracking research in marketing. *Review of Marketing Research., 4*, 123–147. doi:10.1108/S1548-6435(2008)0000004009

Weidlich, J., & Bastiaens, T. (2017). Explaining social presence and the quality of online learning with the SIPS model. *Computers in Human Behavior, 72*, 479–487. doi:10.1016/j.chb.2017.03.016

Wei, J., Chen, J. C., & Adawu, A. (2014). Teaching ESL beginners metacognitive writing strategies through multimedia software. *The CATSOL Journal, 26*(1), 60–75.

Wells, G. (1999). *Dialogic inquiry: Toward a sociocultural practice and theory of education.* Cambridge, UK: Cambridge University Press. doi:10.1017/CBO9780511605895

Wenger, E. (1998). Communities of practice: Learning as a social system. *The Systems Thinker, 9*(5), 2–3.

Wenger, E. (2009). Communities of practice. *Communities, 22*, 57.

Wenger, E. (2011). *Communities of practice: A brief introduction.* Watertoen, MA: Harvard Business Press.

Whitburn, B. (2014). The 'inclusion' of students with vision impairments: Generational perspectives in Australia. *International Journal of Whole Schooling, 10*(1), 1–18. Retrieved from http://hdl.handle.net/10536/DRO/DU:30062408

Wicker, R. B., & Quintana, R. (2000). An innovation-based fluid mechanics design and fabrication laboratory. *Journal of Engineering Eucation, 39*(3), 361–367.

Wiggins, G. (2012). 7 Keys to Effective feedback. *Educational Leadership, 70*(1), 10–16.

Wiliam, D. (2011). *Embedded formative assessment.* Bloomington, IM: Solution Tree Press.

Wiliam, D. (2011). What is assessment for learning? *Studies in Educational Evaluation, 37*(1), 3–14. doi:10.1016/j.stueduc.2011.03.001 PMID:22114905

Wiliam, D., & Thompson, M. (2017). Integrating assessment with learning: What will it take to make it work? In *The future of assessment* (pp. 53–82). New York: Routledge. doi:10.4324/9781315086545-3

Williams, K. C., & Williams, C. C. (2011). Five key ingredients for improving student motivation. *Research in Higher Education, 12*, 1–23.

Willis, M. J., Dann, C. E., Jones, C. M., Lowe, B. J., & Toohey, E. (2012, November). Designing mobile information systems to support WIL experiences. In *Proceedings of the 24th Australian Computer-Human Interaction Conference* (pp. 653-656). ACM. 10.1145/2414536.2414634

Wilson, G., & Stacey, E. (2003). Online interaction impacts on learning: Teaching the teachers to teach online. In G. Crisp, D. Theile, I. Scholten, S. Barker & J. Baron (Eds.), *Interact, Integrate, Impact: Proceedings of the 20th Annual Conference of the Australasian Society for Computers in Learning in Tertiary Education* (pp. 541-551). Retrieved from http://dro.deakin.edu.au/eserv/DU:30009593/stacey-onlineinteractionimpactson-2003.pdf

Wing, L., & Gould, J. (1979). Severe impairments of social interaction and associated abnormalities in children: Epidemiology and classification. *Journal of Autism and Developmental Disorders, 9*(1), 11–29. doi:10.1007/BF01531288 PMID:155684

Winkle, A. (2017). The power of Plickers. *Texas Adult Education & Literacy Quarterly, 21*(2), 8.

Wixon, D. R., Pietras, C. M., Huntwork, P. K., & Muzzey, D. W. (1996). Changing the rules: A pragmatic approach to product development. In D. Wixen & J. Ramey (Eds.), *Field methods casebook for software design* (pp. 57–89). New York: John Wiley & Sons.

Woodcock, P. (2016). Towards dialogue: Audio feedback on politics essays. *European Political Science, 16*(2), 193–205. doi:10.1057/eps.2015.101

Wood, D., Wood, H., & Middleton, D. (1978). An experimental evaluation of four face-to-face teaching strategies. *International Journal of Behavioral Development, 1*(2), 131–147. doi:10.1177/016502547800100203

Woodfine, B. P., Nunes, M. B., & Wright, D. J. (2005). *Constructivist eLearning and dyslexia: Problems of social negotiation in text-based synchronous environments.* Paper presented at the 3rd International Conference on Multimedia and Information & Communication Technologies in Education (m-ICTE2005), Cáceres, Spain. Retrieved from http://www.formatex.org/micte2005/97.pdf

Woodfine, B. P., Nunes, B. P., & Wright, D. J. (2008). Texted-based synchronous e-learning and dyslexia: Not necessarily the perfect match! *Computers & Education, 50*(3), 703–717. doi:10.1016/j.compedu.2006.08.010

World Intellectual Property Organisation. (2012). *Marrakesh treaty to facilitate access to published works for persons who are blind, visually impaired, or otherwise print disabled.* Geneva: Author.

Wotjas, O. (1998, September). Feedback? No, just give us the answers. *Times Higher Education*, 25.

Xiong, Y., & Suen, H. (2018). Assessment approaches in massive open online courses: Possibilities, challenges and future directions. *International Review of Education, 64*(2), 241–263. doi:10.100711159-018-9710-5

Yadav, A., Subedi, D., Lundeberg, M. A., & Bunting, C. F. (2011). Problem-based learning: Influence on students' learning in an electrical engineering course. *Journal of Engineering Education, 100*(2), 253–280. doi:10.1002/j.2168-9830.2011.tb00013.x

Yin, R. K. (2003). *Case study research: Design and methods* (3rd ed.). Thousand Oaks, CA: Sage Publishing.

Yoon, Y. S., Ham, D. H., & Yoon, W. C. (2016). Application of activity theory to analysis of human-related accidents: Method and case studies. *Reliability Engineering & System Safety*, *150*, 22–34. doi:10.1016/j.ress.2016.01.013

Young, Y. F. (2013). Exploring students' language awareness through intercultural communication in computer-supported collaborative learning. *Journal of Educational Technology & Society*, *16*(2), 325–342.

Zhu, J., & Mok, M. (2018). Predicting primary students' self-regulated learning by their prior achievement, interest, personal best goal orientation and teacher feedback. *Educational Psychology*, *38*(9), 1106–1128. doi:10.1080/01443410.2018.1497775

Zhu, Z., & Ji, Q. (2007). Novel eye gaze tracking techniques under natural head movement. *IEEE Transactions on Bio-medical Engineering*, *54*(12), 2246–2260. doi:10.1109/TBME.2007.895750 PMID:18075041

Zichermann, G. (2011). *Gamification has issues, but they aren't the ones everyone focuses on.* Retrieved from http://radar.oreilly.com/2011/06/gamification-criticism-overjustification-ownership-addiction.html

Zientek, L. R. (2007). Preparing high-quality teachers: Views from the classroom. *American Educational Research Journal*, *44*(4), 959–1001. doi:10.3102/0002831207308223

Zimmerman, N. (1990). Self-Regulated Learning and Academic Achievement: An Overview. *Educational Psychologist*, *25*(1), 3–17. doi:10.120715326985ep2501_2

Zuber-Skerritt, O., Fletcher, M., & Kearney, J. (2015). *Professional learning in higher education and communities: Towards a new vision for action research.* London: Palgrave. doi:10.1057/9781137455185

Zubía, J. G., & Alves, G. R. (Eds.). (2012). *Using remote labs in education: two little ducks in remote experimentation* (Vol. 8). Universidad de Deusto.

About the Contributors

Carolyn Berenato is the Chair, Education and Chair Education Specialists Department at Cabrini University. Her research focuses on learning environments; K-12 curriculum and educational environments with a concentration on special education teaching in the areas of art, literacy and mathematics education, specific learning disabilities, dyslexia, pre-service teachers and administration development, and instructional leadership. She teaches undergraduate and masters courses on Literacy, Mathematics, Inclusive Classrooms, Introduction to Special Education, Special Education Practices for School Leaders, and Special Education Law and Policy. Carolyn has also been a literacy coach, assistant middle school principal and elementary school principal with the Upper Darby School District. Dr. Berenato received her B.A. in art history from Rosemont College, her M.S. in education from Saint Joseph's University with Pennsylvania certifications in elementary education K-6, reading specialist K-12, special education K-12, curriculum supervisor K-12, principal K-12, and letter of eligibility for superintendent and her Ed.D. in educational leadership from Saint Joseph's University.

Melissa Cain is a lecturer at the Australian Catholic University in Brisbane teaching courses on Inclusive Education and Arts education. Melissa's doctoral studies centred on the challenges teachers face when implementing culturally diverse Arts education through examining a range of case studies in Australia and Singapore. Her research covers Initial Teacher Education, Inclusive Education, and Creative Pedagogies with a current focus on hearing the voices of students with vision impairment in mainstream schools. Melissa was a school teacher in Australia and South East Asia for 22 years, has managed three large-scale national Learning and Teaching projects, and has produced an international MOOC through EdX and Microsoft. Melissa has received several higher education teaching awards and is the recipient of the Callaway Doctoral Award.

Tyler Cawthray is an academic and Learning Advisor within Library Services at the University of Southern Queensland. His specialisation and research interests are in student learning and development, learning and teaching, and student success. Tyler works with staff and students in the Management, Commerce and Law disciplines, and collaborates with other Education Portfolio staff, to support student learning and development. He provides one-to-one consultations, as well as developing and delivering workshops, developing teaching resources for staff and students, and working with academic staff to identify and address student learning needs.

Subrata Chakraborty has received Ph.D. degree from Monash University, Australia in 2010. He is currently a Senior Lecturer with the School of Information, Systems and Modelling at University of Technology Sydney. He has previously worked in various academic roles at Monash University, Queensland University of Technology, University of Southern Queensland and Charles Sturt University. Dr Chakraborty also worked in the software development sector responsible for developing enterprise scale business systems. His key research interests are in the areas of Education Analytics, Video Coding, Computer Vision, EEG Signals analysis, Decision Support systems and Optimization. He has published in international conferences and journals. Dr Chakraborty is a member of IEEE, ACS (CP), PMI and ACM.

Beverly Dann has worked at the University of the Sunshine Coast for 10 years with a portion of that time leading the Bachelor of Primary Education program. Beverly has recently completed her PhD in Primary Science Education. Her research areas include, science teaching, leadership, assessment and retention of First Year university students. Beverly has a passion for science that is contagious and is evident in her teaching. Her contemporary approach to teaching engages learners of all ages through an inquiry and hands-on approach to teaching and learning. She often incorporates digital technologies in her courses that are transferable to classroom settings.

Ann Dashwood is Associate Professor in Applied Language and Coordinator of the TESOL/ Applied Linguistics specialisation in the School of Linguistics, Adult and Specialist Education at the University of Southern Queensland. As a supervisor of candidates focusing on language learning and language teaching approaches across local and international education, Ann has a special interest in extending intercultural communication awareness and skills of pre-service and post-graduate teachers as they build their discourse capacity in pedagogy. Her research and teaching reflect engagement of teachers and learners in the EAL/D and EFL process of language-based content learning, meaningful feedback and applications of translanguaging.

Nicole Delaney is a Lecturer in Mathematics Curriculum and Pedagogy at the University of Southern Queensland, focusing on early years education. Nicole is a primary trained teacher with a background in software design and development. A thread running through Nicole's work as an educator has been student focus, flexibility and responsiveness in teaching and learning. Nicole has redesigned a number of courses to include digital badges as tools among the practices which facilitate teacher presence and transform student experience in the online space.

Melissa Fanshawe is a Lecturer in Mathematics Curriculum and Pedagogy at the University of Southern Queensland. She has twenty years' experience within Queensland schools as a teacher, advisory teacher, Deputy and Principal. Melissa is a trained teacher for students with vision impairment, gaining her Masters at RIDBC. She is currently completing her PhD in the field of Vision Impairment and enjoys her volunteer roles as an advisor to the board for Vision Australia and the Queensland convenor for the Australian Braille Authority. Melissa is passionate about advocating for standardised accessible formats so students with vision impairment can have independent access to quality education. Melissa has written the initial primary maths courses to address pre-service mindset in teaching mathematics and using digital badges with the aim to minimise anxiety. These have received positive feedback from students, leading to a number of school, faculty and university teaching awards. Melissa was awarded an ASCILITE 2018 innovation award for "Fostering a positive maths mindset for learning through technology-based sense of presence".

Ali Mohamed B. Habibi received his Master's degree in Electronic and Computer Engineering from Griffith University, Brisbane, Queensland, Australia. He is currently working towards his doctoral degree at University of Southern Queensland. Current research interests focus on Engineering and Technology Labs, Internet of Things, Remote laboratories and collaborative learning, and how the Lab learning experience can be improved through a collaborative approach. This includes studying student collaborative behaviors and intelligent remote laboratory systems, and developing new approaches to formative assessment through the use of technology to provide recommendations for the facilitation of online collaboration via remote access laboratories.

Abdul Hafeez-Baig holds a PhD in the domain of healthcare and information systems from USQ, Australia and master's degrees in MIS and MBA from Griffith University, Brisbane, Australia. Since joining USQ in Jan 2004, he has published around 100 referred publications and research grants of 3 million in the domains of health and education regarding technology management. He is very conversant with wireless technology as well as emerging technologies and learning management systems. He teaches information systems concepts to both undergraduate and postgraduate students, including MBA students. He has also supervised numerous post graduate research students. he also has numerous publications in academic and scholarly journals, and has a vast array of scholarly conference papers, all of which have focussed on the domain of information systems. He has extensive experience in the area of information systems, especially relates to the healthcare sector. He is very interested in wireless healthcare applications, systems analysis and design, adoption, the infusion and diffusion of information technology, knowledge management, technology management, digital innovation, M-learning and E-Commerce, outsourcing, networking, healthcare and information technology, and the re-engineering of business processes.

Vineetha Hewagodage is a researcher in the areas of teaching second languages, adult literacy, language testing and refugee studies. She received her Bachelor of Education from the Griffith University, Masters of Education and Doctor of Philosophy from the University of Southern Queensland in Australia. She has a keen interest are in the areas of the importance of the educator-student relationship in learning, the use of digital communication technologies in community-based practice and the scholarship of teaching and learning a second language. Her work has focused on teaching strategies for improving the quality of discussions both within the classroom and in online discussion boards.

Anuradha Mandal is an academic with University of Southern Queensland and has over 5 years of experiences in this sector. She hold Master of Computer Applications (MCA) degree from the University of Pune, India. She has over 10 years of corporate industry experiences in various sectors including telecommunications, data management and banking prior to joining the academia. Her current research focus includes data analytics, education analytics and information systems.

Josh McCarthy is a senior lecturer in Media Arts with specific interests in visual effects, animation, graphic design, photography and digital art. He is the program coordinator of the Graduate Certificate in Compositing and Tracking and the Graduate Certificate in Dynamic Effects and Lighting, and teaches within the Bachelor of Media Arts program in the School of Creative Industries. Josh is also the director of Electricityscape, an Adelaide-based company delivering digital solutions to local and international clients.

James McDowell is a member of the Centre for Technology Enhanced Learning at Lancaster University, where he currently holds a management position with a focus on enhancement of the student experience in the Faculty of Arts and Social Sciences. A Senior Fellow of the HEA (SFHEA) and a Certified Member of the Association for Learning Technology (CMALT), James is an innovative and award-winning researcher-practitioner in the field of TEL, with research interests focussing on video-enhancement of learning and teaching, and the use of educational technologies to promote inclusivity.

Ron Pauley is a Liaison Librarian within Library Services at the University of Southern Queensland. He provides proactive, discipline-based scholarly information and communication services by liaising with clients in Faculties, Schools and Institutes. As Faculty Librarian for Education at the University of Southern Queensland he works with academic staff and students across the university campuses to ensure access to appropriate information resources for programs and courses. He assists staff and students to develop independence in research at undergraduate, postgraduate and doctoral levels while providing customized information literacy (library research skills) training, online tutorial presentations, and individual consultations. He also oversees and maintains the meta-course FYI(Education) and the Education Commons community site. Over 80 separate web pages full of information, demonstrations, video links and tutorials to help you develop your study skills, personal confidence and professional identity.

Alwyn Powell is a Lecturer in Mathematics Curriculum and Pedagogy at the University of Southern Queensland. Alwyn has over 30 years experience in education in the primary, secondary and tertiary sectors. Alwyn has presented on mathematics learning through integrated curriculum and STEAM in primary school throughout Queensland. Alwyn presents practical classroom-ready, hands on workshops, including developing ideas to allow active on-line interaction to assist in reducing anxiety and providing remote teacher access through such programmes as the International On-line Makerspace.

Ahmed Syed is a student of Masters of Business Research at USQ, where he is researching the use of AI in Agriculture. He comes with over 10 years of industry experience as an Innovation Architect for a German IT company i.e. SAP, where he was responsible for driving Innovation and Managed Services adoption. In this role, Ahmed supported the value advisory, design and delivery of innovative solutions. Prior to this Ahmed was in a Digital Transformation strategic advisory role for SAP APJ organization. He assisted clients in achieving their Digital Transformation vision and Best-Run Business capability, focusing on enabling customers to realize their Digital Strategy. Ahmed started the career as a functional consultant and was part of SAP Implementation projects around the globe including Korea, Malaysia and the Middle East.

Anbarasu Thangavelu is an academic and Learning Advisor within Library Services at the University of Southern Queensland. His specialisation and research interests are in student learning and development, learning and teaching, and student success. As a Learning Advisor, he works with staff and students in the Education discipline, and collaborates with other Education Portfolio staff, to support student learning and development. He provides one-to-one consultations, as well as developing and delivering workshops, developing teaching resources for staff and students, and working with academic staff to identify and address student learning needs.

Index

I

L

M

N

O

P

S

T

V

W

www.ingramcontent.com/pod-product-compliance
Lightning Source LLC
LaVergne TN
LVHW062122300926
849161LV00010B/2521
* 9 7 8 1 7 9 9 8 0 4 2 6 0 *